International Scientific Conference Civil Engineering Design and Construction 2024

IOP Conference Series: Materials Science and Engineering
Volume 1323

Varna, Bulgaria
9-13 September 2024

ISBN: 979-8-3313-1145-2
ISSN: 1757-8981

Printed from e-media with permission by:

Curran Associates, Inc.
57 Morehouse Lane
Red Hook, NY 12571

Some format issues inherent in the e-media version may also appear in this print version.

Institute of Physics
Dirac House, Temple Back
Bristol BS1 6BE UK

Phone: 44 1 17 929 7481
Fax: 44 1 17 920 0979

techtracking@iop.org

Additional copies of this publication are available from:

Curran Associates, Inc.
57 Morehouse Lane
Red Hook, NY 12571 USA
Phone: 845-758-0400
Fax: 845-758-2633
Email: curran@proceedings.com
Web: www.proceedings.com

TABLE OF CONTENTS

IOP Conf. Series: Materials Science and Engineering **1323** (2024) 011001 doi:10.1088/1757-899X/1323/1/011001

Preface

The **International scientific conference "Civil Engineering Design and Construction" (DCB)**, which began in 2000 as a national conference organized by the Union of Bulgarian Construction Engineers and the Bulgarian Scientific and Technical Union of Civil Engineering, is held biennially. This conference holds significant importance for Bulgaria, providing a platform for scientists, researchers, and experts to showcase their achievements, share their experiences, and present innovative ideas in various fields of design, construction, and architecture. This international scientific forum has been a platform for in-depth discussions and fruitful meetings since 2006.

The DCB conference was held at Varna Free University "Chernorizets Hrabar" in Bulgaria on September 12-13, 2024. In its thirteenth iteration, is a premier international event organized by the Scientific and Technical Union of Civil Engineering in Bulgaria (NTSSB), with strong supporting from the Federation of Bulgarian Scientific and Technical Unions, Chamber of Engineers in Investment Design, Bulgarian Construction Chamber, Varna Free University, and the "Evrika" Foundation. A key focus of this year's conference was the transformative role of digitalization and Building Information Modeling (BIM) in the construction industry. Participants presented their work on-site, and their presentations were screened in the conference halls.

The conference program featured five sessions, including two plenary sessions and 58 specialized reports. In addition to the primary Bulgarian scientific organizers and representatives from the University of Architecture, Civil Engineering, and Geodesy, the University of Structural Engineering and Architecture "Lyuben Karavelov", Varna Free University "Chernorizets Hrabar", New Bulgarian University, the University of Transport "Todor Kableshkov", European Polytechnical University, the University of Mining and Geology "St. Ivan Rilski", and the Bulgarian Academy of Sciences (BAS), scientists from Northern Ireland, Lithuania, Turkey, and Syria participated in the scientific forum. In addition to thematic discussions on the main topics, a special session for young scientists was organized, supported by the "Evrika" Foundation. Leading companies working in the building and construction sectors were actively presented.

The accepted papers are published in the IOP Conference Series – Materials Science and Engineering and separate volume of Conference Proceedings and copyrighted with an ISSN.

Editors DCB'2024:
Assoc. Prof. Ventseslav Stoyanov, PhD, Eng – Editor in Chief
Assoc. Prof. Lachezar Hrischev, PhD, Eng
Assoc. Prof. Tony Venelinov, PhD

IOP Conf. Series: Materials Science and Engineering **1323** (2024) 011001 doi:10.1088/1757-899X/1323/1/011001

Scientific Committees

IOP Conf. Series: Materials Science and Engineering **1323** (2024) 011002 doi:10.1088/1757-899X/1323/1/011002

Peer review declaration

All papers published in this volume of *IOP Conference Series: Materials Science and Engineering* have been peer reviewed through processes administered by the Editors. Reviews were conducted by expert referees to the professional and scientific standards expected of a proceedings journal published by IOP Publishing

- **Type of peer review:** Single-anonymous
  The review process is single- anonymous, the authors did not know reviewers, but reviewers know the authors.

- **Conference submission management system:** Morressier
  Additional communication to the authors through conference email – editors.conference.dcb@gmail.com

- **Number of submissions received:** 29

- **Number of submissions sent for review:** 29

- **Number of submissions accepted:** 27

- **Acceptance Rate (Number of Submissions Accepted / Number of Submissions Received X 100):** 93.1 %

- **Average number of reviews per paper:** 1.11

- **Average number of improvement requirements per paper:** 2.74

- **Total number of reviewers involved:** 22

- **Any additional info on review process:** Reviewers were asked to judge the manuscripts based on criteria with five level of assessments (very good, good, acceptable, not acceptable, and poor). When at least one of the reviewers confirmed that all assessments were "good", the manuscript was accepted for publication. English language editing and formatting were used in the review process.

- **Contact person for queries:**
  Assoc. Prof. Ventseslav Stoyanov, PhD;
  e-mail: vensy.stoyanov@gmail.com

IOP Conf. Series: Materials Science and Engineering **1323** (2024) 012001 doi:10.1088/1757-899X/1323/1/012001

The effect of concrete maintenance measures on the colour variation in concrete

Mindaugas Daukšys*, Edgaras Fadejevas, Mohamed Wagih Abdou Mohamed Meneaem and Svajūnas Juočiūnas

Faculty of Civil Engineering and Architecture, Kaunas University of Technology, Kaunas, Lithuania

*E-mail: mindaugas.dauksys@ktu.lt

Abstract. This research provides valuable insights into the effects of various maintenance measures on concrete colour variation, helping civil engineers and architects make informed decisions in their projects. During the research, specimens were cast using moulds whose surfaces were covered with different release agents (RA) or without them. To see the impact of measures for concrete maintenance on colour variation, the following ways of coating specimens were chosen: test specimen left in plastic mould (M), specimen wrapped using a flexible polyethene sheet (PS), one test specimen left without any coatings (WC), specimen sprayed with water and wrapped using a flexible polyethene sheet (PSW), specimen coated using a wax-based curing agent (WBA), specimen was coated using siloxane-based water repellent (SBR). The NCS Colour Scan 2.0 instrument was employed to assess colour variation and determine the appropriate colour notation, reducing the influence of subjective evaluator opinion. The research showed that the most popular detected grey colour had the following notation: S 3502-Y. This colour can be described as follows: its visual blackness is 35%, its chromaticity is 2%, and its hue (Y) is yellow. Different measures mainly influenced the hue, which varied in the range from yellow to green. M, PS, PSW, and WBA had the most visible effect on grey colour clouding and marbling. Using those maintenance measures, the visual blackness of the grey colour has reached 50% or even 60%.

1. Introduction

Owners, architects, investors, and developers desire architectural concrete that is consistently coloured and free from discolouration to create an aesthetically appealing appearance. The colour variation of concrete can be influenced not only by the type of cement used, coloured pigments, type of formwork, type of release agent, and ambient air temperature but also by the type of measures used during concrete maintenance, such as wetting with water, covering with polyethene sheet, covering with spreadable and non-spreadable membranes, duration of concrete structure retention in the formwork, the materiality of the formwork surfaces and another. This means that discolouration, a complex issue, depends on various factors and is difficult to control during construction to achieve the highest requirements for formed concrete quality. Therefore, understanding the impact of maintenance measures is crucial for professionals in the field.

Variations in the mix composition, the characteristics of the aggregate and formwork used, the drying conditions of the concrete surfaces, and inadequate finishing techniques can lead to discolouration of concrete surfaces [1]. A lower water-to-cement (w/c) ratio results in a darker surface for the concrete, whereas a higher w/c ratio leads to a lighter surface [2]. The analysis of discolouration on concrete surfaces can be conducted by comparing photos and calculating the percentage of grey and white areas. A more significant percentage of the grey regions indicates the stability of the grey colour [3]. Formwork with a more prominent rough texture absorbs water and releases agents more rapidly, leading to irregularities such as colour variations, clouding, and marbling. The presence of iron oxide and manganese oxide affects the natural grey hue of hardened Portland cement paste [4]. The general impression of colour variation of formed faces of exposed concrete also depends on its tint and tone. It is known that sand used for concrete preparation gives tint, while water amount provides the tone [5].

The quality of the concrete surface is assessed based on various standards that outline the primary criteria for colour, texture, and dimensional tolerances of the formed concrete surface [6-9]. Requirements for the categories of formed concrete surfaces according to colour variation are summarised in Table 1 [10].

Table 1. Categorisation of formed concrete surfaces based on colour uniformity

CIB Report No. 24							
Number of the reference scale photograph	1	2	3	4	5	6	7
NCS notations of reference grayscale photos*	S 1502-B, S 2005-B, S 2005-B20G	S 2502-G, S 4005-B80G, S4502-G	S 6502-Y, S 6010-B90G, S 7005-B80G, S 7005-G20Y	S 8000-N, S 8005-G20Y	S 8500-N	S 8500-N	S 8500-N
ACI 347.3R-13							
Class of surface void ratio		SVR1	SVR2		SVR3	SVR4	
Colour uniformity		CU1 [a, na]	CU1 [a, na]		CU2 [a, na]	CU2 [a], CU3 [na]	
DBV Code of Practise							
Class of surface void ratio		P1	P2		P3	P4	
Colour uniformity		FT1[a, na]	FT2 [a, na]		FT2 [a, na]	FT3 [a], FT2 [na]	
ÖVBB Richtlinie							
Class of surface void ratio	-		4P	3P	2P	1P	
Colour uniformity	-		F1	F1	F1	F2	

* - the match of the colours of reference grayscale photos to the specific notation of the closest colour was done using the colour reader NCS Colour Scan 2.0. Legend: a = absorbent form facing; and na = nonabsorbent form facing.

Table 1 indicates that the most frequently occurring colours (S 2502-Y, S 3502-Y, S 4502-Y) are associated with the second reference grayscale image by CIB, according to visual blackness in colour. According to ACI standards, light and dark colour variations and rust are deemed acceptable for the colour uniformity class CU1, while gradual light and dark discolourations are permissible for the colour uniformity class CU2. Additionally, CU2 requires mostly uniform colour consistency between adjacent placements and layer lines. According to DBV and ÖVBB standards, light and dark discolourations are allowed for the colour consistency class FT1. For the FT2 class,

IOP Conf. Series: Materials Science and Engineering **1323** (2024) 012001 doi:10.1088/1757-899X/1323/1/012001

consistent large areas of light and dark discolouration are acceptable, while minor clouding and discolouration are permissible for the FT3 class.

Literary analysis indicates a need for further research to investigate the influence of various concrete maintenance methods on discolouration. Additional quantitative studies are necessary to grasp the effect of different concrete maintenance measures on colour variations in the resulting surfaces.

This research aimed to examine the impact of different concrete maintenance measures on the changes in colour on the formed concrete surface when different types of release agents are used to cover the surface of moulds for casting.

2. Materials and methodology

Ordinary Portland cement CEM I 52.5 R, sourced from AB "Akmenės cementas" in Lithuania, was utilised for the study and fulfilled the standard LST EN 197-1:2011 requirements. The cement exhibited the following characteristics: a specific surface area of 410 m^2/kg (as determined by the Blaine method), a specific gravity of 3050 kg/m^3, a water demand for standard consistency of 26.5% (as determined by the Vicat method), 0.8 mm in volume stability, an initial setting time of 195 minutes, and a compressive strength of 27.1/54.0 MPa after 2/28 days.

Fine aggregate sand with a 0/4 mm fraction and coarse aggregate gravel with a 4/16 mm fraction sourced from local quarries in Lithuania were employed. The granulometric properties of the fine and coarse aggregate complied with the requirements specified in the standard LST EN 12620:2003+A1:2008. The following properties were observed for the sand with a 0/4 mm fraction: a dry bulk density of 1635 kg/m^3, a specific gravity of 2660 kg/m^3, a water absorption of 0.56%, and a fineness modulus of 2.4 (CF). The gravel particles were round and had a smooth surface. The coarse aggregate was not washed and contained some dust. The gravel with a 4/16 mm fraction exhibited the following properties: a dry bulk density of 1582 kg/m^3, a specific gravity of 2670 kg/m^3, a water absorption of 1.39%, a particle shape index (SI) of 3%, and a particle toughness (by Los Angeles abrasion test) of 24%.

The high-range water-reducing superplasticiser MasterGlenium ACE 430 (Austria), which was employed as a plasticising admixture, exhibited the following characteristics: is a brownish liquid with a density of approximately 1.06±0.02 kg/l and a pH value of 5.5±1.0.

The composition for the concrete mix used in the research was selected from a single concrete mixture producer in Lithuania. This mixture is utilised to cast concrete structures on site when a specific surface appearance is required. The following were the amounts of materials (in kilograms) per cubic meter of the mix: 350 kilograms of Portland cement, 890 kilograms of sand in the 0/4 mm fraction, 980 kilograms of gravel fraction in the 4/16 mm size, and 2.8 kilograms of superplasticiser (0.8% of the cement mass). The water-to-cement ratio was 0.44. The concrete mixture was prepared in a laboratory setting using dry components in a rotating pan mixer Zyklos ZZ 50 HE (Germany). The mixing process was carried out in two stages. Initially, the fine and coarse aggregates and 2/3 of the water were added to the mixer and mixed for approximately one minute. Following this, the mixture was allowed to rest for two minutes. Subsequently, cement, 1/3 of the water, and superplasticiser were added, and the mixture was mixed for approximately two minutes.

After the fresh concrete mix was prepared, the temperature was measured according to the guidelines outlined in LST EN 12350-1:2019. According to LST EN 12350-2:2019, the slump test was conducted, while the density and air content (measured by the pressure method) were determined in line with LST EN 12350-6:2019 and LST EN 12350-7:2019, respectively.

In the laboratory, the test specimens were cast in plastic moulds made from plastic tubes with a diameter of 0.15 meters and a height of 0.3 meters. The moulds were cut to the necessary height, and the bottom of the mould was then secured to a wooden plate. Plastic moulds were coated with two different types of release agents (RA): MasterFinish RL 215 from Austria, which is based on a water-soluble emulsion (RA-WSE), and MasterFinish RL 310 from Austria, which is based on mineral oils (RA-MO). The clean moulds' surfaces were sprayed to apply the release agents, and any excess was removed using a dry absorbent cloth, following the manufacturer's guidelines. The moulds were coated with the RA about 60 minutes before each casting test.

To achieve the desired layer thickness, the concrete mix was manually placed into moulds. After that, it was compacted using a portable electric internal vibrator, the ENAR Dingo (Spain). The vibrator's shaft was flexible, and its poker head had a diameter of 25 mm and a frequency of 13.750 vibrations per minute. Compaction commenced when the poker head penetrated the fresh concrete layer a few centimetres. The poker head was held in a fixed vertical position during the compaction of fresh concrete, and the vibration time lasted approximately 16 seconds.

Three separate scenarios were analysed in the course of the research: first, test specimens were cast in plastic moulds that were not coated with RA, and aggregates were either heated or not heated before mix preparation; second, test specimens were cast in plastic moulds coated with RA-WSE, and aggregates were either heated or not heated before mix preparation; third, test specimens were cast in plastic moulds coated with RA-WSE, and aggregates were either heated or not heated before mix preparation.

The casted test specimens were kept in plastic moulds in the laboratory at 20±2°C temperature and humidity of 42±5% for two days. After demolding, the surface of concrete specimens was coated using different measures for concrete maintenance, and specimens were left in the same conditions. Considering concrete maintenance measures used for concrete maintenance at the construction site for the research were chosen as follows: MasterKure 216 WB (Austria) wax-based curing agent for newly laid concrete, which inhibits evaporation of water during the setting period, Sikagard®-700 S (Switzerland) siloxane-based water-repellent impregnation, and flexible polyethene sheet (Lithuania). The wax-based curing agent (WBA) had the following properties: white liquid, density at about 0,98±0.02 kg/l, while the siloxane-based water repellent (SBR) had such properties: colourless liquid, density at about 0,80 kg/l, flash point at about 30°C. Polyethene sheet (PS) had the following properties: flexible, translucent/waxy, and chemical-resistant.

The following progress of coating specimens was used in the research to see the impact of measures for concrete maintenance on concrete colour variation. Table 2 presents notation and descriptions of measures used for concrete maintenance to make it easier to understand.

Table 2. Notation and description of measures used for concrete maintenance

Notation	Description of measures used for concrete maintenance
M	test specimen left in plastic mould
PS	specimen wrapped using a flexible polyethene sheet
WC	test specimen left without any coatings
PSW	specimen sprayed with water and wrapped using a flexible polyethene sheet
WBA	specimen coated using a wax-based curing agent
SBR	specimen was coated using siloxane-based water-repellent

The NCS Colour Scan 2.0 instrument (Sweden) was used to analyse the formed concrete surface's colour variation. The portable colour reader is designed to match the colours of the analysed concrete surface with the data stored in its memory and provide an automatic colour notation or code. This notation or code is based on the NCS system, but it can be converted to the RAL system notation if required. The colour notation is described in this way: the first two digits signify the percentage of visual darkness in the colour, while the next two digits indicate the colour's chromaticity or intensity. The letter-number combination following the hyphen indicates the colour's hue, specifying which chromatic attributes are present and their relationship. The NCS Colour Scan 2.0 reader is capable of measuring areas of 2, 4, and 8 mm in size, and it automatically extracts four dominant colours. Additionally, the size and flatness of the measurement area are suitable for the selected test specimens.

The test specimens' photos were taken using a Nikon D3200 camera with 24 megapixels and a Nikon DX AF-S Nikkor 18-55 mm lens. Additionally, a Quantuum flashlight was mounted separately. The camera's main parameters were an f-number of 8, a shutter time of 1/160 s, a flash intensity of 15, and a sensitivity of ISO 100.

3. Results and discussions

The composition of the concrete mix remained unchanged during the entire research process. Table 3 presents fresh concrete properties determined for each batching. From Table 3, it can be seen that the properties of fresh concrete depend on the mixture temperature. When aggregates were heated before mix preparation, the mixture air content increased from 3.4% to 4.8%, density decreased from 2398 kg/m^3 to 2374 kg/m^3, and slump decreased from 210 mm to 50 mm compared to the mixture without aggregates heating. B-1 series slump values can be attributed to the slump class of S5, while B-2 series – to the slump class S2.

Table 3. Fresh concrete properties

Series	Air content, %	Density, kg/m^3	Slump value, mm	Mixture temper., °C	Tem., °C	Hum., %	Notes
B-1-1	3.5	2383	210	22.2	21.0	37.1	Aggregates were not heated before mix preparation
B-1-2	3.2	2412	210	24.4			
Average	3.4	2398	210	23.3			
B-2-1	4.8	2375	50	38.0	23.2	42.6	Aggregates were heated before mix preparation
B-2-2	4.7	2373	30	40.0			
Average	4.8	2374	40	39.0			

After two days of curing, the test specimens' surfaces were treated differently for each batch, using different measures for concrete maintenance. Table 4 presents a view of test specimens prepared according to three distinct scenarios when the concrete specimens' surfaces were treated using the following measures: M, PS, WC, PSW, WBA, and SBR. In such a way, prepared specimens were left for one week in laboratory conditions. Photos in Table 4 reveal that the surfaces are initially darker when measures like PSW, WBA, and SBR are used for concrete maintenance. After one week, the impact of measures on concrete discolouration was determined after one test specimen was removed from the plastic mould and two specimens from the flexible polythene sheet for each batch. After two hours, a visual inspection of the concrete specimens' colour variations was performed, and images were captured for documentation.

Table 5 shows the view of test specimens coated using different measures for concrete maintenance when a 23°C temperature mixture for specimen casting was used. In comparison, Table 6 shows the view of test specimens when a 39°C temperature mixture was used. From visual analysis, it can be noticed that ordinary Portland cement CEM I of 52.5R gave the concrete specimens a grey colour. Whereas the sand fraction of 0/4 mm that gave the concrete tint was delivered from the same quarry, and the water that gave the concrete tone amount was constant for all batches, the measures used for concrete maintenance could have affected the variation of grey colour. It means that different measures used for concrete maintenance caused the variation of the natural grey colour of concrete when plastic mould was coated with various types of RA or without RA during the research.

Table 4. The view of test specimens coated using different measures for concrete maintenance: A 23°C temperature mixture was used for specimen casting (a); a 39°C temperature mixture was used for specimen casting (b). Explanation of used measures for concrete maintenance (from left to right): M, PS, WC, PSW, WBA and SBR (notations described in Table 3)

a)		b)	
Specimens cast in moulds whose surfaces were not coated with RA			
a)		b)	
Specimens cast in moulds whose surfaces were coated with RA-WSE			
a)		b)	
Specimens cast in moulds whose surfaces were coated with RA-MO			

Grey colour clouding and marbling were noticed on the specimen surface when measures for concrete maintenance like M, PS, PSW, and WBA were used. The mentioned maintenance measures affect water evaporation from concrete specimens during curing, i.e., they stop water evaporation. Table 6 showed that PS and PSW had the most visible clouding and marbling effect on concrete when the specimens were cast in plastic moulds without RA; M, PS, PSW and WBA had the most visible impact when RA-WSE was used; and PSW and WBA had the most visible effect when RA-MO was used. It was noticed that PS and PSW had the most visible clouding and marbling effect on concrete when the specimens were cast in plastic moulds without RA and using RA-WSE; M, PS, PSW and WBA had the most visible impact when RA-MO was used. The formed concrete surface has many air pores, making it difficult to see the impact of clouding and marbling. It is worth mentioning that visual analysis also revealed that concrete specimens cast using a 39°C temperature mixture are characterised by slighter colour variation than specimens cast using a 23°C temperature mixture. Standards specifying the main requirements for the formed concrete

IOP Conf. Series: Materials Science and Engineering **1323** (2024) 012001 doi:10.1088/1757-899X/1323/1/012001

surface allow discolouration in large areas and minor clouding in small areas, depending on the surface categories.

Table 5. The view of test specimens after curing using different measures for concrete maintenance. A 23°C temperature mixture was used for specimen casting. Notations described in Table 3

The measures used for concrete maintenance					
M	PS	WC	PSW	WBA	SBR

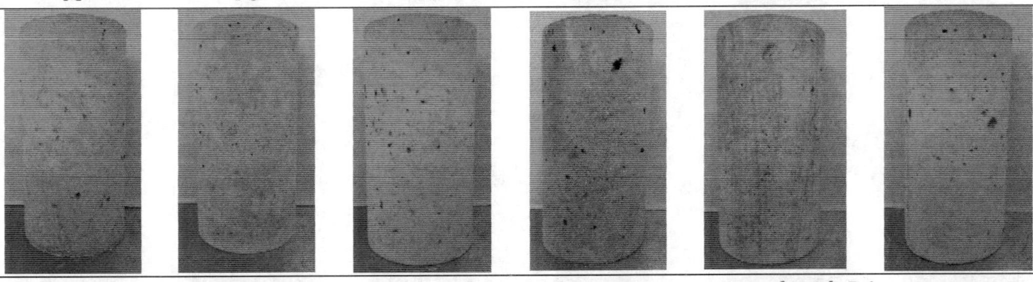

Specimens cast in moulds whose surfaces were not coated with RA

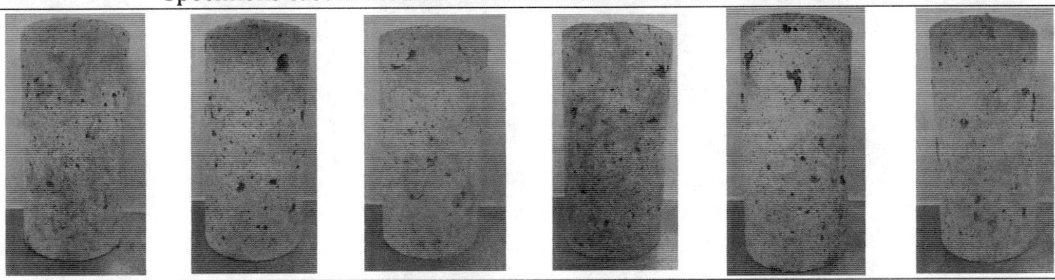

Specimens cast in moulds whose surfaces were coated with RA-WSE

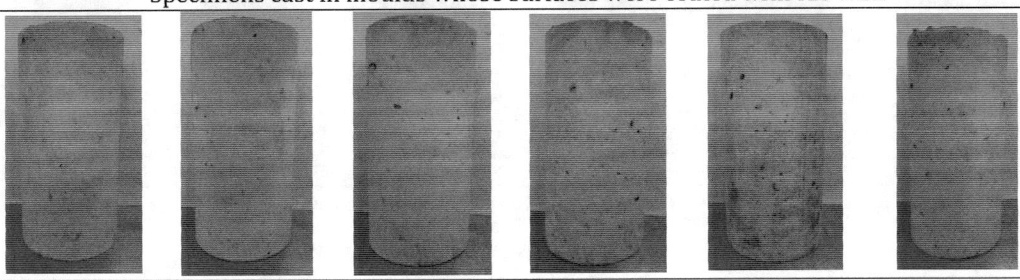

Specimens cast in moulds whose surfaces were coated with RA-MO

Visual analysis depends on the evaluator's subjective opinion, so the NSC colour reader was used for colour variation inspection to avoid this. In this way, the surface of test specimens was analysed using an NCS Colour reader to detect the colour difference when different measures for concrete maintenance were used. This allows us to determine the colour code of the analysed concrete surfaces.

IOP Conf. Series: Materials Science and Engineering **1323** (2024) 012001 doi:10.1088/1757-899X/1323/1/012001

Table 6. The view of test specimens after curing using different measures for concrete maintenance. A 39°C temperature mixture was used for specimen casting. Notations described in Table 3

The measures used for concrete maintenance					
M	PS	WC	PSW	WBA	SBR
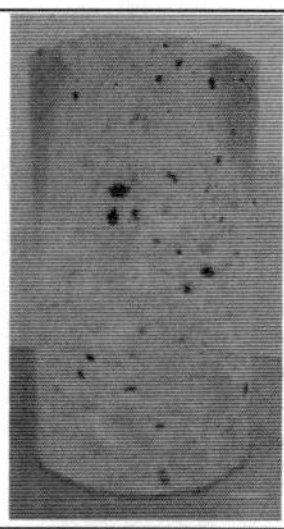	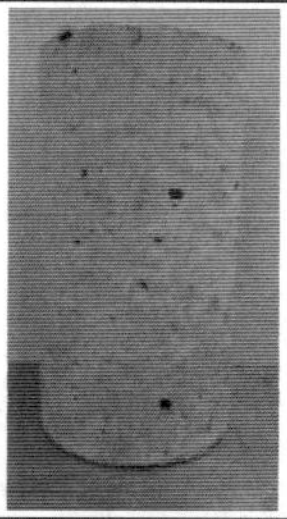		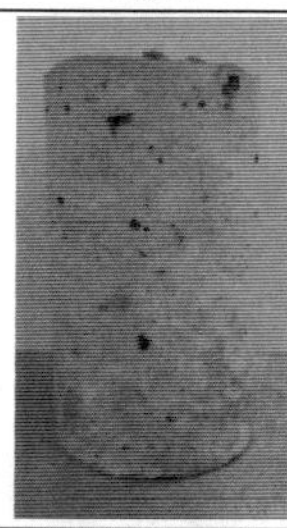		
Specimens cast in moulds whose surfaces were not coated with RA					
	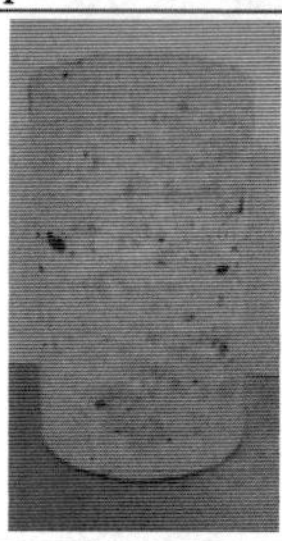	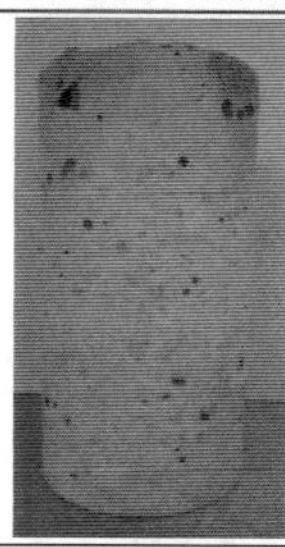	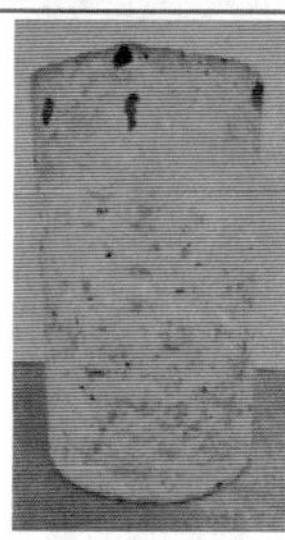	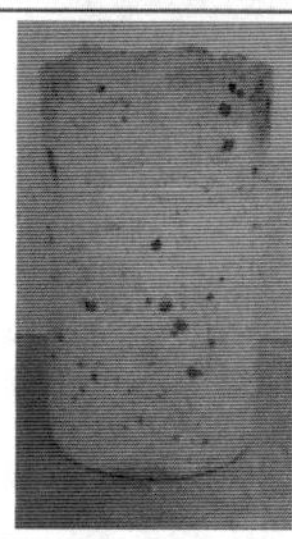	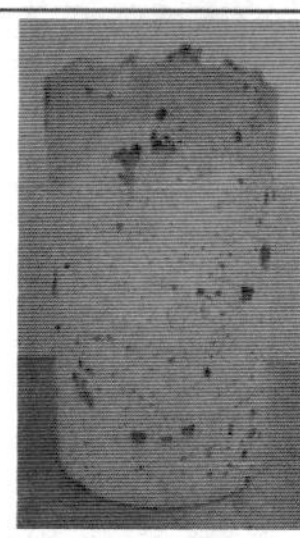
Specimens cast in moulds whose surfaces were coated with RA-WSE					
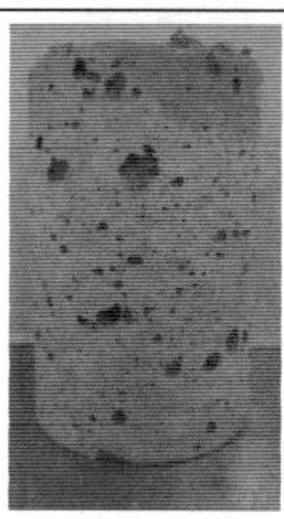	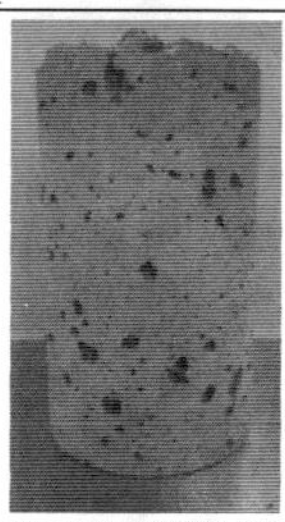	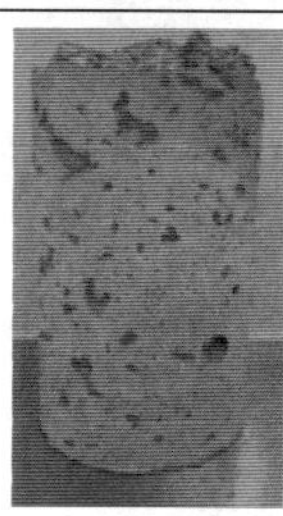	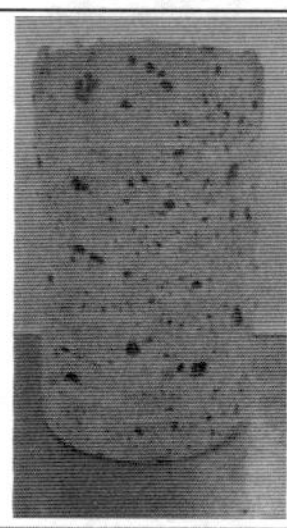	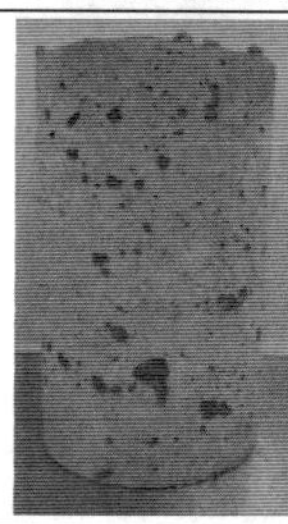	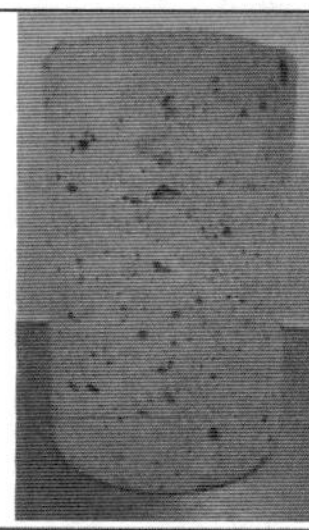
Specimens cast in moulds whose surfaces were coated with RA-MO					

Table 7 presents the notation of colours of test specimens coated using different measures for concrete maintenance when a 23°C temperature mixture for specimen casting was used. In comparison, Table 8 presents the notation of colours of test specimens when a 39°C temperature mixture was used. The findings of the colour notation revealed the variation in colour of the examined concrete surfaces, which are not always perceptible to the human naked eye. Tables 7 and 8 revealed that the most popular detected grey colour had the following notation: S 3502-Y. This grey colour can be described as follows: its visual blackness is 35%, its chromaticity is 2%, and its hue (Y) is yellow. This means that this colour, with a visual blackness of 35%, can be chosen as the primary colour of concrete. It was noticed that the grey colour of concrete varied from darker to lighter, and almost all specimens had a yellow hue. There were a few cases when the grey colour had a green hue. Depending on the measure type used for concrete maintenance, visual blackness, compared to the primary colour, varied from 25% to 50% when a mixture of 23°C temperature was used and from 25% to 60% when a mixture of 39°C temperature was used.

Table 7. Depending on the used measure for concrete maintenance, colour variation was obtained on the test specimen surface. A 23°C temperature mixture was used for specimen casting. Notations described in Table 3

Colour notation	The measures used for concrete maintenance					
	M	PS	WC	PSW	WBA	SBR
S 2502-Y	■	■	■		■	■
S 3000-N			■			
S 3005-Y50R					■	
S 3502-Y	■	■	■	■	■	■
S 4005-Y20R					■	
S 4005-G80Y					■	
S 4502-Y	■			■	■	
S 5005-G80Y	■					
Specimens cast in moulds whose surfaces were not coated with RA						
S 2502-Y		■	■		■	
S 3000-N		■	■			
S 3500-N			■			
S 3502-Y	■	■	■	■	■	■
S 4000-N				■		
S 4005-Y20R	■				■	
S 4005-G80Y				■		■
S 4502-Y	■			■	■	
S 5005-G80Y		■				
S 5502-Y				■		
Specimens cast in moulds whose surfaces were coated with RA-WSE						
S 2502-Y	■	■	■	■	■	■
S 3000-N		■		■		
S 3005-G80Y					■	■
S 3502-Y	■	■	■	■		■
S 4005-Y20R				■	■	
S 4005-G80Y						■
S 4502-Y		■			■	
S 5005-Y20R					■	
Specimens cast in moulds whose surfaces were coated with RA-MO						

The visual analysis concluded that specimens cast using a 39°C temperature mixture were characterised by slighter colour variation than specimens cast using a 23°C temperature mixture. Using a mixture of 39°C for specimen casting, the colour changes in the NCS colour deck in 5 steps depending on the measure used for concrete maintenance. Using the mixture of 23°C, the colour changes even in 7 steps depending on the measure used for concrete maintenance.

Table 8. Depending on the used measure for concrete maintenance, colour variation was obtained on the test specimen surface. A 39°C temperature mixture was used for specimen casting. Notations described in Table 3

Colour notation	The measures used for concrete maintenance					
	M	PS	WC	PSW	WBA	SBR
S 2502-Y	■		■		■	■
S 3000-N			■			
S 3005-Y50R	■		■		■	
S 3502-Y	■	■	■	■	■	■
S 4005-Y20R			■	■		
S 4502-Y	■	■		■		
S 5005-Y20R	■					
Specimens cast in moulds whose surfaces were not coated with RA						
S 2502-Y	■	■	■		■	■
S 3005-Y50R	■		■		■	■
S 3005-G80Y					■	
S 3502-Y	■	■	■	■	■	■
S 4005-Y20R	■		■		■	■
S 4502-Y	■			■		
Specimens cast in moulds whose surfaces were coated with RA-WSE						
S 2502-Y	■	■	■		■	■
S 3005-Y50R					■	■
S 3502-Y	■	■	■	■		■
S 4005-Y20R	■	■			■	■
S 4005-G80Y	■	■				
S 4502-Y				■		
S 5005-Y50R				■		
S 6005-Y20R				■		
Specimens cast in moulds whose surfaces were coated with RA-MO						

Standards that outline the primary requirements for the colour and texture of formed concrete surfaces require preparing mock-ups or large-scale concrete elements to determine the impact of the release agent, form-facing materiality, mixture composition, and other uses. This means that the impact of the scale effect on test specimens can extend to the overall perception of colour variation. The distance used to evaluate the formed concrete surface quality also affects the impression of colour variation. Some standards also declare that colour variation analysis should be done after one month of curing. The colour variation analysis using a colour reader was also done after four weeks. The test specimens were kept in the laboratory conditions at a temperature of 20±2°C and humidity of 42±5%. It was noticed that the colour variation became more stable during the rest of the curing period. The visual blackness of darker colours decreased from 60% to 45%, which means that the grey colour became lighter. Civil engineers and architects can make rational decisions in their projects by knowing the impact of various maintenance measures on concrete colour variation and considering specific requirements for formed surface categories.

4. Conclusions

The most visible grey colour clouding and marbling effect was noticed when the specimens were cast in moulds whose surfaces were covered using RA-WSE; the mixture temperature was 23°C and were used such measures: specimens left in plastic mould (M), specimens wrapped using a flexible polyethene sheet (PS), specimens sprayed with water, wrapped using a flexible polyethene sheet (PSW), and specimens coated using a wax-based curing agent (WBA). Meanwhile, M, PS, PSW and WBA had the most visible effect on concrete clouding and marbling when the specimens were cast in plastic moulds using RA-MO and the mixture temperature was 39°C. Visual analysis shows that specimens cast using a 39°C temperature mixture are characterised by slighter colour variation than specimens cast using a 23°C temperature mixture.

Maintenance measures like leaving the test specimen in a plastic mould (M), wrapping it using a flexible polyethene sheet (PS), spraying it with water and wrapping it using a flexible polyethene sheet (PSW), and coating it using a wax-based curing agent (WBA) affect water evaporation from concrete specimens during curing, i.e., they stop water evaporation. Using those maintenance measures, the visual blackness of the grey colour has reached 50% or even 60%.

The most popular detected grey colour had the following notation: S 3502-Y. This colour can be described as follows: its visual blackness is 35%, its chromaticity is 2%, and its hue (Y) is yellow. This colour, with a visual blackness of 35%, was chosen as the primary colour of concrete. There were a few cases when the grey colour had a green hue.

The overall perception of colour variation was checked only after one week when different measures were used for concrete maintenance. Later, after four weeks, the colour variation became more stable during the rest of the curing period. The visual blackness of darker colours decreased from 60% to 45%, which means that the grey colour became lighter.

References

[1] National Ready Mixed Concrete Association 2007 Discolouration Concr. in pract. 23 p.
[2] American Concrete Institute 1984 Guide to residential cast-in-place concrete construction No. 81-40 505 p.
[3] Garg S, Nim KS, Bajpai KK, Misra S 2019 Enhancement in the quality of near surface concrete using some formwork liners Constr. Build. Mater. 207 722–733.
[4] Püttbach E 1987 Pigments for the colouring of concrete Betonw.+Fertigteiltech. 53 2 1–9.
[5] Toogood E. 2020 Webinar: Achieving cast in situ visual concrete The concrete centre.
[6] CIB Working Commission W29 1973 Tolerances on blemishes of concrete Report No. 24 Concr. Surf. Finish. 1–8.
[7] American Concrete Institute 2014 Guide to Formed Concrete Surfaces 347.3R-13 1–21.
[8] PERI GmbH 2018 Best Practice for Architectural Concrete 1–112.
[9] Österreich Vereingung für Beton- und Bautechnick 2002 ÖVBB Richtlinie „Sichtbeton – Geschalte Betonflächen" 1–31.
[10] Savukaitis G, Daukšys M, Juočiūnas S, Kriptavičius D 2021 The influence of new and used formwork coated with different release agents on the appearance of the formed concrete surface J. Build. Eng. 42 102807.

IOP Conf. Series: Materials Science and Engineering **1323** (2024) 012002 doi:10.1088/1757-899X/1323/1/012002

Impact of Impurities on the Properties of Recycled Aggregate for Use in Civil Engineering Work and Road Construction. Part 1: Technical Assessment

Roumiana Zaharieva [1*], Boyan Petrov[1], Daniel Evlogiev[1], Magdalena Kostadinova[1] and Borislav Simonov[1]

[1] Department of Building Materials and Insulations, University of Architecture, Civil Engineering and Geodesy, Sofia, Bulgaria

*E-mail: zaharieva_fce@uacg.bg

Abstract. Construction and demolition waste (CDW) is a significant source of mineral resources. Recycled concrete aggregate fraction 0/63 mm is among the most used materials. The present study aims to evaluate the influence of impurities such as asphalt, bricks, soil, and lightweight materials on the properties of recycled aggregate. Two types of blended recycled materials are composed – one, representative for recycled aggregate from CDW generated at road projects and another type, representative for recycled materials from buildings rehabilitation and demolition. The main characteristics of blended recycled aggregate with different amounts of impurities are being investigated according to EN 13 242. The maximum dry density and optimum water content according to EN 13286-2 and the California Bearing Ratio (CBR) according to EN 13286-47 are also determined. The source of CDW is found to have a significant impact on the performance of the recycled material. The rule of mixtures and some testing methods are not always appropriate for the recycled materials. Based on the results, some recommendations regarding the management of CDW and the appropriate use of recycled aggregates in civil engineering and road construction are made.

1. Introduction

Road construction is among the biggest consumers of mineral resources worldwide [1]. At the same time, road rehabilitation and reconstruction generate a huge quantity of construction and demolition waste (CDW) – in Bulgaria, in 2011 it amounted to 33% [2]. The scarcity of natural materials, the environmental limitations related to new quarries exploitation, and the high transportation costs lead to the significant recycling share of the roads generated CDW. Waste asphalt is recovered via cold or hot asphaltic pavement recycling [3], [4], thus contributing to the circular economy of the construction sector. Recycled concrete aggregates (RCA) have been also largely implemented for many years in various unbound materials for embankments, road sub-base, and base courses [5], [6]. Thus, the road sector has created good practices for sustainable CDW management. The requirements of the Waste Framework Directive [7] have boosted the recycling of CDW to achieve the target material recovery of 70%. The road applications became attractive for the recovery not only of road works generated CDW, but also of CDW from other sources. However, when derived from building rehabilitation, reconstruction,

IOP Conf. Series: Materials Science and Engineering **1323** (2024) 012002 doi:10.1088/1757-899X/1323/1/012002

and demolition, the valuable RCA might contain significant amounts of non-hazardous impurities such as clay bricks, roof tiles and cladding tiles, glass, plastics, timber, gypsum, insulating materials, etc. [8]. The impurities might affect the performance of RCA and the road sector might not accept those aggregates. There are different studies on the impact of ceramic impurities – according to *Poon and Chan*, the brick aggregate reduces the sensitivity of recycled material from moisture content [9] the maximum dry density might be achieved at a different from the optimum moisture content, determined by the Proctor test [10]. Moreover, although the ceramic decreases the resistance to fragmentation, the fines generated during the compaction lead to a better consolidation of the embankment, and the bearing capacity assessed by the California Bearing Ratio (CBR) finally increases [11]. The opposite is claimed by [9]– it seems that the CBR is noticeably reduced with a higher brick aggregate content of the recycled material because of both the lower density of the bricks and the poorer interlocking between the bricks and concrete particles.

The selective demolition and the separate collection of CDW is the best way (in terms of costs and resources to be used in recycling) to overcome that major drawback of recycled materials – the presence of impurities. However, it is not possible to always apply selective demolition techniques because of the increased costs, time restrictions, peculiarities of the buildings, etc.

The present paper covers the technical aspects of the impurities impact assessment on the properties of the recycled aggregate for use in civil engineering work and road construction. The aim is to outline the possibilities and limitations of the application of mixed recycled materials. The environmental challenges related to the presence of impurities are discussed in the second part of the study.

2. Materials and methods

Recycled materials fraction 0/63 mm from 2 sources of CDW are studied, based on performed inspection of several construction sites:

- Roads and facilities such as parking, aprons, other pavements, concrete ditches, and culverts, etc., where the impurities in concrete CDW are from asphaltic pavements and soils. A significant presence of natural crushed stone CDW might also be observed.
- Buildings with reinforced concrete structures or masonry walls, where apart from concrete and bricks the CDW might contain a significant amount of other materials.

Recycled materials constituents are designated as per EN 933-11 [12] (Table 1). The individual constituents are obtained by a laboratory recycling of selected CDW.

Three types of recycled material fraction 0/63 mm are investigated. Type 1 consists of single-component materials: concrete aggregate only (Rc100), bricks aggregate only (Rb100) and ground asphalt (Ra100). This type serves as a basis to assess the impurities impact on the performance of the blended recycled materials (Table 2). The blends of Type 2 contain recycled concrete with different amounts of impurities such as natural stone (10% and 20%), ground asphalt (10%, 20% and 30%) and soil (0.5%). Type 2 aggregate would be representative for the recycled materials from the roads and pavements. Type 3 blends aim to reproduce recycled materials from buildings rehabilitation and demolition CDW. They contain up to 70% of recycled bricks, 0.5% of soil, and quite a big amount of floating materials – 10 cm^3/kg which is less than 1% and depends on the floating materials particle densities. The source of floating materials are wooden elements, joinery, plastic pipes, and various insulation materials. Type 3 blends would

IOP Conf. Series: Materials Science and Engineering **1323** (2024) 012002 doi:10.1088/1757-899X/1323/1/012002

result from commingled demolition and of rehabilitation works. The performance of recycled aggregate is compared to that of natural stone NA (limestone) fraction 0/63 mm.

Table 1. Constituents of recycled materials and CDW source description

Constituents of recycled materials	*Designation of constituents as per EN 933-11*	*Origine of used CDW*
Concrete, concrete products, mortar Concrete masonry units	Rc	Recycled concrete from a reinforced concrete structure of a demolished multistore parking, Sofia, Bulgaria
Unbound aggregate, natural stone Hydraulically bound aggregate	Ru	Unbound natural aggregate fraction 0/63 mm, Studena quarry near Sofia, Bulgaria
Clay masonry units (i.e. bricks and tiles) Calcium silicate masonry units Aerated non-floating concrete	Rb	Recycled masonry of dense clay bricks with a limited amount of masonry mortar, from a demolished building near Sofia, Bulgaria
Bituminous materials	Ra	Ground asphalt from road surface
Floating materials	FL	Recycled materials fraction 0/31.5 mm from building carpentry (30 vol.%) and lightweight building plastics (70 vol.%)
Other: Cohesive (i.e. clay and soil) Miscellaneous: metals (ferrous and non-ferrous), non-floating wood, plastic and rubber Gypsum plaster	X	Lumps of sandy clay

Table 2. Blends of recycled materials

		Type 1 (reference)			Type 2 (from roads)			Type 3 (from buildings)		
Constituents		Rc100	Rb100	Ra100	Rc90 Ra10	Rc70 Rcu80 Ra20	Rc50 Rcu70 Ra30-X0.5	Rc90 Rb10-X0.5 FL10	Rc70 Rb30-X0.5 FL10	Rc30 Rb70-X0.5 FL10
Concrete and mortar	Rc, %	100			90	70	50	90	70	30
Natural stone	Ru, %					10	20			
Bricks	Rb, %		100					9.5	29.5	69.5
Asphalt	Ra, %			100	10	20	29.5			
Wood and plastics	FL, cm^3/kg							10	10	10
Clay	X, %						0.5	0.5	0.5	0.5

IOP Conf. Series: Materials Science and Engineering **1323** (2024) 012002 doi:10.1088/1757-899X/1323/1/012002

The European norm EN 13242 + A1 [13] specifies the properties of recycled aggregates obtained by processing recycled materials for hydraulically bound and unbound materials for civil engineering work and road construction. The main characteristics of Types 1, 2 and 3 aggregates were determined using the requested standardised methods.

To assess the performance of recycled materials and their ability to be used in base and sub-base courses, two additional tests were performed – Proctor compaction test according to EN 13286-2 [14] for reference density and optimum water content and test for the determination of California bearing ratio and linear swelling, according to EN 13286-47 [15].

3. Results and discussion

3.1 Geometrical properties of recycled materials

The grading of the studied aggregate (EN 933-1) [16] is presented in Figure 1 and Figure 2.

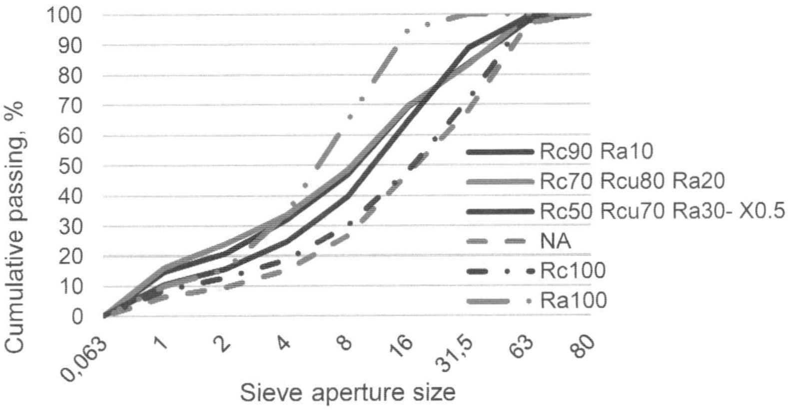

Figure 1 Comparison of grading curves between recycled materials of Type 1 (reference) and Type 2 (from roads), as described in Table 2.

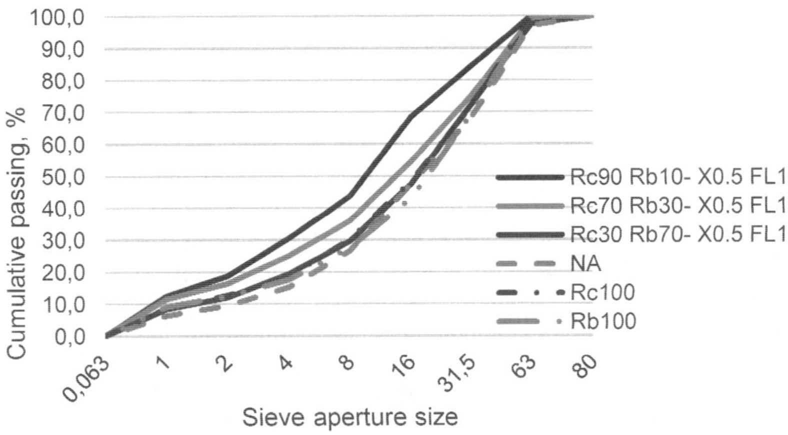

Figure 2 Comparison of grading curves between recycled materials of Type 1 (reference) and Type 3 (from buildings), as described in Table 2.

IOP Conf. Series: Materials Science and Engineering **1323** (2024) 012002 doi:10.1088/1757-899X/1323/1/012002

The crushing of different types of CDW is different and the grain size of the recycled materials can vary widely, which also affects the grading of the blends. Type 2 blended recycled aggregates are finer than the natural aggregate fraction 0/63 mm (NA) and recycled concrete aggregate of (Rc100) because of the presence of ground asphalt where the dominating fraction is 2/16 mm – Figure 1, but the rule of mixtures cannot be applied. The deviations within Type 3 blends (Figure 2) are relatively smaller than those of Type 2, but the rule of mixtures cannot be applied either. The impurities might then lead to additional deviations in the grading of recycled materials.

The coefficient of uniformity $Cu=d_{60}/d_{10}$ of the granular mixes is a representative parameter for their compatibility. For the majority of the recycled aggregate mixes the Cu values are around 16 (Table 3), which might be assessed as satisfactory [17], [18]. However, the very high value (Cu above 30) for the intermediate mix of Type 3 (Rc50 Rb30- X0.5 FL10) shows that quite large grain size deviations are possible for recycled materials.

Table 3 Geometrical properties of recycled materials and quality of the fine fraction

type	Cu (Coefficient of uniformity)	FI		SI		Fines content	SE (Sand equivalent)	MB (Methylene blue value)
	-	%	class	%	class	%	%	g/kg
NA	7.9	8	FI20	0	SI20	5	71	0.75
Rc100	15.8	4	FI20	0.3	SI20	6	76	0.5
Rb100	15.8	29	FI35	0.03	SI40	24	86	1
Rc90 Ra10	16.0	6	FI20	0.1	SI20	5	75	0.6
Rc70 Rcu80 Ra20	16.0	9	FI20	0.13	SI20	7	82	0.5
Rc50 Rcu70 Ra30 X0.5	16.0	9	FI20	0.3	SI20	8	82	0.5
Rc90 Rb10- X0.5 FL10	16.0	8	FI20	0.1	SI20	8	72	0.5
Rc50 Rb30- X0.5 FL10	31.5	12	FI20	0.1	SI20	9	70	1
Rc30- Rb70 X0.5 FL10	15.8	24	FI35	0.03	SI40	23	70	1.5

The shape of the grains of the aggregates is of great importance to the ability of the materials to be compacted. Flat and elongated particles are to be avoided because the compaction becomes more difficult and the compaction degree cannot be achieved. In terms of the flakiness index (FI, as per EN 933-3) [19], apart from the blend with predominant bricks fraction (Rc30- Rb70 X0.5 FL10), all other blends do not differ much from the natural aggregate NA and recycled concrete aggregate Rc100 – the percentage of flat particles is between 4% and 12% (Table 3), which allows classifying the recycled materials as of category FI20, according to the EN 13242 [13]. The reason for the higher flakiness index (FI35) of the blend with 70% bricks is namely the bricks aggregate, where the share of flat particles is 29%. The National Annex to EN 13242 [20] would allow all studied blends to be used in base courses of recycled materials even for heavy-load roads, for which the maximum FI shall be 35.

IOP Conf. Series: Materials Science and Engineering **1323** (2024) 012002 doi:10.1088/1757-899X/1323/1/012002

The shape index (SI), (determined as per EN 933-4) [21] of bricks and blend Rc30- Rb70 X0.5 FL10 is also twice higher (SI40) than that of other blends, Rc100 and NA, all having SI20 (Table 3). Again, according to [18], all recycled materials and blends can be used in base courses of heavy-load roads. The rotary crushers can produce particles having shapes closer to the cubic, so the industrially recycled materials with a higher share of bricks might have suitable SI and FI.

3.2 Fines content and quality of fine fraction

The limits in the fines content result from certain requirements for the mechanical and hydraulic performance of aggregate (e.g. drainage, swelling and stiffness) [22]. According to [20], for the heavily loaded pavements, the maximum fines content shall be 7% for the base courses of non-fractioned recycled material and 12% for the sub-base of non-fractioned material, when tested according to [16]. All recycled materials of Type 2 and Type 3, except the blend with 70% of bricks (Rc30- Rb70 X0.5 FL10) (Table 3), meet the requirements for the sub-base, but only Rc100, Rc90 Ra10 and Rc70 Rcu80 Ra20 meet the requirements for the base courses. Restrictions for the base course of lightly and medium-loaded roads are smaller, so the recycled concrete materials with impurities of less than 30%, can still be used there [20].

The sand equivalent (determined acc. to EN 933-8:2012) is used to qualify aggregates for applications where sand is desirable but fines and dust are not. The National Annex to EN 13242 [20] requires for the base course of heavy load roads to use recycled materials with sand equivalent of at least 35%. The values of all recycled materials for the sand equivalent are superior – they vary between 69.8% to 85.9% and for some of the blends are even higher than the value for NA (71.1%) – Table 3. The Bulgarian limits look quite low, but the measured SE values of recycled aggregates would satisfy the requirements of more exigent technical specifications (e.g. Greek, French), where the limits values for the unbound base course and subbase might reach 60% [23].

The Methylene Blue (MB) test is a widely used method for detecting the presence of clay minerals in aggregates. The values of MB as per EN 933-9 of the studied blends vary between 0.5 and 1.5 (Table 3). The higher values of 1 and 1.5 are representative of aggregates with a bigger share of brick aggregates (30%, 70% and 100% of the Rb component) [23]. The Bulgarian Technical specification for road construction [18] and the National Annex to EN 13242 [20] do not specify any limit values for MB of non-fractioned aggregate, but only sand equivalent values. However, according to Nikolaides A. et al [23], the sand equivalent test is not a reliable test to determine whether or not any aggregate material is suitable for pavement construction. As for the determination of suitability of a non-limestone aggregate, a Methylene Blue (MB) test must be performed and in case the SE criterion is not satisfied, the acceptance or rejection should be based on the methylene blue criterion.

3.3 Physical and mechanical properties

The significant variations in the apparent particle densities of the recycled aggregate (between 2690 kg/m^3 for NA, through 2330 kg/m^3 for concrete aggregate and 2000 kg/m^3 for brick aggregate) determine the large variety of the loose bulk density of the granular mixes, both measured as per EN 1097-6 [24] – Table 4. However, the rule of mixtures cannot be applied –the densities of mixtures are systematically lower than the calculated values, because of the heterogeneity of the recycled aggregate.

Table 4 Physical properties of the different aggregates

Type	Loose bulk density	Apparent particle density	WA_{24}	MS Frost resistance	Maximum dry density	Optimal moisture content
	kg/m³	kg/m³	%	%	kg/m³	%
NA	1670	2690	2.1	3.1	2250	2.6
Rc100	1410	2330	5.2	7.7	1950	9.4
Rb100	960	2000	16.0	73	1770	13.9
Ra100	1340	2440	2.4	-	-	-
Rc90 Ra10	1460	2130	5.0	7.5	1990	10.8
Rc70 Rcu80 Ra20	1530	2140	4.8	6.8	2000	8.7
Rc50 Rcu70 Ra30 X0.5	1490	2290	4.2	5.9	1990	7.6
Rc90 Rb10- X0.5 FL10	1390	2300	6.3	8.7	1940	10
Rc50 Rb30- X0.5 FL10	1260	2170	8.1	12	1900	11.1
Rc30- Rb70 X0.5 FL10	1130	2050	12.8	44.2	1760	11.6

The porosity of the cement paste in concrete aggregate leads also to approximately two and a half times bigger water absorption (WA_{24}) of Rc100, compared to that of the natural aggregate NA and asphalt aggregate Ra100, which does not exceed 2.4%. The water absorption of ceramic is the most pronounced and reaches 16% for Rb100. Logically, the WA_{24} of each aggregate mix is located between those extreme values (Table 4), but the rule of mixtures does not seem to be applicable. Apart from the influence of the materials themselves, the WA_{24} might also be affected by the variations in the recycled aggregate grading – the WA_{24} value is calculated as a weighted average of the water absorptions measured on the 3 fractions (0/4 mm, 4/31.5 mm, and 31.5/63 mm), but the different mixes have different shares of these three fractions.

Water absorption of aggregate impacts their frost resistance. The frost resistance has been assessed by the Magnesium sulfate test as per EN 1367-2 [25], often applied as alternative testing to the direct freezing-thawing cycles of aggregate. The Magnesium sulfate test results are summarised in Table 4. There is a huge variation in the mass loss values (MS) – it is the most expressed for bricks aggregate (Rb100) overpassing 70%, while for natural aggregate (NA) is only around 3%, and for concrete aggregate (Rc100) is more than 7.5%. These results confirm the findings in [9].The explanation of differences consists in the very high open porosity of bricks, determining a very big absorption of testing solution during the immerse phase, leading to an immense crystalline pressure on the ceramic walls of the voids during the drying stage. When the crystalline pressure exceeds the ceramics tensile strength, the cracks and disintegration of brick particles occur. The same mechanism is valid for concrete particles too, but an additional sulphate-induced destruction of cement paste in concrete aggregate increases the MS values [26]. The ground asphalt does not exhibit high water absorption and, as expected, the soundness of blends of asphalt and concrete aggregates even slightly increased. For the blends of Type 3, the results are also logical – with the increase of brick particles, the soundness decreases. All blends, having a mass loss not exceeding 25% would be classified as compliant with the Bulgarian requirements for unbound base courses of heavy-loaded roads. The specifications for base courses and subbases of not heavy load roads allow the use of aggregates with a soundness value of up to 35%. Thus, only the blend with predominant brick content (Rc30- Rb70 X0.5 FL10) is excluded from the possibilities to be used in base-courses [18].

The resistance to fragmentation as per EN 1097-2:2020 [27] is a critical parameter of the mechanical performance of aggregates. The Los Angeles value (LA) is representative of the aggregate toughness and its ability to withstand various forms of abrasion, including crushing degradation, and disintegration. The determined LA values are illustrated in Figure 3.

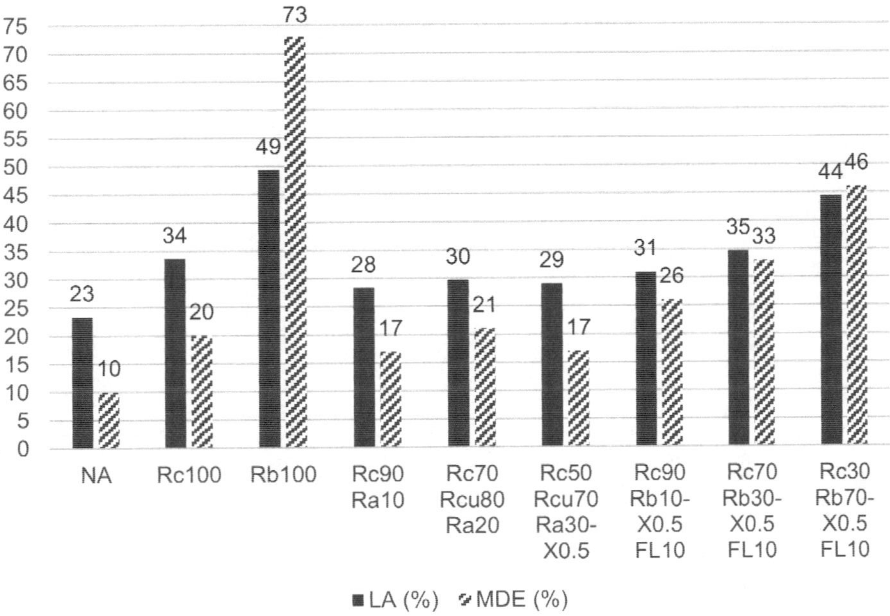

Figure 3 Resistance to fragmentation (LA) and resistance to wear (MD_E) of the different aggregates

They are calculated as a weighted average of LA values, measured for three standard fractions (4/8 mm, 8/11.2 mm, 11.2/16 mm) of each tested material. All recycled materials are less resistant to fragmentation than the reference natural aggregate (NA). The majority of recycled blends and Rc100 are of categories LA30 and LA35 and can be used in all unbound road bases. Because of the very friable recycled bricks (LA is 50 for Rb100), the blend with more than 30% recycled bricks (Rc30- Rb70 X0.5 FL10) is of category LA45 and is suitable rather for base courses of not heavily loaded roads [18], [20]. The recycled asphalt of a share between 10% and 30% in the aggregate mixes does not practically impact the LA of concrete dominant recycled aggregate.

The same trends in aggregate performance are observed with the resistance to wear, determined as per EN 1097-1 and evaluated by the Micro-Deval coefficient (MD_E) [28]. However, the negative effect of bricks is even more pronounced. The MDe of Rb100 is 7 times higher than MD_E of NA and more than 3.5 times than that of Rc100 - Figure 3. The bigger difference is due to the bigger number of rotations (10000 min^{-1}) compared to the LA test and the presence of water in MD_E test. So, the MD_E seems to be more sensitive to the mechanical performance of recycled aggregate than the LA. However, in the Bulgarian technical specifications, there are no MD_E limits.

The moisture-density curves (Proctor curves) are presented in Figure 4. They are an indication of the sensitivity of the density concerning the variations of moisture content for the materials [9].

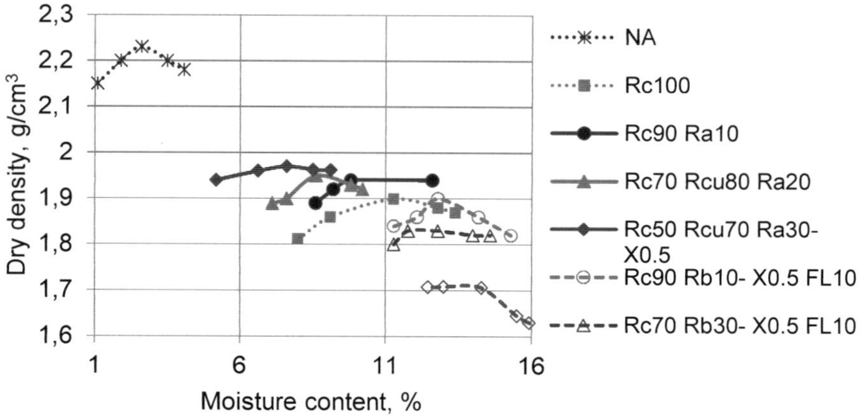

Figure 4 Proctor compaction curves of the different recycled aggregate

The shape of the curves and the obtained values for the maximum dry density (MDD) and optimum moisture content (OMC) differ considerably. The curves of recycled materials are generally flatter, meaning the MDD could be achieved even when there is a certain deviation from the OMC of aggregate. The differences in the MDD and OMC are mainly attributed to the physical properties of aggregate particles, but also to the relatively finer grading of recycled aggregates of Type 2. It was found that the MDD of the recycled concrete aggregate (Rc100) and blends with asphalt is about 12±1% lower than the MDD of NA (Table 4). For the mix with a significant brick content (Rc30- Rb70 X0.5 FL10), the MDD might be even below 1800 kg/m^3 (i.e. is reduced by more than 20%, compared to NA's MDD). As a result of the high water absorption of the crushed clay brick particles, the incorporation of more than 10% of recycled bricks increased significantly the OMC – up to 11.6%, while for the concrete aggregate (Rc100) and mixes with dominant concrete aggregate, the OMC is around 9% (9.3±1.2 %), still much higher than 2.6% for NA (Table 4).

The values of CBR, by which the bearing capacity of the base courses in a compacted state is directly evaluated, are presented in Figure 5. Apart from one blend of Type 2, containing 30% of ground asphalt (Rc50 Rcu70 Ra30 X0.5), all other values are above the threshold of 80%, required for the base course of heavy-loaded roads [18]. However, the CBR of 70% is enough to allow the use of that blend in roads with medium loads, where CBR shall be at least 50%. Despite the lower mechanical performance (in terms of LA and MDe) of Rc100 compared to NA, the CBR values of the two aggregates are practically identical, because of the good interlocking of concrete particles. The presence of the ground asphalt leads to CBR reduction by 8%, 17% and 53% for the blends Rc90 Ra10, Rc70 Rcu80 Ra20 and Rc50 Rcu70 Ra30 X0.5. The CBRdecrease in those blends is mainly due to the lower bonding between the bitument coated particles. Another reason might be related to the variations in the blends grading (Figure 1) and to the peculiarity of CBR testing being done on a fraction with a maximum size of 31.5 mm, which significantly increases the percentage of recycled asphalt in the sample.

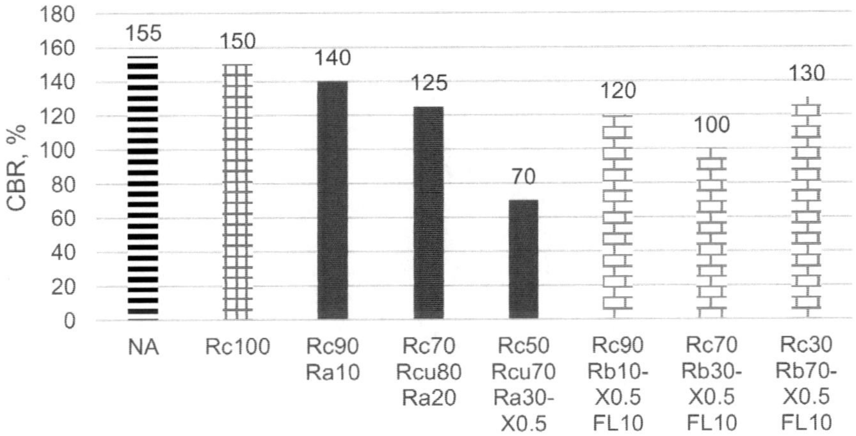

Figure 5 CBR of the different materials

A certain CBR decrease (of the order of 20±10%) of Type 3 aggregates compared to NA and Rc100 is observed. However, no relationship between the bricks content and the CBR values can be established, as stated in [4]. The highest CBR within Type 3 blends is determined for the mix Rc30- Rb70 X0.5 FL10 having predominant brick particles (70%), followed by the mix with 20% and the lowest CBR is for the blend with 30% of bricks. Our results confirm the findings of other researchers reporting relatively high CBR values when recycled bricks are present in the blends [11]. An additional contribution to the good mechanical behaviour of compacted blends of concrete and bricks aggregate might be the residual hydraulic activity of cement paste in concrete particles and pozzolanic activity of the brick fines [29].

4. Conclusions and future work

The results of the laboratory testing undertaken in this research have shown overall that the incorporation of impurities such as bricks up to 30% and ground asphalt up to 20% into crushed concrete has "low to minimal affect" on the physical and mechanical properties of the original material and does not compromise the ability of recycled aggregates to be used in civil engineering work and road construction. Some blends of Type 2 (CDW originating from roads and facilities) and Type 3 (CDW originating from buildings) were demonstrated to satisfactorily meet the current Bulgarian specification requirements for various purposes, including the most exigent such as the unbound base course of heavy-load roads. The other impurities such as timber, plastics, soil do not harm the aggregate if kept at low content, e.g. less than 0.5% for soil and non-floating timber, rubber, etc. and not more than 10 cm^3/kg of lightweight impurities (plastics, floating timber).

When the ground asphalt share is more than 20% but does not exceed 30%, the aggregates are still acceptable for road base courses of light and medium-load roads.

The maximum recommended brick impurities limit is also 30%, because of the lower performances of blends with a predominant brick fraction (70%) and mainly because of the unsatisfactory frost resistance, assessed by $MgSO_4$ test.

IOP Conf. Series: Materials Science and Engineering **1323** (2024) 012002 doi:10.1088/1757-899X/1323/1/012002

Therefore, a coarse on-site sorting of CDW would be sufficient to ensure a composition of CDW that would allow good-quality recycled materials with multiple possible uses in civil engineering works and road construction.

The performed work revealed that the rule of mixtures is not always able to predict the recycled aggregate blends properties. Some of the tests are not adapted to the peculiarities of the recycled aggregate, either. For instance, the $MgSO_4$ test is too severe for concrete recycled aggregates because $MgSO_4$ induces additional chemical corrosion of the old cement paste. Many tests such as Los Angeles test, Micro-Deval test, CBR are not performed on the entire 0/63mm fraction, but on specified sub-fractions. This approach might be relevant for uniform rock materials as in natural aggregates, but not for aggregates where the finer fractions differ a lot from the coarser fractions, both in terms of composition and performance.

In the future, within the project MOBICCON-PRO (2022-2025), funded within HORIZON-CL-2022-TWIN- TRANZITION-01-10 programme of the European Union, the findings will be verified by investigating whether the industrially recycled aggregate from various sources of CDW will satisfy the requirements of the technical specifications. The recommendations for uses of recycled materials will be demonstrated and validated in real-scale projects.

The use of recycled aggregates, especially in unbound applications, can be hindered by their unsatisfactory environmental performance. In addition to the chemical requirements of EN 13242, additional restrictions may be imposed by national legislation. In the second part of this article, the influence of impurities will be discussed, as well as the environmental risks because of lack of additional environmental criteria to the recycled aggregate.

Acknowledgments

The research presented in the paper was carried out in the framework of the Project 101091679 — MOBICCON-PRO, funded within HORIZON-CL-2022-TWIN- TRANZITION-01-10 programme of the European Union. The authors acknowledge the financial support of the project.

References

[1] Mankelow J., et all, 2021. Collation of the results of the 2019, Aggregate Minerals Survey for England and Wales. British Geological Survey Open Report, OR/21/024.
[2] Ministry of Environment and waters of Republic of Bulgaria, "National Strategic Plan on CDW management 2010-2020," Ministry of Environment and waters, Sofia, 2011.
[3] Mollenhauer K., et al, 2023, Cold recycled asphalt for pavements with optimized resource and energy efficiency: proposal for harmonized mix and pavement design, Transportation Research Procedia, Volume 72, Pages 3031-3038.
[4] He J., et all, 2024, Quantifying the agglomeration effect of reclaimed asphalt pavement on performance of recycled hot mix asphalt, Journal of Cleaner Production, Volume 442, 141044.
[5] A. Rao, K.N. Jha, S. Misra, 2007, Use of Aggregates from Recycled Construction and Demolition Waste in Concrete, Resources Conservation and Recycling, vol. 50, pp. 71-81.
[6] V.W.Y. Tam, 2009, Comparing the implementation of concrete recycling in the Australian and Japanese construction industries, J. Clean. Prod., 17 (7), pp. 688-702.
[7] Directive 2008/98/EC of the European Parliament and of the Council of 19 November 2008 on waste and repealing certain Directives, Waste Framework Directive, OJ L 312, 22.11.2008, pp. 3–30
[8] Vieira C., Pereira P., 2015, Use of recycled construction and demolition materials in geotechnical applications: A review, Resources, Conservation and Recycling, Volume 103, pp. 192-204.
[9] Poon C., Chan D., 2006, Feasible use of recycled concrete aggregates and crushed clay brick as unbound road sub-base, Construction and Building Materials, Volume 20, Issue 8, pp. 578-585.
[10] European Committee of Standardisation, EN 13286-2:2011 Unbound and hydraulically bond mixtures – Part 2: Test methods for laboratory reference density and water content – Proctor compaction.

[11] Aatheesan T. et al, 2009, Crushed brick blends with crushed concrete for pavement sub-base and drainage applications, Australian Geomechanics Journal, Vol 44, No 2, pp. 62-72.
[12] European Committee of Standardisation, EN 933-11:2009 Tests for geometrical properties of aggregates - Part 11: Classification test for the constituents of coarse recycled aggregate.
[13] European Committee of Standardisation, EN 13242:2002+A1:2007 Aggregates for unbound and hydraulically bound materials for use in civil engineering work and road construction.
[14] European Committee of Standardisation, EN 13286-2:2011 Unbound and hydraulically bound mixtures - Part 2: Test methods for laboratory reference density and water content - Proctor compaction.
[15] European Committee of Standardisation, EN 13286-47:2022 Unbound and hydraulically bound mixtures - Part 47: Test method for the determination of California bearing ratio, immediate bearing index and linear swelling.
[16] European Committee of Standardisation, EN 933-1:2012 Tests for geometrical properties of aggregates – Part 1: Determination of particle size distribution – Sieving method.
[17] Dimter S. et al, Impact of different wood ash fractions on the load-bearing capacity of sand mixtures, Proceedings of CETRA 7th International Conference on Road and Rail Infrastructure, May 2022, Croatia, pp. 449-456.
[18] Ministry of Regional Development and Public Works of Bulgaria, "Technical specification of the Road Infrastructure Agency," Ministry of Regional Development and Public Works, Sofia, 2014.
[19] European Committee of Standardisation, EN 933-3:2012 Tests for geometrical properties of aggregates - Part 3: Determination of particle shape - Flakiness index.
[20] Bulgarian Institute for Standardization, BDS EN 13242:2002+A1:2007/NA: 2017 Aggregates for unbound and hydraulically bound materials for use in civil engineering work and road construction, Sofia, 2017.
[21] European Committee of Standardisation, EN 933-4:2008 Tests for geometrical properties of aggregates - Part 4: Determination of particle shape - Shape index.
[22] Hatipoglu M., et all, 2019, Effects of Fines Content on Hydraulic and Mechanical Performance of Unbound Granular Base Aggregates, Journal of Transportation Engineering, Part B: Pavements, Volume 156, Issue 1.
[23] Nikolaides A, et all, 2007, Sand equivalent and methylene blue value of aggregates for highway engineering, Foundations of Civil and Environmental Engineering, Volume 10.
[24] European Committee of Standardisation, EN 1097-6:2022 Tests for mechanical and physical properties of aggregates - Part 6: Determination of particle density and water absorption.
[25] European Committee of Standardisation, EN 1367-2:2009 Tests for thermal and weathering properties of aggregates - Part 2: Magnesium sulfate test.
[26] Chini AR, Kuo SS, Armaghani JM, Duxbury JP. Test of recycled concrete aggregate in accelerated test track. J Transp Eng 2001; 127(6): 486–92.
[27] European Committee of Standardisation, EN 1097-2:2020 Tests for mechanical and physical properties of aggregates - Part 2: Methods for the determination of resistance to fragmentation.
[28] European Committee of Standardisation, EN 1097-1:2011 Tests for mechanical and physical properties of aggregates - Part 1: Determination of the resistance to wear (micro-Deval).
[29] Likes L. et al, Recycled concrete and brick powders as supplements to Portland cement for more sustainable concrete, Journal of Cleaner Production 364 (2022) 132651.
[30] Vegas I. , Ibañez J. et al. 2011, Pre-normative research on the use of mixed recycled aggregates in unbound road sections, Construction and Building Materials, Volume 25, Issue 5, pp. 2674-2682, ISSN 0950-0618.

IOP Conf. Series: Materials Science and Engineering **1323** (2024) 012003 doi:10.1088/1757-899X/1323/1/012003

Impact of Impurities on the Properties of Recycled Aggregate for Use in Civil Engineering Work and Road Construction. Part 2: Environmental Assessment

Roumiana Zaharieva[1*], Daniel Evlogiev[1], Borislav Simonov[1], Magdalena Kostadinova[1] and Boyan Petrov[1]

[1] Department of Building Materials and Insulations, University of Architecture, Civil Engineering and Geodesy, Sofia, Bulgaria

*E-mail: zaharieva_fce@uacg.bg

Abstract. Recycled concrete aggregate fraction 0/63 mm is used as unbound materials in several construction works, for which the environmental soundness is of a great importance. The present study aims to evaluate the influence of impurities such as asphalt, bricks, soil, and lightweight materials on the environmental properties of recycled concrete aggregate. The chemical characteristics are determined as per EN 13242+A1 requirements. The environmental performance is evaluated based on a batch leaching test according to EN 12457-1 (L/S =2) and on the content of organic materials such as PAHs, PCBs, C10-C40. The two sets of tests were found to give conflicting results: no problematic chemical characteristics were found even at a high percentage of impurities in the recycled concrete aggregate, while the environmental performance was not satisfactory; the recycled aggregates were found not inert because of the significant amounts of sulphates, chlorides and total dissolved solids, sometimes heavy metals and, when blended with ground asphalt, organic compounds such as hydrocarbons. Thus, backfills and road facilities can be a source of soil and groundwater pollution if a pre-demolition audit, identification of contaminated CDW, decontamination activities and selective demolition are not implemented. Some improvements in the regulatory framework are proposed.

1. Introduction

Recycled concrete aggregate from construction and demolition waste (CDW) has been used for many years as a replacement for natural aggregate in civil engineering works and road construction because of its excellent mechanical properties. Reclaimed asphalt is also being used in pavements [1]. According to [2], the main outputs of recycled aggregate include substitution of virgin aggregates in unbound applications, such as the construction of roads, parking areas, and embankments. During the last two decades more and more transportation authorities have included recycled aggregate in their technical specifications and design manuals [3], including in Bulgaria [4, 5].

The use of recycled aggregate has multiple benefits in the sustainability context – savings of natural resources and landfills capacities, recovery of huge amount of CDW, reduced transportation-related environmental impacts, etc. However, due to the presence of hazardous substances in construction materials and contamination during the exploitation of buildings, the recycled aggregate from CDW might have some negative impacts on the environment; particularly the leaching of chemical pollutants have been a critical concern [1, 2, 6]. Vegetation die-off due to high-pH and high concentration of metals leached from recycled concrete aggregate are reported. Furthermore, the leaching of organic compounds from recycled asphalt is also a potential consideration. The leached chemical pollutants can potentially cause human health impacts via direct exposure or consumption when mixed with ground and surface water resources [1]. Leaching from recycled materials is a complex process and can be influenced by a range of factors, which can be categorised as physical, chemical and external.

The magnitude of this risk depends on the source of CDW, in situ identification of hazardous components, implementation of selective demolition, recycling technology, different uses – in bound or unbound materials, water content (denoted by liquid to solid ratio L/S), pH, DOC in seeping water, particle size and in compaction degree, presence of coatings and impermeable layers, etc. [1, 7].

The European standard EN 13242 +A1 [8] considers recycled aggregate as equal to natural ones, but sets some additional requirements to be met, for example, to declare the category of recycled aggregate according to EN 933-11 [9] and to control some properties more often, including those related to the presence of harmful substances. However, only a few harmful substances are listed in EN 13242+A1 [8], mainly related to the sulphates (acid-soluble sulphates and water-soluble sulphates) and total sulphur content. For the unbound uses of aggregate, the control of impurities affecting the setting and hardening rates of hydraulically bound mixtures is not relevant. The risk of releasing heavy metals (by leaching) or other hazardous substances is only broadly considered and no test methods are specified. The standard refers to the national regulations of the member countries. In some countries with advance CDW management, such as Denmark, the stress is put on the pre-demolition audit, decontamination of buildings, and characterisation of CDW (hazardous or non-hazardous), thus ensuring high quality, including environmental, of the recycled materials [10]. In the UK, the quality protocol for inert waste has enabled the greater use of recycling aggregates by providing a consistent and thorough methodology for establishing when it ceases to become a waste [11]. In Austria, the following environmental parameters of aggregates shall be tested: pH value, electric conductivity, contents of chromium (total), copper, ammonium, nitrite, sulphates, index of carbon hydrides, sum of 16 polyaromatic hydrocarbons [7]. Depending on the values of those parameters, four classes of environmental compatibility (A+, A, B and C) of recycled aggregates are specified and recommendations of use are made, according to the application form and field of the aggregates [7] – Table 1.

In Bulgaria, the National Annex to EN 13242 [4] which regulates the uses of aggregates according to their technical performances is not more specific than the European standard in terms of the hazardous compounds – neither for natural, nor for recycled aggregates, and does not prescribe any additional requirements. The Technical Specification issued in 2014 by the Road Infrastructure Agency [5], contains a part about thresholds for sulphates, as well as limits for the water-soluble chlorides.

At the same time, the main Ordinance on CDW management [12] allows pre-treated inert CDW to be used in backfilling. According to the Commission Decision 2011/753/EU [13] backfilling is defined as a recovery operation where suitable waste is used for reclamation purposes in excavated areas or for engineering purposes in landscaping and where the waste is a substitute for non-waste materials. There is no specification of the type of waste, other than it

IOP Conf. Series: Materials Science and Engineering **1323** (2024) 012003 doi:10.1088/1757-899X/1323/1/012003

has to be suitable, meaning appropriate for the purpose and without causing environmental harm.

Table 1. Environmental Engineering Applicable Fields (Minimum Requirements), as per [7]

Form of application	Hydro-geological delicate area	Hydro-geological less delicate area	Within the non-hazardous waste landfill body
Unbound without cover layer	Quality class A^+	Quality classes A^+, A	Quality classes A^+, A, B, C
Unbound with cover layer or in bound form with or without cover layer	Quality classes A^+, A	Quality classes A^+, A, B	Quality classes A^+, A, B, C
Aggregate	Quality classes A^+, A, B	Quality classes A^+, A, B	Quality classes A^+, A, B, C

In the Bulgarian legislation, the pre-treatment is understood as crushing and screening according to the design grading requirements for the backfilling material. The pre-treated CDWs remain waste and are not construction products in terms of the Regulation EC 305/2011 [14], as the recycled aggregates are. The inert character shall be proven by testing only if CDW originates from potentially contaminated sites as specified in Annex 10 to [12]. The limits of hazardous compounds and the degree contamination, which might be acceptable for the inert waste, are specified only in the Ordinance for Landfills [15]. There, Table 1 in Appendix 1, provides a list of waste, which might be accepted in landfills for inert waste without testing, among which are the CDW of codes 17 01 01 (concrete), 17 01 02 (bricks), 17 01 03 (tiles and ceramics), 17 01 07 (mixtures of concrete, bricks, tiles and ceramics other than those mentioned in 17 01 06), 17 02 02 (glass) and 17 05 04 (soil and stones other than those mentioned in 17 05 03) as per the European List of Waste [16], when they have not been contaminated or treated by hazardous substances. The common practice in Bulgaria does not include an appropriate system for the traceability of CDW and practically all CDWs of concrete, bricks, tiles, glass and rock materials are considered inert and can be used "for engineering purposes in landscaping and where the waste is a substitute for non-waste materials" [17, 18].Despite some significant differences between the landfills for inert waste, where some requirements for the degree of the bottom's impermeability exist [19], and the backfills where there are no such requirements, the limits for landfills in [15], if tested at all, are applicable as environmental criteria to prove the inert character of CDW which will be used for backfilling. Another contradiction is that when the backfilling is executed with recycled aggregate conforming to EN 13242 [8], there is no need to prove the inert character of recycled aggregate; thus, the environmental criteria might not be satisfied.

It has been concluded that CDW is a highly heterogeneous material, and the very limited information about material characteristics and potential leaching properties of the CDW being recycled in construction works offers a poor basis for assessing the environmental consequences of the recycled aggregates utilisation [6].

The present study aims to evaluate the influence of non-hazardous impurities such as bricks, soil, non-tar-containing asphalt and lightweight materials on the environmental properties of recycled concrete aggregate. Consequently, the technical criteria might be amended by an environmental assessment when the most appropriate unbound uses of recycled aggregates are to be defined.

2. Materials and methods

2.1. Preliminary studies

Two main sources of CDW are considered:

- CDW generated from road rehabilitation and demolition of reinforced concrete facilities;
- CDW generated from the rehabilitation and demolition of buildings having bearing structures from reinforced concrete or masonry.

A preliminary study was performed on the constituents of the crushed CDW from 12 various sources – 3 from building demolition, 4 from building refurbishment, 1 from road rehabilitation, 1 from facility reconstruction, 2 from facilities demolition and 1 from facilities rehabilitation. Three of the crushed CDWs were sourced from a recycling plant near Sofia. The other CDWs were sampled from construction sites. When the CDW sample was too coarse, it was additionally crushed in a laboratory to a size below 31.5 mm. The classification of the CDW constituents was performed according to EN 933-11 on the fraction 4/31.5 mm (Table 2).

Table 2. Components of the sampled CDW, determined as per EN 933-11 on a fraction 4-31.5mm.

Cource of CDW	Building refurbishment	Building refurbishment	Building refurbishment	Road rehabilitation	Multi-storey parking demolition	Building refurbishment	Building Demolition	Building Demolition	Building demolition	Dam wall rehabilitation	Railway sleepers removal	Concrete pavement demolition
Samle No	112-1	112-2	112-3	112-4	112-5	22-2	16663	16664	16665	124-2	124-3	124-4
Rc - Concrete and mortar, %	44.7	87.3	12.2	-	100	100	50.5	64.6	50.8	100	100	73.7
Ru - CA, stone and CA with cement mortar, %	-	-		-	-	-	44.3	19.9	26.6	-	-	18.4
Rb - Bricks and ceramics, %	55.3	12.7	87.8	-	-	-	3.8	12.1	12.8	-	-	0.4
Ra - Bitumen materials, %	-		-	100	-	-	0.9	1.9	2.1	-	-	4.8
Rg – glass, %	-		-	-	-	-	0	0	0	-	-	-
FL - Floating materials, cm^3/kg	2.3	1.8	1.5	-	-	-	10.0	5.7	6.1	-	-	1.1
X - Clay, soil, metals, non floating wood, plastic materials and rubber, gypsum, %	-	-	-	-	-	-	0.5	0	0.4	-	-	0.3

Based on the investigations of the Bulgarian demolition and refurbishment practices [20], it is assumed that no pre-demolition audit has been performed and only some of the hazardous components of the CDW (such as asbestos-containing materials) have been removed before the construction or demolition work, but not potentially contaminated paint, plasters, coatings, sealings, etc. [10, 21]. It seems the common practice in Bulgaria does not imply selective demolition because of higher costs, although the selective demolition has been introduced as compulsory by the national legislation for more than 10 years [17]. A rough sorting of CDW is done in situ, mainly for separating the valuable CDW such as metals.

It has been found that the CDW originating from roads and facilities consists mainly of concrete. When asphalt pavement is present, usually the asphaltic layer is ground and removed separately, but still some asphaltic impurities might be present (up to 10%). In some cases, however, no separate collection is performed and then the asphalt impurities might be nearly 30%. When a deep reconstruction is performed, or removal of the pavement is done, the CDW might contain a certain amount of natural stone from the base and sub-base courses, but not exceeding 20%, as well as a marginal quantity of soil, because the natural aggregate is usually processed separately – Table 2.

The composition of CDW from buildings is the most diverse. Even in those with a reinforced concrete load-bearing structure, there are impurities of bricks, up to 30%, due to the typical for Bulgaria construction of enclosures and partition walls. In buildings with a load-bearing structure made of masonry, the bricks are the dominant fraction, but there are impurities up to 30% of masonry mortars, plasters, renders and concrete elements. Visually, the amount of lightweight impurities such as plastics (pipes, insulating materials) and timber (carpentry) can be assessed as quite high, but due to their low density, the mass share is less than 0.5%. When tested and expressed as recommended per EN 933-11 [9], the share of floating constituents does not overpass 10 cm^3/kg.

2.2. Materials

Fraction 0/63 mm has been chosen as a representative for the recycled aggregate to be used in an unbound state in civil engineering work and road construction [22, 23]. As it has been proven in Part 1 of the present paper [24], the technical assessment revealed that when the impurities such as bricks, ground asphalt, plastics, and timber do not exceed 30%, the performance of recycled aggregates might be decreased, but there are still convenient fields of application of those aggregates, at least for backfilling. Moreover, since Bulgarian legislation allows backfilling to be done with inert CDW, without further recycling and conformity assessment as aggregate, it is this fraction that is used most in backfilling and landscaping.

Based on the classification results of CDW, three types of laboratory samples of recycled concrete aggregates fraction 0/63 mm were produced (Table 3), combining different amounts of impurities, thus aiming to cover "worst case scenarios" when the separate collection of CDW is not done or only partially done:

- Type 1 consists of single-component CDW: concrete aggregate only (Rc100), brick masonry aggregate only (Rb100) and ground asphalt (Ra100). This type serves mainly as a basis of the aggregate blends of Types 2 and 3. However, the aggregate Rc100 would be representative of the demolition and repair works of reinforced concrete facilities (e.g. dam wall) and elements (e.g. railway sleepers).
- Type 2 contains recycled concrete with impurities of natural stone (10% and 20%), ground asphalt (10%, 20% and 30%) and soil (0.5%). Type 2 aggregate would be representative of the recycled materials from roads and pavements;
- Type 3 contains up to 70% of recycled bricks, 0.5% of soil, and quite a big amount of floating materials – 10 cm^3/kg. The source of floating materials are wooden elements, carpentry, plastic pipes, and various insulation materials. Type 3 blends would result from the demolition and rehabilitation works of different buildings.

The floating materials (FL) were composed of laboratory crushed CDW from building carpentry (30 vol.%) and lightweight building plastics (70 vol.%) to a fraction 0/31.5 mm. The other constituents marked with X are represented by lumps of sandy clay.

IOP Conf. Series: Materials Science and Engineering **1323** (2024) 012003 doi:10.1088/1757-899X/1323/1/012003

Table 3 Recycled aggregates composition

Constituents as per EN 933-11 [9]		Type 1 (reference)			Type 2 (from roads)			Type 3 (from buildings)		
Description	Designation and share	Rc100	Rb100	Ra100	Rc90 Ra10	Rc70 Rcu80 Ra20	Rc50 Rcu70 Ra30- X0.5	Rc90 Rb10- FL10 X0.5	Rc70 Rb30- FL10 X0.5	Rc30 Rb70- FL10 X0.5
Concrete, concrete products, mortar Concrete masonry units	Rc, %	100			90	70	50	90	70	30
Unbound aggregate, natural stone Hydraulically bound aggregate	Ru, %					10	20			
Clay masonry units (i.e. bricks and tiles) Calcium silicate masonry units Aerated non-floating concrete	Rb, %		100					9.5	29.5	69.5
Bituminous materials	Ra, %			100	10	20	29.5			
Floating materials (wood and plastics)	FL, cm^3/kg							10	10	10
Other: Cohesive (i.e. clay and soil) Miscellaneous: metals (ferrous and non-ferrous), non-floating wood, plastic and rubber, Gypsum plaster	X, %						0.5	0.5	0.5	0.5

2.3. Test methods

The chemical characteristics of aggregates as per EN 13242+A1 [8] (acid-soluble sulfates, total sulfur content and water-soluble sulfates) were determined. The tests were carried out according to BDS EN 1744-1+A1 [25].

The risk of releasing heavy metals and other environmental performance were evaluated based on a batch leaching test according to EN 12457-1 [26] with a "Liquid to Solid Ratio equal to 2" (L/S=2). For the unbound uses of recycled materials, where they are usually in a compacted state and might even be covered by more impermeable layers (asphalt, concrete), the

IOP Conf. Series: Materials Science and Engineering **1323** (2024) 012003 doi:10.1088/1757-899X/1323/1/012003

ratio L/S=2 was assessed as more representative than the ratio L/S=10, usually applied in Bulgaria for the waste characterization [2, 27]. Only for a few samples, for comparison purposes only, the tests were doubled as the ratio L/S =10 [28]. The following elements contents were measured as per EN ISO 11885 [29]: Arsenic – As, Barium – Ba, Cadmium – Cd, Chromium – Cr, Copper – Cu, Mercury – Hg, Molybdenum – Mo, Nickel – Ni, Lead – Pb, Antimony – Sb, Selenium – Se, Zinc– Zn. The fluorides and sulphates were determined as per EN ISO 10304-1 [30]. The other relevant environmental parameters were determined as follows: dissolved organic carbon (DOC) and total organic carbon (TOC) as per EN 1484 [31], total soluble solids as per EN 15216 [32], polycyclic aromatic hydrocarbons (PAH) as per EN 15527 [33] and nonpolar hydrocarbons C10-C40 and C10-C22 as per EN 14039 [34]. Concerning PAHs, since it has been reported that the PAHs in non-asphalt containing CDW did not seem to be critical [6], only Ra100 and blends of Type 2 were tested.

All tests were performed in accredited laboratories and according to relevant methods, but with a different precision, still satisfactory as required. The content of elements and substances was expressed in mg/kg of the dry mass.

For the assessment of the environmental risks, the limit values for inert and non-hazardous waste specified by the Bulgarian Ordinance of Landfills [15] are applied. The test results, for the harmful substances content, are also discussed in the context of the Danish CDW management basis [10], EN 13242 requirements [4] and Austrian guidelines for recycled aggregate [7]. Some of the applicable limits are summarized in Table 4.

3. Results and discussion

3.1. Assessment of the inert character of the selected CDW

The leaching test results on the sampled CDW are presented in Table 5. The results are further discussed in the context of the Ordinance for Landfills [15] and the values are compared to the limits given in Table 4.

Only one CDW sample (112-1) fails, in terms of sulphates content, the threshold for non-hazardous waste, but since the TDS value remains within the limits, after conducting additional tests it might still be assessed as non-hazardous. The same sample, originating from building refurbishment works (removal of painted render and repair of an external brick wall which has suffered frost damages), has a relatively high content of Mo and fluorides, both more than for inert, but still within the limits for non-hazardous waste.

Two more samples (112-2 and 22-2), both originating from buildings refurbishment, do not have inert character because of both higher sulphates and TDS contents. The plaster on masonry walls might be the main source of sulphates in these two samples, as well as in 112-3, where the sulphates are also above the limit of 560 mg/kg. Surprisingly, higher sulphates content was also measured in the CDW sample from the parking demolition (112-5) where no plaster was present. This CDW has been chosen (prior to the environmental testing) as a basis for the blends of Type 2 and Type 3.

A significant content of chlorides was found in the CDW sample from the removal of concrete tiles pavement (124-4). Most probably, the chlorides originate from de-icing salts, traditionally used during the winter periods on the sidewalks.

Table 4. Limit values for the content of harmful substances in inert and non-hazardous CDW and in recycled materials

Environmental parameter/Harmful elements and substances	According to the Bulgarian Ordinance for Landfills [15] at L/S = 2 l/kg (L/S=10 l/kg)*		According to EN 13242 [4]	According to the Austrian Guidelines for recycled aggregate [7]	
	For inert CDW [mg/kg]	For non-hazardous CDW with low organic content [mg/kg]	[mg/kg]	For Quality class A+, CDW originating from Civil Engineering Works	For Quality class C, CDW originating from Buildings
pH value	-	-	-	7.5÷12.5	7.5÷12.5
Electric conductivity	-	-	-	150 mS/m	250 mS/m
Arsenic – As	0.1 (0,5)	0.4 (2)	-	-	-
Barium – Ba	7 (20)	30 (100)	-	-	-
Cadmium – Cd	0.03 (0,04)	0.6 (1)	-	-	-
Chromium – Cr (total)	0.2 (0,5)	4 (10)	-	0.3 mg/kg	1.5 mg/kg
Chromium – Cr-VI	-	-	-	-	-
Copper – Cu	0.9 (2)	25 (50)	-	0.5 mg/kg	5 mg/kg
Mercury – Hg	0.003 (0,01)	0.05 (0.2)	-	-	-
Molybdenum – Mo	0.3 (0,5)	5 (10)	-	-	-
Nickel – Ni	0.2 (0,4)	5 (10)	-	-	-
Lead – Pb	0.2 (0,5)	5 (10)	-	-	-
Antimony – Sb	0.02 (0,06)	0.2 (0.7)	-	-	-
Selenium – Se	0.06 (0,1)	0.3 (0.5)	-	-	-
Zinc– Zn	2 (4)	25 (50)	-	-	-
Ammonium - N	-	-	-	1 mg/kg	30 mg/kg
Nitrite - N	-	-	-	0.5 mg/kg	8
Chlorides	550 (800)	10000 (15000)	-	-	-
Fluorides	4 (10)	60 (150)	-	-	-
Sulphates	560 (1000)	10000 (20000)		1500 mg/kg	6000 mg/kg
Sulphates – SO_4 (water soluble)			2000	-	-
Sulphates – SO_4 (acid soluble)	-	-	2000	-	-
Sulphur - S	-	-	10000	-	-
Dissolved organic carbon - DOC	240 (500)	380 (800)	-	-	-
Total dissolved solids - TDS	2500 (4000)	40000 (60000)	-	-	-
Index of carbon hydrides	-	-	-	1 mg/kg	40 mg/kg
Total organic carbon - TOC	30000	Not specified	-	-	-
Polychlorinated biphenyls - PCBs	1	Not specified	-	-	-
Total Petroleum Hydrocarbons (TPH) (C10 ÷ C40)	500	Not specified	-	-	-
Polycyclic aromatic hydrocarbons – PAHs (total)	700	Not specified	-	4 mg/kg	25 mg/kg

*For the value in brackets, the eluate has to be produced according to EN 12457-4 [28]

Two out of three samples from building demolition were also characterized as non-inert because one of them (16663) has a bigger amount of Cr (total) and Sb, while in the second (16665) the TDS exceeded the threshold of 2500 mg/kg.

Thus, only 4 out of 12 samples met the requirements for inert CDW, in terms of their leaching values. However, since the CDW sample 112-4 originated from the rehabilitation of a road with asphaltic concrete pavement and the major CDW component was ground asphalt (Table 4), the content of organic compounds was also measured. It was found that the total petroleum hydrocarbons (C10 ÷ C40) content of 1500 mg/kg exceeded 3 times the threshold for inert waste (500 mg/kg) as per [15]. The other assessed parameters (DOC, TOC and PAHs) do not exceed the limits, but it shall be mentioned that the PAH's limit of 700 mg/kg was extremely high, compared to other regulations [6]. PCBs have not been identified.

Table 5. Leaching test results for the different CDW samples

Source of CDW	Building refurbishment	Building refurbishment	Building refurbishment	Road rehabilitation	Multi-storey parking demolition	Building refurbishment	Building Demolition	Building Demolition	Building demolition	Dam wall rehabilitation	Railway sleepers removal	Concrete pavement demolition
Samle No	112-1	112-2	112-3	112-4	112-5	22-2	16663	16664	16665	124-2	124-3	124-4
Arsenic – As, [29], mg/kg	<0.1	<0.1	<0.1	<0.1	<0.1	<0.5	0.043	<0.02	<0.002	<0.1	<0.1	<0.1
Barium – Ba, [29], mg/kg	<1	<1	<1	<1	<1	<1	0.5	2.6	4.1	<1	<1	<1
Cadmium – Cd, [29], mg/kg	<0.03	<0.03	<0.03	<0.03	<0.03	<0.04	<0.002	<0.002	<0.002	<0.03	<0.03	<0.03
Chromium – Cr (total), [29], mg/kg	<0.2	<0.2	<0.2	<0.2	<0.2	<0.2	**0.22**	0.12	0.055	<0.2	0.025	<0.2
Copper – Cu, [29], mg/kg	<0.5	<0.5	<0.5	<0.5	<0.5	<0.5	0.098	0.021	0.079	<0.5	<0.5	<0.5
Mercury – Hg, [35], mg/kg	<0.01	<0.01	<0.01	<0.01	<0.01	<0.01	<0.001	<0.001	<0.001	<0.01	<0.01	<0.01
Molybdenum – Mo, [29], mg/kg	**2.3**	<0.3	<0.3	<0.3	<0.3	<0.5	0.06	0.04	0.023	<0.3	<0.3	<0.3
Nickel – Ni, [29], mg/kg	<0.2	<0.2	<0.2	<0.2	<0.2	<0.2	<0.004	<0.004	<0.004	<0.2	<0.2	<0.2
Lead – Pb, [29], mg/kg	<0.2	<0.2	<0.2	<0.2	<0.2	<0.5	0.031	0.077	0.033	<0.2	<0.2	<0.2
Antimony – Sb, [29], mg/kg	<0.02	<0.02	<0.02	<0.02	<0.02	<0.05	**0.04**	0.017	0.019	<0.02	<0.02	<0.02
Selenium – Se, [29], mg/kg	<0.06	<0.06	<0.06	<0.06	<0.06	<0.1	<0.02	<0.02	<0.02	<0.06	<0.06	<0.06
Zinc– Zn, [29], mg/kg	1.5	<0.5	<0.5	<0.5	<0.5	<0.5	0.011	0.007	0.005	<0.5	<0.5	<0.5
Chlorides, [30], mg/kg	267	868	<100	<100	<100	<100	13.9	58	10.8	<100	<100	**1629**
Fluorides, [30], mg/kg	**12**	<4	<4	<4	<4	-	0.37	0.43	0.4	<4	<4	<4
Sulphates – SO_4 [30], mg/kg	**11235**	**681**	**741**	<100	**1116**	**2495**	119	18.6	10.6	238	<100	159
Total dissolved solids - TDS, [32], mg/kg	**18939**	**9818**	2250	307	2389	**2581**	1608	2280	**2968**	328	808	1907

So, only 3 of the selected CDW satisfy the requirements of the Bulgarian legislation to be used in backfilling. However, since none of the 12 CDW samples originates from potentially contaminated sites, in general, according to the Ordinance on CDW management [12], none will be subject to compulsory testing to prove that the CDW is inert.

3.2. Environmental assessment of recycled aggregates.

When the recycling of mineral CDW includes only crushing and screening, there are no means of converting non-inert CDW into recycled materials with no environmental impacts. The leaching test results of the recycled materials are presented in Table 6.

The recycled aggregates Rc100 and Ra100 (Table 6) of Type 1 are chemically identical to the CDW of samples 112-3, 112-4 and 112-5 respectively (Table 5), because the aggregates fraction 0/63 mm results from the mechanical laboratory recycling of those CDW. As it has been proven in item 3.1., they are not inert – Rc100 and Rb100 because of the bigger amount of leached sulphates, while Ra100 because of the much higher TPH (C10-C40) content than the limit value for inert waste.

Table 6. Environmental parameters of the recycled aggregates

Substance, [test method], units	Type 1 (reference)			Type 2 (from roads)			Type 3 (from buildings)		
	Rc100	Rb100	Ra100	Rc90 Ra10	Rc70 Rcu80 Ra20	Rc50 Rcu70 Ra30-X0.5	Rc90 Rb10-X0.5 FL10	Rc70 Rb30-X0.5 FL10	Rc30 Rb70-X0.5 FL10
Tests performed according to EN 12457-1 [26]									
Arsenic – As, [29], mg/kg	<0.1	<0.1	<0.1	<0.1	<0.1	<0.1	<0.1	<0.1	<0.1
Barium – Ba, [29], mg/kg	<1	<1	<1	<1	<1	<1	<1	<1	<1
Cadmium – Cd, [29], mg/kg	<0.03	<0.03	<0.03	<0.03	<0.03	<0.03	<0.03	<0.03	<0.03
Chromium – Cr (total), [29], mg/kg	<0.2	<0.2	<0.2	<0.2	<0.2	<0.2	<0.2	<0.2	<0.2
Copper – Cu, [29], mg/kg	<0.5	<0.5	<0.5	<0.5	<0.5	<0.5	<0.5	<0.5	<0.5
Mercury – Hg, [35], mg/kg	<0.01	<0.01	<0.01	<0.01	<0.01	<0.01	<0.01	<0.01	<0.01
Molybdenum – Mo, [29], mg/kg	<0.3	<0.3	<0.3	<0.3	<0.3	<0.3	<0.3	<0.3	<0.3
Nickel – Ni, [29], mg/kg	<0.2	<0.2	<0.2	<0.2	<0.2	<0.2	<0.2	<0.2	<0.2
Lead – Pb, [29], mg/kg	<0.2	<0.2	<0.2	<0.2	<0.2	<0.2	<0.2	<0.2	<0.2
Antimony – Sb, [29], mg/kg	<0.02	<0.02	<0.02	<0.02	<0.02	<0.02	<0.02	<0.05	<0.02
Selenium – Se, [29], mg/kg	<0.06	<0.06	<0.06	<0.06	<0.06	<0.06	<0.06	<0.1	<0.06
Zinc– Zn, [29], mg/kg	<0.5	<0.5	<0.5	<0.5	<0.5	<0.5	<0.5	<0.5	<0.5
Chlorides, [30], mg/kg	<100	<100	<100	<100	<100	<100	<100	200	<100
Fluorides, [30], mg/kg	<4	<4	<4	<4	<4	<4	<4	<4	<4
Sulphates – SO_4 [30], mg/kg	**1116**	**741**	<100	**1050**	**807**	**606**	**1158**	**1125**	**784**
Total dissolved solids - TDS, [32], mg/kg	2389	2250	307	1895	1610	930	2064	2111	899
Tests performed according to EN 12457-4 [28]									
Fluorides, [30], mg/kg	-	-	<20	<20	<20	<20			<20
Sulphates – SO_4 [30], mg/kg			30	**2200**	**1900**	420			**1400**
Total dissolved solids - TDS, [32], mg/kg			<1500	**4700**	3700	<1500			3400
Dissolved organic carbon – DOC, [31], mg/kg	-	-	17	49	35	<10	-	-	47
Total organic carbon – TOC, [36], mg/kg	-	-	17	49	35	10	-	-	47
Sum 16 EPA PAHs, [33], mg /kg	-	-	67.5	5.4	10.5	44.8	-	-	-
TPH C10÷C40, [34], mg/kg	-	-	**1500**	280	**540**	**550**	-	-	<40
Tests required by EN 13242 [8]									
Sulphates – SO_4 (acid soluble), [25], mg/kg	-	-		5300	3900	2900	6100	5100	2500
Sulphur – S, [25]	-	-	-	2500	1800	1600	2600	2300	1500
Sulphates – SO_4 (water soluble), [25], mg/kg	-	-	-	1200	900	700	1400	1300	1200

The sulphates in Rc100 and Rb100 have contaminated all blends of Type 2 (because of the concrete content) and Type 3 (because of both concrete and brick content). The other components in the recycled materials (Ra, Ru, X and FL) have practically contributed to the reduction of sulphates content in the aggregates, but not below the threshold of 560 mg/kg [15]. It shall be noted that according to Article 10 of the Waste Framework Directive [18], the waste shall be subject to separate collection and shall not be mixed with other waste or other materials with different properties.

However, when tested according to [25] and evaluated as per EN 13242 [8], all blended recycled aggregates meet the requirements for the water-soluble sulphates of the best category of recycled aggregate SS0.2, for which the maximum sulphates content is specified to 0.2% (i.e. 2000 mg/kg). The content of acid-soluble sulfates determines the recycled materials of category AS0.8. The category of total sulphur content will be S1, because for all blends the values are below 1%. These categories allow the use of recycled materials in all unbound applications specified by the Bulgarian Technical Legislation [4, 5], including the most demanding such as a base course of heavily-loaded roads and the most vulnerable to leaching such as road shoulders and parking areas.

If based only on the sulphates content, the recycled aggregates of Type 2 and Type 3 would be classified as environmental quality Class A, according to the Austrian Guidelines, i.e. not suitable only for unbound use without a cover layer in hydro-geological delicate areas, defined as water-source preservation areas and areas with set conditions regarding water management [7]. One blend of each type, where the concrete content does not exceed 50% (Rc50 Rcu70 Ra30- X0.5 and Rc30 Rb70- X0.5 FL10) would even be classified as Class A+, i.e. would have no application restriction. It must be mentioned that the environmental class limits in [7] are based on eluate with L/S ratio equal to 10 [28].

However, the sulphates content values above 1000 mg/kg determine the non-inert character of the waste as per the Bulgarian Ordinance on Landfills [15] when L/S = 10 is applied(Table 4). It means, that whatever the L/S ratio is, only one blend (Rc50 Rcu70 Ra30- X0.5) with less than 50% concrete, heavily polluted by sulphates, will pass the requirements for backfilling if the inert character is to be tested. This blend, however, will fail the assessment because of the higher amount of TPH (C10 ÷ C40) than that specified in [15], due to the significant ground asphalt content (30%). The same negative assessment will be made for the blend with 20% of ground asphalt (Rc70 Rcu80 Ra20). Only 10% of ground asphalt does not lead to an inacceptable increase of total petroleum hydrocarbons (C10 ÷ C40) – Table 6.

The measured values for the TOC and DOC (Table 6) remained within the limits specified in [15] for inert waste (Table 4). The DOC content was between 10 and 50 mg/kg, which confirms the findings of other researchers for the mineral fraction of the CDW, reporting DOC as 5 to 100 mg/kg [6]. The TOC values are identical to the DOC values, meaning the total organic carbon consists of dissolved organic carbon. The presence of small amounts of light organic impurities (FL of 10 cm^3/kg) does not significantly affect the TOC and DOC content in the recycled aggregate.

Concerning PAHs, the values of 67.5 mg/kg and 44.8 mg/kg, measured in samples with 100% and 30% of ground asphalt, characterize the aggregates as environmentally incompatible, because they overpass the limit of 25 mg/kg, specified for the lowest Class C by the Austrian Guidelines [7]. The other two blends, with 10% and 20% of ground asphalt, have PAHs content below 12 mg/kg, which would classify them as of environmental quality class A, i.e. not acceptable only for unbound use without a cover layer in hydro-geological delicate areas. The

Danish CDW management basis is more severe – it allows CDW with total (based on the sum of 16 congeners) PAHs content below 4 mg/kg, and specifies limits of 0.3 mg/kg for some congeners, too – Benz(a) pyrene (PAHs) and Dibenzo(a,h) anthracene (PAHs) [10]. None of those limits are respected by the tested samples containing ground asphalt – Table 7. It can be concluded that, to avoid any risk for the environment, the asphalt content in unbound recycled aggregates shall be kept below 10%. The Bulgarian Landfills Ordinance stipulates the PAHs limit value for inert waste to 700 mg/kg [15], which seems to be unjustified and inappropriate, as the limit of 200 mg/kg applicable in Finland for utilization of soil in non-residential areas has been seen as extremely high [6].

Table 7. Specific PAHs congeners content in selected aggregates

PAH's congeners	Recycled aggregates designation			
	Ra100	Rc90 Ra10	Rc70 Rcu80 Ra20	Rc50 Rcu70 Ra30- X0.5
Benz(a) pyrene, [33], mg/kg	4.3	0.41	0.74	2.6
Dibenzo(a,h) anthracene, [33], mg/kg	0.81	<0.05	0.09	0.27

4. Conclusions

4.1. Impact of impurities on the environmental performance of recycled aggregate

The regulatory implication [15, 19] that in the absence of external contamination the concrete CDW is inert was not upheld.

It has been proven that the source of CDW has a critical impact on the environmental performance of recycled aggregates, even when CDW were sampled from presumably non-polluted sites and consisted mainly of concrete and bricks.

The limited presence of non-hazardous impurities such as bricks (up to 30%), soil (up to 0.5%), non-tar-containing asphalt (up to 10%) and lightweight materials (up to 10 cm^3/kg, determined as per EN 933-11) does not affect the environmental properties of recycled concrete aggregate, but rather contributes to the decrease of its degree of contamination, if the concrete CDW has been polluted. However, as prescribed in the Article 10 of the Waste Framework Directive [18], the "waste ... shall not be mixed with other waste or other materials with different properties".

There are some risks for the environment even when the recycled aggregates for unbound materials for use in civil works and road construction are assessed as technically conform for the intended uses.

4.2. Improvements required in the regulatory framework

It has been found that the requirements in the national legislation [12] only inert CDW to be used in backfilling are often more severe than those for recycled aggregate, usually obtained only by crushing and screening of the same CDW, i.e. without changing the environmental characteristics. The majority of the studied recycled aggregates were found non-inert because of the significant amounts of sulphates, chlorides and total dissolved solids, sometimes heavy metals and, when blended with ground asphalt, harmful organic compounds. However, the acceptable PAH's limits of 700 mg/kg for inert waste in the Bulgarian legislation [15] are too high and create a significant risk for soil's and water's pollution when the CDW are used in backfilling or recycled aggregate for unbound applications. Several times lower PAH's limits have been specified in other countries [6, 37].

The basic standard EN 13242 [8] and its Bulgarian National Annex [4], setting out the requirements for recycled aggregates, do not seem to perform well when an environmental assessment of aggregates must be made. The chemical characteristics under control of those standards, related mainly to the sulphates content, have determined all studied recycled aggregates as suitable for all unbound uses, while the assessment according to other guidelines [7, 10] has revealed some restrictions to be applied. For example, the presence of asphalt in the recycled aggregate creates a risk of polluting the environment with hydrocarbons (C10-C40) and PAHs. The excessive, according to [7], sulphate content, shall limit some uses of those recycled aggregates in hydro-geological delicate areas.

The risk of releasing heavy metals (by leaching) or other hazardous substances is only broadly considered by EN 13242 and no test methods are specified [8]. In the lack of appropriate national amendments on these issues, the waste legislation was implemented. Two problems have been outlined – appropriate testing methods (e.g. L/S ratio) and which limits to be applied to the recycled aggregates (e.g. for the inert or non-hazardous waste).

It is recommended to amend the Bulgarian Annex to EN 13242 [4] by specifying relevant environmental characteristics, methods of testing and appropriate limits for the relative uses of the recycled aggregates. Those limits might also be applied to the pre-treated CDW which is suitable to be used in backfilling.

4.3. Implications for the management of CDW in Bulgaria

The pre-demolition audit shall be mandatory to identify the contaminated CDW. Among the commonly recognised pollution mechanisms (paints, coatings, industrial processes), the sulphates and chlorides contamination by de-icing salts shall also be considered;

In order to keep some unavoidable impurities in relatively low limits, the selective demolition and on-site sorting should be applied to the greatest extent possible, given the technology and costs of demolition and refurbishment.

The recycling entities shall have information on the source of CDW. The CDW heterogeneity shall be taken into account in relation to decisions regarding utilisation of the aggregates [6]. When some potential risks of contamination exist, a more sophisticated recycling technology might be implemented – for example, the recycled aggregate might be washed-out to eliminate excessive water-soluble sulphates, chlorides, fluorides, etc. Such technology will also contribute to the removal of lightweight impurities.

Acknowledgements

The research in this study was supported by project "MOBile and Innovative Circularity for CONstruction PROducts (MOBICCON-PRO)" co-funded by the European Union's Horizon Europe programme under grant agreement ID 101091679.

References

[1] Hung, V. Q. et. al.,2024, Leaching characteristics of metals from recycled concrete aggregates (RCA) and reclaimed asphalt pavements (RAP),Heliyon, https://doi.org/10.1016/j.heliyon.2024.e30407 (accessed on 12.07.2024)

[2] Butera, S. et al., 2015, Construction and demolition waste: Comparison of standard up-flow column and down-flow lysimeter leaching tests, Waste Management, 43, 386-397, https://doi.org/10.1016/j.wasman.2015.04.032

[3] Gupta N. et al, 2017, Recycled concrete aggregate as road base: leaching constituents and neutralization by soil Interactions and dilution, Waste Management, 72, https://doi.org/10.1016/j.wasman.2017.11.018
[4] Bulgarian Institute for Standardization, BDS EN 13242:2002+A1:2007/NA:2017 Aggregates for unbound and hydraulically bound materials for use in civil engineering work and road construction. Bulgarian National Annex, Sofia (in Bulgarian)
[5] Ministry of Regional Development and Public Works of Bulgaria, "Technical specification of the Road Infrastructure Agency," Sofia, 2014 (in Bulgarian).
[6] Butera, S. et al., 2014, Composition and leaching of construction and demolition waste: Inorganic elements and organic compounds. Journal of Hazardous Materials, 276, 302-311, https://doi.org/10.1016/j.jhazmat.2014.05.033
[7] Austrian Construction Materials Recycling Association & Austrian Recycled Construction Materials Quality Assurance Association, 2009, Guideline for Recycled Construction Material, 8th Edition, Vienna.
[8] European Committee for Standardization, EN 13242:2002+A1:2007 Aggregates for unbound and hydraulically bound materials for use in civil engineering work and road construction
[9] European Committee for Standardization, EN 933-11:2009 Tests for geometrical properties of aggregates - Part 11: Classification test for the constituents of coarse recycled aggregate
[10] The Zealand Network for Construction and Construction Waste, 2020, Management basis for construction and construction waste, Forvaltningsgrundlag for bygge- og anlægsaffald, Sjællandsnetværket for Bygge- og Anlægsaffald, (in Danish), ISBN 978-87-972448-0-7, https://www.odsherred.dk/media/53lfc34i/forvaltningsgrundlag_bygge_og_anlaegsaffald.pdf (accessed 12.07.2024)
[11] Delloite SA, 2016, Construction and Demolition Waste management in United Kingdom, V2 – September 2015 (revised 27/1/16), https://ec.europa.eu/environment/pdf/waste/studies/deliverables/CDW_UK_Factsheet_Final.pdf, (accessed 12.07.2024)
[12] The Council of Ministers of the Republic of Bulgaria, 2017, Ordinance on the construction waste management and the use of recycled building materials, State Gazette, 98, 08.12.2017 (in Bulgarian)
[13] Commission Decision of 18 November 2011 establishing rules and calculation methods for verifying compliance with the targets set in Article 11(2) of Directive 2008/98/EC of the European Parliament and of the Council (notified under document C(2011) 8165) (2011/753/EU), OJ L 310, 25.11.2011, p. 11–16.
[14] Regulation (EU) No 305/2011 of the European Parliament and of the Council of 9 March 2011 laying down harmonised conditions for the marketing of construction products and repealing Council Directive 89/106/EEC, OJ L 88, 4.4.2011, p. 5–43.
[15] The Council of Ministers of the Republic of Bulgaria, Ordinance No. 6 on the conditions and requirements for the construction and operation of landfills and other facilities and installations for the recovery and disposal of waste, State Gazette, 80, 13.09.2013, last amended State Gazette, 01.05.2021 (in Bulgarian).
[16] Commission Decision of 3 May 2000 replacing Decision 94/3/EC establishing a list of wastes pursuant to Article 1(a) of Council Directive 75/442/EEC on waste and Council Decision 94/904/EC establishing a list of hazardous waste pursuant to Article 1(4) of Council Directive 91/689/EEC on hazardous waste (notified under document number C(2000) 1147) (Text with EEA relevance) (2000/532/EC)Text with EEA relevance, http://data.europa.eu/eli/dec/2000/532/2023-12-06 (accessed on 18 03 2024)
[17] The Council of Ministers of the Republic of Bulgaria, Waste Management Act, State Gazette, 53, 13.07.2012, last amended State Gazette ,108, 30.12 2023 (in Bulgarian)
[18] Directive 2008/98/EC of the European Parliament and of the Council of 19 November 2008 on waste and repealing certain Directives, Waste Framework Directive, OJ L 312, 22.11.2008, p. 3–30
[19] Council Directive 1999/31/EC of 26 April 1999 on the landfill of waste, OJ L 182 16.7.1999, Consolidated text: http://data.europa.eu/eli/dir/1999/31/2018-07-04 (accessed on 12. 07. 2024)
[20] Zaharieva, R. et al, 2024, Pre-demolition audit as a tool for appropriate CDW management – a case study of a public building, Proceedings of 16th Int. Scientific Conf. “Civil and Environmental Engineering”, Slovakia, E3S Web of Conferences 550, 01045 (2024), https://doi.org/10.1051/e3sconf/202455001045
[21] Zaharieva, R., Y. Kancheva, 2020, Hazardous Construction Waste - Types and Sources, Yearbook of the University of Architecture, Civil Engineering and Surveying, 53 (2), 441-456, Sofia (in Bulgarian)
[22] Vieira C., Pereira P., 2015, Use of recycled construction and demolition materials in geotechnical applications: A review, Resources, Conservation and Recycling, 103, 192-204.
[23] Vegas I.et al., 2011, Pre-normative research on the use of mixed recycled aggregates in unbound road sections, Construction and Building Materials, 25 (5), 2674-2682, ISSN 0950-0618
[24] Zaharieva R. et all, Impact of Impurities on the Properties of Recycled Aggregate for Use in Civil Engineering Work and Road Construction. Part 1: Technical Assessment, XIII International Scientific Conference “Civil Engineering

Design and Construction", September 12 - 13, 2024, Varna, Bulgaria VFU " (under print in the same volume of IOP Conf. Ser.)

[25] European Committee for Standardization, EN 1744-1:2009+A1:2012 Tests for chemical properties of aggregates - Part 1: Chemical analysis

[26] European Committee for Standardization, EN 12457-1:2003 Characterisation of waste - Leaching - Compliance test for leaching of granular waste materials and sludges - Part 1: One stage batch test at a liquid to solid ratio of 2 l/kg for materials with high solid content and with particle size below 4 mm (without or with size reduction)

[27] Allegrini, E. et al., 2015, Life cycle assessment and residue leaching: The importance of parameter, scenario and leaching data selection. Waste Management (2015), 38, 474-485, http://dx.doi.org/10.1016/j.wasman.2014.12.018, (accessed on 12. 07. 2024)

[28] European Committee for Standardization, EN 12457-4:2003 Characterisation of waste - Leaching - Compliance test for leaching of granular waste materials and sludges - Part 4: One stage batch test at a liquid to solid ratio of 10 l/kg for materials with particle size below 10 mm (without or with size reduction)

[29] European Committee for Standardization, EN ISO 11885:2009 Water quality - Determination of selected elements by inductively coupled plasma optical emission spectrometry (ICP-OES) (ISO 11885:2007)

[30] European Committee for Standardization, EN ISO 10304-1:2009 Water quality - Determination of dissolved anions by liquid chromatography of ions - Part 1: Determination of bromide, chloride, fluoride, nitrate, nitrite, phosphate and sulfate (ISO 10304-1:2007)

[31] European Committee for Standardization, EN 1484:2019-04 Water analysis - Guidelines for the determination of total organic carbon (TOC) and dissolved organic carbon (DOC)

[32] European Committee for Standardization, EN 15216:2008 Characterization of waste - Determination of total dissolved solids (TDS) in water and eluates

[33] European Committee for Standardization, EN 15527: 2008-09 Characterization of waste - Determination of polycyclic aromatic hydrocarbons (PAH) in waste using gas chromatography mass spectrometry (GC/MS)

[34] European Committee for Standardization, EN 14039:2005 Characterization of waste - Determination of hydrocarbon content in the range of C10 to C40 by gas chromatography

[35] European Committee for Standardization, EN ISO 17852:2008 Water quality - Determination of mercury - Method using atomic fluorescence spectrometry (ISO 17852:2006)

[36] European Committee for Standardization, EN 1484: 1997-08 Water Analysis - Guidelines for the Determination of Total Organic Carbon (Toc) and Dissolved Organic Carbon (DOC)

[37] Government of Spain Ministry of Agriculture, Food and Environment, Order AAA/661/2013, of April 18, amending Annexes I, II and III of Royal Decree 1481/2001, of December 27, regulating the disposal of waste by landfill (in Spanish), Orden AAA/661/2013, de 18 de abril, por la que se modifican los anexos I, II y III del Real Decreto 1481/2001, de 27 de diciembre, por el que se regula la eliminación de residuos mediante depósito en vertedero, https://www.boe.es/eli/es/o/2013/04/18/aaa661 (accessed 12.07.2024).

IOP Conf. Series: Materials Science and Engineering **1323** (2024) 012004 doi:10.1088/1757-899X/1323/1/012004

E-waste as a potential precursor for geopolymers

Aleksandar Nikolov[1*], Liliya Tsvetanova[1], Borislav Barbov[1], Bogdan Ranguelov[2] and Mihail Tarasov[1]

[1] Institute of Mineralogy and Crystallography, Bulgarian Academy of Sciences, Sofia, Bulgaria
[2] Institute of Physical Chemistry, Bulgarian Academy of Sciences, Sofia, Bulgaria

*E-mail: y8sashko@yahoo.com

Abstract. The rapid growth in electronics production and the shortening product lifespans have resulted in a growing problem of electronic waste (e-waste). E-waste is a complex mixture of organic and inorganic materials, making its recycling a challenge. This study explores the potential of powdered e-waste as a geopolymer precursor and investigates the geopolymerization of reactive phases in e-waste. Research techniques, including XRD, XRF, SEM/EDX, and FTIR, were used to characterize the e-waste and obtained geopolymers. Significant contents of reactive SiO_2, Al_2O_3 and CaO components were detected in the e-waste which can participate in the formation of geopolymer gel. The present study elucidate the feasibility of utilizing e-waste as a geopolymer precursor, offering a potential for sustainable waste management and resource recovery in the electronics recycling industry.

1 Introduction

In recent decades, the production of electronics has grown exponentially, and the useful life of products has become shorter. This leads to the generation of more and more waste from used electronics, the so-called "e-waste", or WEEE (waste from electrical and electronic equipment). Electronics is a major consumer of precious metals and its recycling is essential. Unfortunately, the recycling of electronics is complicated because they are composed of organic and inorganic materials – most often organic polymer matrix reinforced with glass fibres. The recycling process usually involves manual dismantling, coarse crushing, grinding or pulverization, followed by physical (dry) and/or chemical separation (wet). Depending on the type of electronics, e-waste contains approximately 30% metals and 70% nonmetals [1]. The waste residue after metal separation most often contains materials such as "textolite", glasstextolite, polyamide, "getinax", glass/carbon fibre and other polymer composite. The waste residue cannot be used to make new electronics with the same properties, which leads to the need of another solution for the utilization of e-waste.

A common practice for utilization of waste products is to add them into inorganic binder materials for building purposes or road construction [2]. A number of studies have been carried out, mostly by Indian scientists, to replace part of the sand in the composition of geopolymer or Portland cement mixtures [3, 4]. The results showed an increase in the compressive and tensile strength with sand replacement up to 20 % with e-waste [5, 6]. Excessive replacement results in a drastic drop in strength, because of loss on adhesion between binder matrix and the organic substance in e-waste.

The aim of the present study is to evaluate the potential of local e-waste as a geopolymer precursor and elucidate the geopolymerization of the reactive inorganic phases in the e-waste. The study is focused on the characterisation of a local e-waste and syntesized geopolymer based on the e-waste. The organic components and precious metals presented in the e-waste stays relatively inert and do not participate actively in the geopolymer process hence are not of the scope of the paper.

IOP Conf. Series: Materials Science and Engineering **1323** (2024) 012004 doi:10.1088/1757-899X/1323/1/012004

2. Materials and methods

The geopolymer precursor in the present study was powdered WEEE provided from local Bulgarian recycling plant which extract valuable metals from e-waste. The present e-waste was produced by recycling electronics, mostly printed circuit boards. After manual disambling of precious components, the printed crciut boards were crushed and milled in industrial steel ball mill. The copper components were collected by dry separation based on density differences. This recycling process generates powdered e-waste (EW) approximatly 150 tonnes annualy with tendency of increasing.

The geopolymer activator solution was prepared prior by mixing liquid sodium metasilicate (water glass) with molar ratio (SiO_2/Na_2O=2.98) and density – 1.42 g/cm^3, potassium hydroxide on pellets (Sigma-Aldricht) and tap water.

The powder XRD patterns were recorded on an Empyrean (Malvern-Panalytical) diffractometer using CuKα radiation at 40 kV and 30 mA. The chemical composition was analyzed on pressed tablets using a Rigaku Supermini 200 wavelength dispersive X-ray fluorescence spectrometer, capable of analyzing all elements from fluorine. FTIR spectra were recorded on a Tensor 37 spectrometer (Bruker) with a 4 cm^{-1} resolution using KBr pellets. The micromorphology and local elemental composition of the e-waste and geopolymer fragments and their powdered samples were investigated with a JEOL 6390 scanning electron microscope and point standardless EDX analysis. Samples for SEM/EDX examination were coated with gold. The loss on ignition (LOI) was detemined by calcination of 20 g e-waste at 550 °C in muffle furnace.

3. Results

3.1 Characterization of the e-waste (EW)

SEM examination of the e-waste reveals that the material is well disintegrated into two main parts - glass fibers (up to 1 mm long and up to 10 μm in diameter) and shapeless pieces (20-300 μm) of organic material (Fig. 1 a, b). According to EDX analysis (white box inset on the right side of Fig. 1b), the fibers are characterized by a high content of Si, Al and Ca, consistent with "E-glass" fibers commonly used in electronics due to its very high electrical resistivity [7].

Figure 1. SEM images of the e-waste (EW) with magnification: a) x100; b) x500. In the white boxes – standardless elemental EDX analyses (wt.%)

The apparent presence of C and Br in the fiber composition is due to the influence on the analysis of the neighboring organic polymer material, consisting mainly of C, O and Br (white box inset on the left side of Fig. 1b). Bromine is typically used as a fire retardant additive in electrical plastics.

The loss on ignition (550°C) of the e-waste was determined to 30.1 wt.%. This value represents the content of organic and volatile matter in the e-waste. Chemical composition (XRF analysis) of the initial EW and the product of its calcination at 550 °C (EW550) are shown in Table 1. In addition to SiO_2, CaO, Al_2O_3 and Br, which are the main components, the raw e-waste contains also in quantities from a few percent to tenths of a percent

CuO (metal Cu – residual after dry separation), BaO, MgO, TiO_2, FeO, SO_3, K_2O, Na_2O and some other minor components. About 25 wt. % (shortage of XRF analysis total to 100%) of the e-waste can be attributed to light elements of organic material, such as carbon, not detected by XRF.

XRF analysis of EW550 coupled with LOI value (Table 1) shows that bromine content decreased sharply after calcination due to its volatilization along with organic material. Table 1 also shows that there is a noticeable difference in the content of CaO, Al_2O_3, CuO, MgO in the original and treated e-waste, which can hardly be explained by the calcination process and is rather related to the heterogeneous composition of the e-waste.

Table 1. Chemical composition of initial e-waste (EW) and product of its calcination at 550 °C (EW550), determined by XRF analysis, in wt. %

Sample	SiO_2	CaO	Al_2O_3	Br	CuO	BaO	MgO	TiO_2	FeO	SO_3	K_2O	Na_2O	Other*	LOI	Sum
EW	34.19	22.04	9.34	5.42	1.11	0.50	0.49	0.42	0.36	0.38	0.26	0.14	0.35	-	75.00
EW550	35.26	17.37	11.19	0.06	2.36	0.59	1.32	0.33	0.27	0.49	0.18	0.28	0.20	30.10	100.00

* Other components are ZrO_2, SrO, ZnO, Cl

The phase composition of the raw electronic waste (EW) is characterized by the presence of mainly low-ordered phases, associated with a wide amorphous halo between 15 and 35° 2θ in XRD patterns (Fig. 2). These phases relate to glass fiber and organics. Other phases of the e-waste are residual metal copper and various organic phases: s-triazine and polyglycine (Fig. 2). After calcination at 550 °C the organic phases oxidized to graphite and copper to tenorite (CuO).

The FTIR spectra of e-waste showed bands related to the plastics [8] and hydroxyl groups which after calcination disappeared (Fig. 3). The peaks at 2964 cm^{-1}, 2930 cm^{-1}, and 2872 cm^{-1} show the presence of CH_3, C–H, CH_2 vibrations, respectively. The characteristic peak at 1747 cm^{-1} refers to C–H, while peaks at 1608 cm^{-1} and 1510 cm^{-1} represents C=C stretching vibrations. The peak corresponding to 737 cm^{-1} is attributed to aromatic ring vibration. The absorption band (EW) at 3450 cm^{-1} is attributed to the O-H stretching vibration. The bands about 1018-1020 cm^{-1} and 451-453 cm^{-1} observed at e-waste (EW and EW550) are correlated to T-O asymmetric stretching mode and T-O bending mode, respectively (where T is Si or Al). The well define bands at 1452 cm^{-1} at sample EW550 is related to BO_3 bending and 1468 cm^{-1} at sample EW is CH_2 bending.

3.2 Geopolymer synthesis and characterisation

The high content of reactive SiO_2, Al_2O_3 and CaO found in e-waste, as well as the good openness of the carriers of these components - fibers, make the studied e-waste promising precursors for the production of geopolymers.

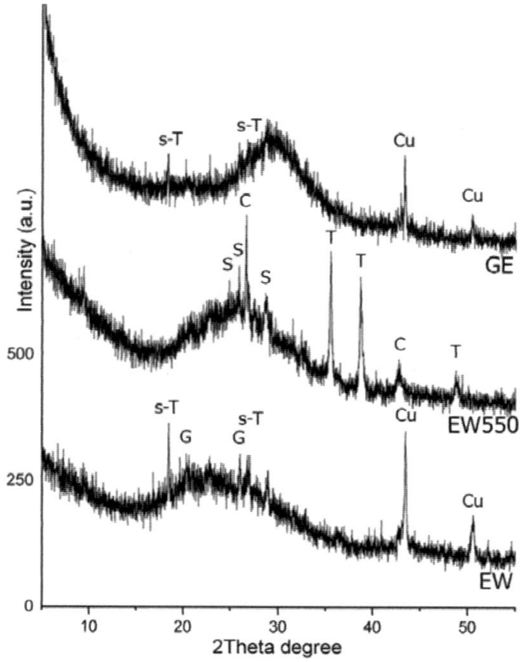

Figure 2. XRD patterns of initial e-waste (EW), e-waste calcined at 550 °C (EW550) and geopolymer based on e-waste (GE). Legend: C – graphite/carbon, Cu – copper, S – SiO_2, T – tenorite, s-T- s-triazine, G – polyglycine

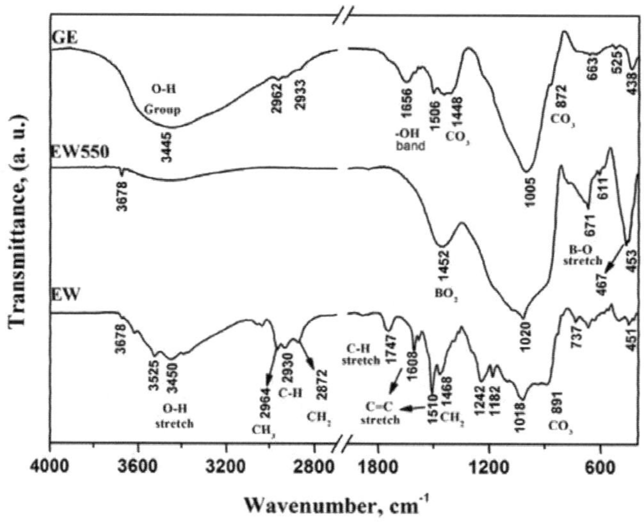

Figure 3. FTIR spectra of initial e-waste (EW), e-waste calcined at 550 °C (EW550) and geopolymer based on e-waste (GE)

The activator solution was prepared one day prior geopolymer syntesis by mixing 67 g sodium water glass, 18.7g KOH pellets and 18.5 tap water. The obtained activator solution was characterised by following molar ratio: H_2O/M_2O = 12.5; SiO_2/M_2O = 0.9, where M is Na or K. The geopolymer synthesis was performed by mixing the raw e-waste with the alkali activator solution at water to solid ratio equal to 0.50 (wt.%). The fresh mixture was homogenized and poured in steel moulds, wrapped in polyethylene and cured for 24 hours in oven at 80 °C, followed by demoulding and maturing one month at laboratory conditions (25 °C, relative humidity ~ 65%). The XRD results showed that amorphous halo at 15–35 2θº shifted to 25–40 2θº which is characteristic for geopolymers (Fig. 2). There were no significant changes of the peaks of organic phases and metalic copper hence that they did not react at geopolymerisation. The results from FTIR showed shift of the T-O asymmetric stretching band from 1018 cm^{-1} to 1005 cm^{-1} indicating reconfiguration of the silica and alumina after geopolymerization. The position of this band is related to Si/Al ratio of the formed geopolymer gel and lowering the ratio indicates higher inclusion of Al in the geopolymer gel [9, 10]. The bands in the regions around 3445 cm^{-1} and 1656 cm^{-1} at geopolymer based on e-waste correspond to stretching vibration and bending vibrations of OH^- groups, respectively. The bands about 1448 cm^{-1} and 872 cm^{-1} are related to asymmetric stretching vibration and out plane bending of CO_3 groups. Carbonate phases presented are product of natural carbonation of the free alkali in geopolymer with contact to atmospheric carbon dioxide.

The SEM images reveal that the formed geopolymer gel was highly homogeneous (Fig. 4). The dissolution of E-glass fibres in alkaline solutions was report recently [11]. The high alkalinity of the activator and high temperature (80 °C) promotes the dissolution of the fibres in e-waste. Thus, the geopolymer gel was a product of metasilicate derived from activator solution, and dissolved fibres. The extend of dissolution of e-waste fibres vary to the initial diameter of the fibers. Numerous fibres appeared after geopolymerization with hollow central part which could be due to incomplete dissolution of the glassy fibre and/or inert organic fibre covering. The presence of cracks and cylindrical pores around the remnant particles is indicative for significant geopolymer gel shrinkage.

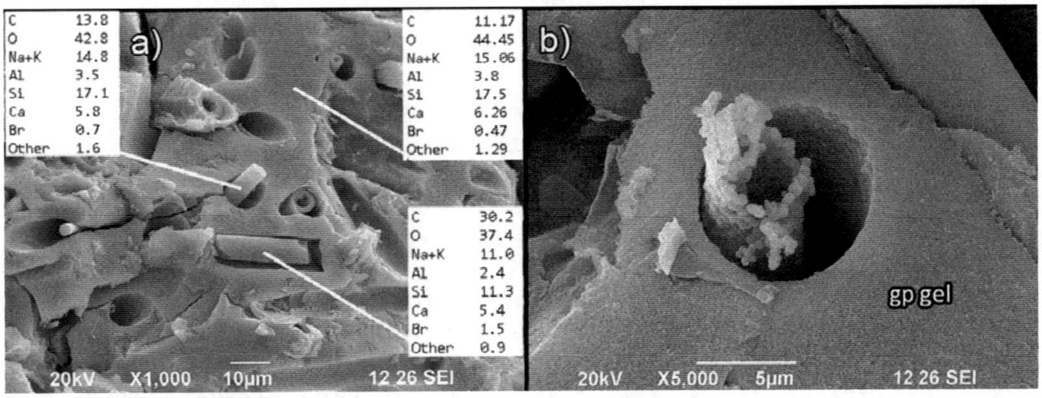

Figure 4. SEM images of geopolymer based on e-waste (GE) with magnification: a) x1000; b) x5000. In the white boxes – standardless elemental EDX analyses (wt.%)

The examination of the e-waste showed that material consist of organic matrix, glass fibres and residual copper. The organic content represent the matrix of the base of the circuit boards with commonly used polymers such as epoxy, polyamides, polyethylene terephthalates, polyethylene naphthalates, etc. The organic polymers contains bromine which is commonly added as flame retardant. Calcination at 550°C significantly reduced the bromine content which makes the calcination of the e-waste delicate operation since bromine is considered toxic. The organic matter and residual copper stayed inhert at the geopolymer syntesis. The reactive component in the e-waste were the glass fibres which were characterised by amorphous structure. The high content of SiO_2, Al_2O_3, and CaO reacted througth the geopolymerization. The resulted geopolymer gel was characterised by homogenous structure and sigificant shrinkage which appeared as cracks and porosity around organic particles and redemnant fibres.

4. Conclusions

The study elucidate the complex composition of e-waste, predominantly consisting of inorganic glass fibres, organic polymers, fire-retardants (bromine compounds) and residual copper. The prelimenary experiments showed that e-waste is promising geopolymer precursor. The reactive compound in the e-waste was the inorganic glass fibres. The fibres dissolves at different extend at the action of the alkali activator to take part of the formed geopolymer gel.

In conclusion, this study provides valuable insights into the feasibility of utilizing e-waste as a geopolymer precursor, offering a potential for sustainable waste management and resource recovery in the electronics recycling industry. Further research is needed to assess the mechanical properties, long-term durability and environmental implications of e-waste based geopolymers.

Acknowledgements
The research was funded by project № KP-06-N77/9 from 04.12.2023, financed by National Science Fund Bulgaria. The authors acknowledge the technical support: XRD - project PERIMED BG05M2OP001-1.002-0005 /29.03.2018 (2018–2023), SEM – project INFRAMAT (National Roadmap for Research Infrastructure) financed by Bulgarian Ministry of Education and Science, CoE "National center for/of mechatronics and clean technologies" (Project BG05M2OP001–1.001–0008) supported by the ERDF within the Bulgarian OP "SESG 2014–2020". Special thanks to Stela Atanasova-Vladimirova (IPC–BAS) for the SEM images.

References

[1] Donadkar M and Solanke S (2016) Review of E-waste material used in making of concrete Int. J. Sci. Technol. Eng. 2(07) p 66-69
[2] Sachdeva A and Sharma U (2020) An overview of utilization of e-waste plastic in road construction ICSCEP 2019 p 101-109
[3] Raut S, Dhapudkar R, and Mandaokar M (2018) Experimental study on utilization of E-waste in cement concrete Int. J. Eng. Sci.(IJES) 5(5) p 82-86
[4] Verma P, Dhurvey P and Gour C (2023) Utilization of E-Waste as Coarse Aggregates in Geopolymer Concrete Sustain. Approaches Strategies E-Waste Manage. Util. IGI Global p 224-238
[5] Sundar M and Raj S (2017) Study on Characteristics of Geopolymer Concrete with E-Waste. J. Mech. Civ. Eng., 14(3) p 21-27
[6] Partheeban P, Kalaiyarrasi A and Narayanan L (2021) Performance evaluation of geopolymer concrete using E-waste and M-sand. Res. Eng. Struct. Mater. 7(2) p 183-198
[7] Jones F and Huff N (2018) The structure and properties of glass fibers, in Handb. Prop. Text. Tech. Fibres. Elsevier p 757-803
[8] Barnwal A, Mir S, Dhawan N (2020) Processing of discarded printed circuit board fines via flotation J. Sustainable Metall.6 p 631-642
[9] Nikolov A, Nugteren H and Rostovsky I (2020) Optimization of geopolymers based on natural zeolite clinoptilolite by calcination and use of aluminate activators Constr. Build. Mater. 243 p 118257
[10] Nikolov A (2022), Sodium aluminate activated geopolymers based on natural zeolite clinoptilolite. Rev. Bulg. Geol. Soc. 83(3) p 27-30
[11] Lee S, et al. (2019) Dissolution Technology Development of E-Glass Fiber for Recycling Waste of Glass Fiber Reinforced Polymer J. Korean Ceram. Soc. 56(6) p 577-582

IOP Conf. Series: Materials Science and Engineering **1323** (2024) 012005 doi:10.1088/1757-899X/1323/1/012005

Microstructural analysis of phases in cement mortars with quartz and limestone filler

Ventseslav Stoyanov[1,2], Vilma Petkova[3*] and Ekaterina Serafimova[4]

[1] Technology and Construction Management Department, University of Structural Engineering and Architecture „Lyuben Karavelov", Sofia, Bulgaria
[2] Safety Management and Prevention Department, Academy of Ministry of Interior, Sofia, Bulgaria
[3] Institute of Mineralogy and Crystallography, Bulgarian Academy of Sciences, Sofia, Bulgaria
[4] Engineering Ecology Department, University of Chemical Technology and Metallurgy, Sofia, Bulgaria

*E-mail: vilmapetkova@gmail.com

Abstract. The objects of the study were cement mortars prepared from white Portland cement, natural river sand, marble powder and a high range water reducing admixture. Microstructural analysis was carried out by powder X-ray diffraction, scanning electron microscopy and differential thermal analyses, enhanced with evolved gas analysis. Analysis of the results showed that stable structure formed at early stages, with minimal changes in phase composition and quantity during subsequent aging in water. The mortars, in which the sand was completely replaced by marble powder with fine additive, was characterized by stability of composition, which gives reason to conclude for the preparation of cement mortars with high durability.

1. Introduction

Cement-based materials are one of the most commonly used materials in construction works [1]. Basically, cement mortars are mixture of paste and fine aggregates – the paste being water and cement and the fine aggregates being natural, crushed or recycled sand. Cement paste can be considered the "glue" that binds the fine aggregates together to form a mortar. These fine aggregates are considered to be inert, and the most often requirements for their use are particle size distribution and absence of fine particles (e.g. clay content) and harmful impurities [2]. The performance of mortars is improved by the addition of fine additives and chemical admixtures. The fine additives are divided into reactive additives with pozzolanic activity and inert additives - usually quartz, limestone or dolomite [3]. It is assumed that inert additives only fill the spaces remaining between the particles of fine aggregates and thus increase the compressive strength without participating in chemical reactions [3, 4]. Sometimes it is even better to convert potentially reactive additives to inert so that they can be used in a regulated manner in cement compositions, which is particularly important for the incorporation of waste materials [5, 6].

The term "inert material" refers to a material that does not react with chemicals in its environment, and there has been recent discussion as to whether inert additives and aggregates should be considered chemically inert in the hydration process of cementitious materials. The chemistry of the alkali-silica reaction in cement mortars and concretes has been best studied in this field and has already had review articles and monographs published [7-9]. Limestone has long been shown to react with the aluminate cement minerals to form carboaluminates, but not all researchers accept the consequences of these reactions on new phases and compressive strength of cement compositions [10-12]. Recently, research in this area has intensified, especially with the possibility of utilizing a variety of construction wastes and assigning specific applications to these mortars [3, 13-16]

In our previous work, we have found that the study of cement mortars for exterior architectural elements is particularly promising. For this purpose, we used cement compositions prepared from white Portland cements, marble powder (metamorphosed limestone or dolomite) and chemical admixture [17]. These compositions require a precise composition because they are subject to a variety of weathering influences in their life cycle.

This manuscript aims to analyze the formed phases in cement mortars, containing fine marble additives and aggregates, as well as the influence of a high-water-reduced-admixture, by the methods of thermogravimetric and differential thermal analyses and Fourier transform infrared spectroscopy.

2. Materials and methods

2.1 Mortar constituents and sample preparation

Using mechanical mixing, we prepared three types of mortar by combining distilled water and white Portland cement CEM I 52.5 N, produced by Devnya Cement (Bulgaria) with mineral composition (in wt%): Ca_3SiO_5 – 72.13, Ca_2SiO_4 – 15.28, $Ca_3Al_2O_6$ – 5.23 and $Ca_4Al_2Fe_2O_{10}$ (C4AF)—0.61 and compressive strength at 28 days of curing —56.2 N/mm^2. The fine aggregate for reference mortar was natural river sand with particle size distribution ranges between 0.08 and 2.00 mm and fully compliant with CEN Reference sand EN 196-1 [18]. The studier mortars were prepared by marble powder (AIAS S.A. White Marble Products, Greece) with dolomitic consistence of about 95 wt%. The maximal size of fine aggregate particles was 2.00 mm, half of which were fine additives (< 0.125 mm); the content of fines (< 0.063 mm) was 5.0 wt%. Polycarboxylate-based admixture Sika ViscoCrete 5-800 was used as a high range water reducer (HRWR) (density 1.07 g/cm^3 at 20°C), which was added 2 % of the mass of cement. The amount of distilled water added was tailored to achieve uniform workability of the fresh mortars. When HRWR was used, the fresh mortar had a self-leveling effect.

Table 1. Mortar (sample) composition and their labelling

Mortars	Composition of mortar	Cement to aggregate ratio	Water to cement ratio	Samples (days of curing)	
				28 days	120 days
C-S	Cement + Sand	1:3	0.50	C-S028	C-S120
C-M	Cement + Marble powder	1:2	0.60	C-M028	C-M120
C-MH	Cement + Marble powder + HRWR (2.00 wt% of cement)	1:2	0.40	C-M$_H$028	C-M$_H$120

The mortars, designated as C-S, C-M, and C-MH, have their compositions shown in **Table 1**. These mortars were prepared according to the procedure described in EN 196-1. After hardening of 1 day the prisms of mortars were demoulded and submerged in water at 20°C for periods of 28 and 120 days of curing, when they were tested for compressive strength. Small parts were taken from the broken prisms and then samples were prepared for testing with methods of microstructural analysis. The samples labelled as C-S028, C-M028, C-M_H028, C-S120, C-M120, and C-M_H120, respectively (**Table 1**). Detailed information regarding the chemical composition of the raw materials and the sampling procedure is provided in [17].

2.2 Experimental methods
Powder X-ray diffraction (PXRD) patterns were collected at room temperature using a step-scan regime with a step size of 0.05° over a 2-theta range of 5° to 75°. The measurements were conducted on a D2 Phaser diffractometer (Bruker AXS, Germany) The experiments were performed in Bragg-Brentano geometry, utilizing Ni-filtered Kα radiation (λ = 0.15418 nm) at an accelerating voltage of 10 kV and a current of 30 mA. The phase composition was identified using the ICDD database [19].

Scanning electron microscopy (SEM) was employed to examine the morphology of the samples. The surface was analysed in secondary electron emission mode using a digitized SEM Philips PH Model 515, equipped with secondary electron image (SEI) detectors. The microscope operated at an acceleration voltage of 30 kV and a magnification of 40,000x. After being dried for 12 hours at 60 ± 5 °C, fragments of the broken samples, each with an approximately flat surface area of 10×10 mm, were coated with a thin layer of gold.

The thermal properties of the materials were studied using thermogravimetric and differential thermal analyses (TG-DTG-DTA). Measurements were conducted with a SETSYS2400 thermal analyser (SETARAM, France) and an OmniStar mass spectrometer for qualitative analysis of the evolved gases (EGA). The temperature range was 20-1100°C, with a heating rate of 10°C/min. The mass of samples was 15 ± 1 mg the gas used was static air.

3. Results and discussion

3.1 Morphological and X-ray phase analysis

The provided data, consisting of PXRD patterns and corresponding SEM micrographs, offers valuable insights into the mineralogical composition of the samples (Figure 1). By analyzing these data, we can identify the primary solid phases present and categorize them into two distinct groups (**Table 2**):

i) minerals derived from the source materials (white Portland cement, river sand and marble powder) – Belite, Albite, Anorthite, Mg-rich calcite, Quartz, Dolomite belite, albite, anorite, Mg-rich calcite, quartz, dolomite;
ii) hydrated minerals formed during the hydration process – Portlandite, Hillebrandite, Scolecite, Ettringite, Hemi- and Monocarboaluminate.

Dolomite, the primary mineral in the marble powder, plays a role in phase formation, as evidenced by the formation of carboaluminates in mortars with high aggregate content. Despite its low chemical reactivity, the dolomite participates in the formation of Portlandite, Ettringite and Hillebrandite (CSH gel) in all mortars, regardless of water to cement ratio and other solution parameters (**Table 1**). Conversely, river sand results in the formation of sulfoaluminates and Scolecite (CSAH gel). The HRWR water reducer does not significantly affect phase formation due to its low content and organic nature, it influences the rheological properties of the C-M_H mortar (28 and 120 days of water curing).

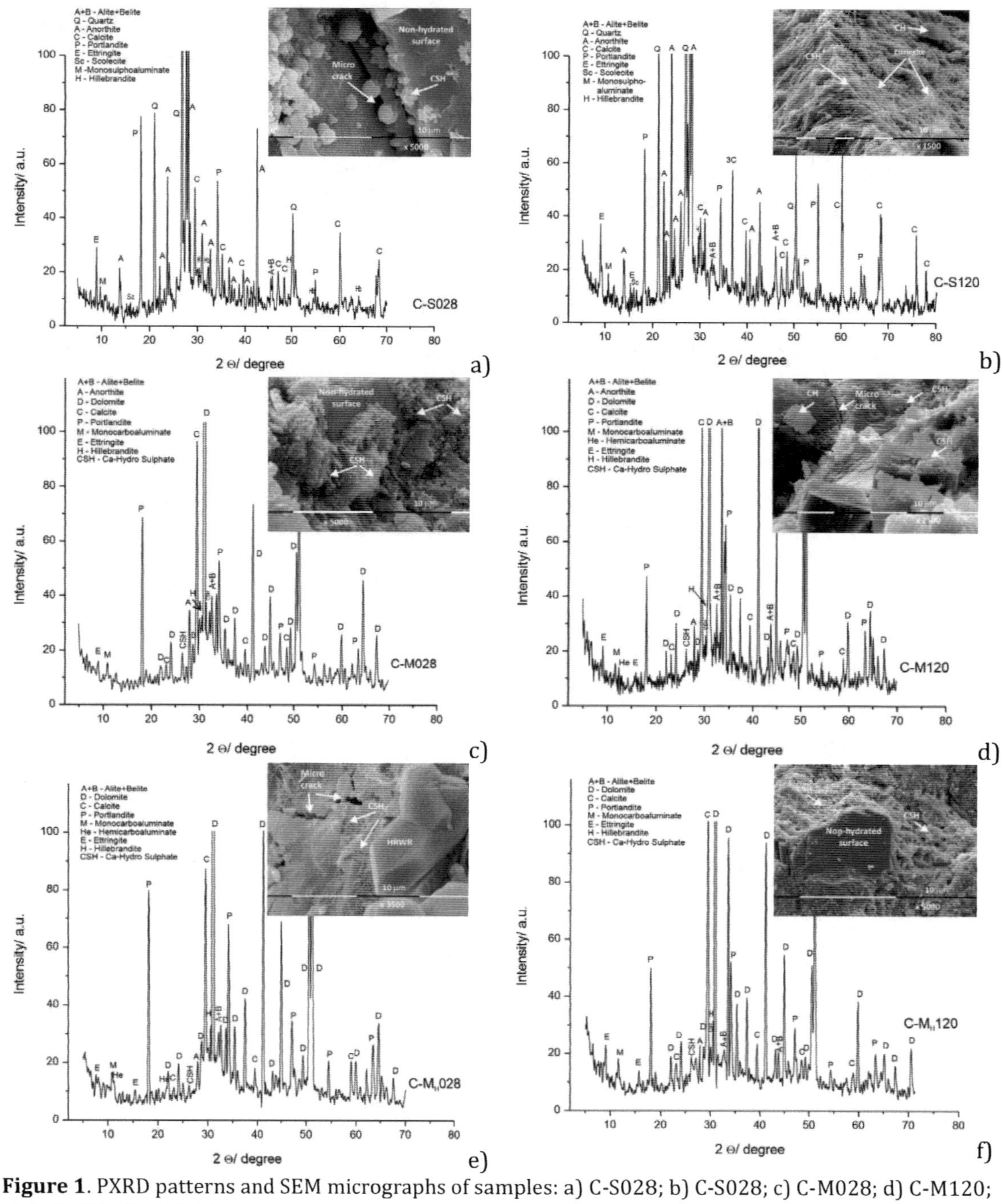

Figure 1. PXRD patterns and SEM micrographs of samples: a) C-S028; b) C-S028; c) C-M028; d) C-M120; e) C-M$_H$028; f) C-M$_H$120.

The low water-to-cement ratio is a primary driver of phase formation during hydration, as evidenced by the presence of unhydrated minerals in the mortar constituents (**Table 2**, **Figure 1**). This water deficiency is confirmed by the morphological analysis of the specimens performed by scanning electron microscopy (inset micrographs in **Figure 1**). Mortars prepared with marble powder exhibit distinct areas of unhydrated material, microcracks, and pores (**Figures 1c-1f**). Portlandite formation is characterized by flat, two-dimensional crystals, particularly in the reference mortar C-S (C-S028 and C-S120). The fine network of small crystals represents CSH/CSAH phases, which are primarily identifiable through SEM imaging due to their small size and X-ray amorphous nature. The elongated, needle-like crystals of ettringite are prominent on the micrographs, especially after 120 days of water curing."

Table 2. PXRD patterns and SEM – identified phases

No	Characteristics of minerals	Sample	Results from PXRD
1.	Relict minerals from the raw cement	All	Belite (C_2S)– $2CaO.SiO_2$ Albite (C_3S)–$(Na,Ca)Al(Si,Al)_3O_8$ Anorthite (CAS_2) – $CaO.Al_2O_3.2SiO_2$
2.	Phases from additives	C-M028, C-M120, C-M_H028 and M_H120 C-S028 and C-S120	Dolomite – $CaCO_3.MgCO_3$ Mg-rich Calcite – $CaCO_3$ Quartz, 46-1045 – SiO_2
3.	New phases forming CSH/CSAH gel:	All	Portlandite (CH) – $Ca(OH)_2$
3.1.	- containing CO_3^{2-}	C-M120 and C-M_H120	Monocarboaluminate, $3CaO.Al_2O_3.CaCO_3Ca(OH)_2.5H_2O$
		C-M120 and C-M_H 028	Hemicarboaluminate – $3CaO.Al_2O_3.0.5CaCO_3.0.5Ca(OH)_2.5.5H_2O$
3.2.	- containing SO_4^{2-}	All	Ettringite – $6Ca(OH)_2.Al_2(SO_4)_3.26H_2O$
		C-S028 and C-S120	Monosulphoaluminate– $Ca_4Al_2SO_{10}.12H_2O$
3.3.	- hydroxyl silicates formation from CaO, Al_2O_3, SiO_2	All	Hillebrandite– $Ca_2SiO_3(OH)_2$
		C-M028, C-M120 and C-M_H 120	CSH – Calcium hydroxyl silicate
		C-S028 and C-S120	Scolecite - $CaAl_2Si_3O_{10}.3H_2O$

3.2 Thermal analysis

Simultaneous thermogravimetric-differential thermal-differential scanning calorimetry (TG-DTG-DSC) analysis coupled with evolved gas analysis (EGA) was conducted on the C-S, C-M, and C-M_H series samples (**Figures 2 and 3**). EGA, using mass spectrometry, was employed to accurately quantify mass losses during the thermal decomposition of minerals from the white cement mortar, aggregates, and newly formed hydrated phases (CSH/CSAH gels).

The thermal analysis revealed four distinct temperature ranges for sample decomposition (**Figure 2**). Based on the known solid phases, these ranges were attributed to:

1) dehydration of crystal water (at 40-200°C) ,
2) dehydration of $Ca(OH)_2$ (at 420-470°C),
3) dehydration of CSH phases (at 520-700°C),
4) decarbonation of $MgCa(CO_3)_2$ and $CaCO_3$ (at 700-850°C).

All samples exhibited mass losses (ML) within specific ranges: 1.0-4.0% in the first range (28-120 days), 1.3-0.7% in the second range, 0.5-4.1% in the third range, and 2.3-3.3% in the fourth range (for sample C-S). However, for samples C-M and C-MH, ML values were significantly higher, reaching 27.6-29.2% (28-120 days). Despite the relatively low ML values in most samples, the observed trends suggest ongoing hydration processes. This is likely attributed to a decrease in portlandite content and a concomitant increase in water incorporated into the structure of CSH phases and crystal water.

The total ML increased from 12.20% (C-S028) to 40.70% (C-M120) due to the increased ML of water in the 1st and 3rd ranges and decarbonation in the 4th range. The ML ratio of Portlandite to CSH phase in samples C-S028 and C-S120 was approximately 1:1.

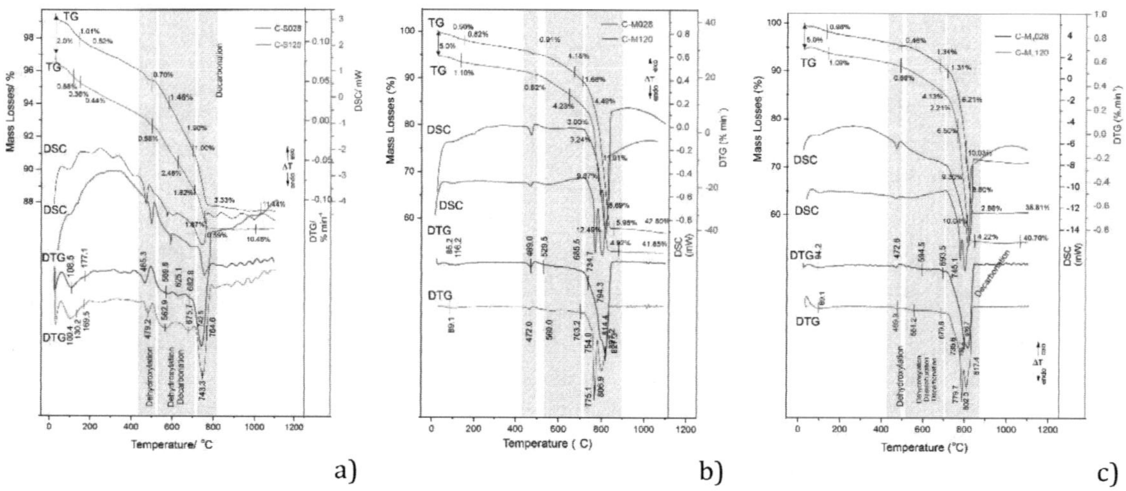

Figure 2. Results from thermal decompotion of samples: a) C-S028 and C-S120; b) C-M028 and C-M120; c) C-M_H028 and C-M_H120.

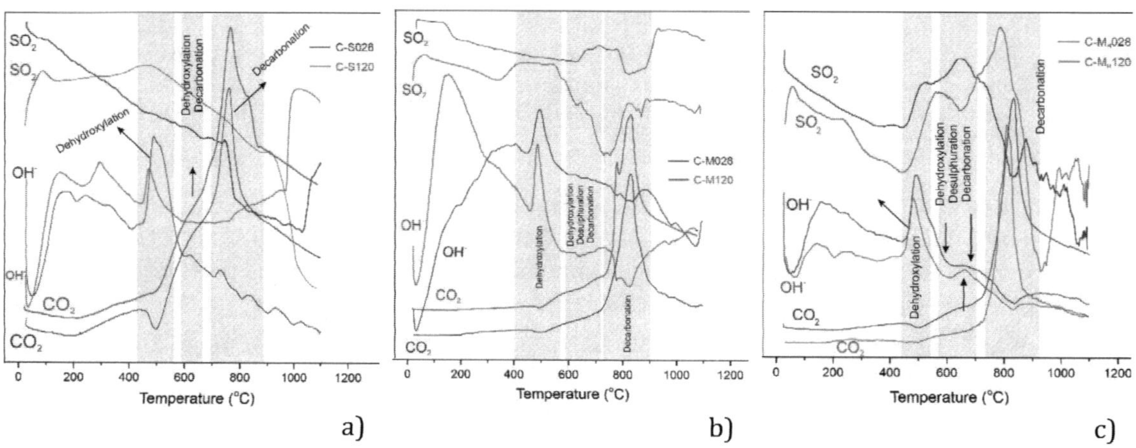

Figure 3. Results from Evolved Gas Analysis (EGA) of samples: a) C-S028 and C-S120; b) C-M028 and C-M120; c) C-M_H028 and C-M_H120.

The substitution of marble powder for sand as an aggregate resulted in a slight increase in portlandite hydration water content in C-M028 and C-M_H028, although this trend was reversed with prolonged hydration (120 days of curing). The water content within CSH phases increased from 0.94% in C-M28 to 1.41% in C-M120 and from 1.08% in C-M_H028 to 4.07% in C-M_H120. The HRWR additive facilitated the formation of a highly dense structure in sample C-MH, as evidenced by the distinct exothermic peak observed in the differential thermal analysis (DTA) curve of C-MH028 (**Figure 2c**) within the 420-470°C temperature range.

Decarbonization occurred at higher temperatures in samples C-M and C-MH compared to C-S, likely attributable to the denser structure, particularly in C-MH. These findings align with our previous research on the physical and mechanical properties of these samples, including density, pore volume and compressive strength [17].

Conclusion

Thermogravimetric analysis (TGA), differential thermal analysis (DTA), X-ray powder diffraction (PXRD), and scanning electron microscopy (SEM) collectively revealed the formation of new hydration phases in the studied cement mortars. Compaction of the structure led to an increase in CSH phases, predominantly evidenced by elevated mass losses. The reduced mass loss associated with crystal water dehydration is likely attributed to the water-reducing additive.

Comparative analysis of 28- and 120-day cured samples indicates that a stable structure formed at early stages, with minimal changes in phase composition and quantity during subsequent aging in water. The combination of white cement, marble powder, and fine particles, potentially augmented by water-reducing admixtures, suggests a promising approach for achieving high-durability cement mortars.

References

[1] Gagg CR. Cement and concrete as an engineering material: An historic appraisal and case study analysis. Engineering Failure Analysis. 2014 May 1;40:114–40.

[2] Li Y, Ren S, editors. 6 - Building Decorative Concrete and Mortar. In: Building Decorative Materials [Internet]. Woodhead Publishing; 2011, p. 92–109.

[3] Kaminskas R, Cesnauskas V, Kubiliute R. Influence of different artificial additives on Portland cement hydration and hardening. Construction and Building Materials. 2015 Oct 1;95:537–44.

[4] Tasdemir C. Combined effects of mineral admixtures and curing conditions on the sorptivity coefficient of concrete. Cement and Concrete Research. 2003 Oct 1;33(10):1637–42.

[5] López-Zaldívar O, Lozano-Díez RV, Verdú-Vázquez A, Llauradó-Pérez N. Effects of the addition of inertized MSW fly ash on calcium aluminate cement mortars. Construction and Building Materials. 2017 Dec 30;157:1106–16.

[6] Petrov BY, Zaharieva RA. Recycling of concrete construction and demolition waste in alternative binders: Part 1 – Technical study. IOP Conf Ser: Mater Sci Eng. 2020 Oct;951(1):012008.

[7] Figueira RB, Sousa R, Coelho L, Azenha M, de Almeida, JM, Jorge PAS, Silva CJR. Alkali-silica reaction in concrete: Mechanisms, mitigation and test methods. Construction and Building Materials. 2019 Oct 20;222:903–31.

[8] Struble L, Skalny J, Mindess S. A review of the cement-aggregate bond. Cement and Concrete Research. 1980 Mar 1;10(2):277–86.

[9] Glasser FP. Chemistry of the alkali-aggregate reaction. In: The Alkali-Silica Reaction in Concrete. CRC Press; 1991.

[10] Farran J. Contribution minéralogique à l'étude de l'adhérence entre les constituants hydratés des ciments et les matériaux enrobes, Thèse, 1957, 100 p. (in French)

[11] Feldman RF, Ramachandran VS, Sereda PJ. Influence of CaCO3 on the Hydration of 3CaO·Al2O3. Journal of the American Ceramic Society. 1965;48(1):25–30.

[12] Soroka I, Stern N. Calcareous fillers and the compressive strength of portland cement. Cement and Concrete Research. 1976 May 1;6(3):367–76.

[13] Tasong WA, Cripps JC, Lynsdale CJ. Aggregate-cement chemical interactions. Cement and Concrete Research. 1998 Jul 1;28(7):1037–48.

[14] Lawrence P, Cyr M., Ringot E. Mineral admixtures in mortars effect of type, amount and fineness of fine constituents on compressive strength. Cement and Concrete Research. 2005 Jun 1;35(6):1092–105.

[15] Vogt C. Ultrafine particles in concrete: Influence of ultrafine particles on concrete properties and application to concrete mix design. KTH, School of Architecture and the Built Environment (ABE); 2010.

[16] Aqel M, Panesar DK. Physical and chemical effects of limestone filler on the hydration of steam cured cement paste and mortar. Revista ALCONPAT. 2020 Aug;10(2):191–205.

IOP Conf. Series: Materials Science and Engineering **1323** (2024) 012005 doi:10.1088/1757-899X/1323/1/012005

[17] Petkova V, Stoyanov V, Pelovski Y. TG–DTG–DTA in studying white self-compacting cement mortars. J Therm Anal Calorim. 2012 Aug 1;109(2):797–806.
[18] EN 196-1:2016 - Methods of testing cement - Part 1: Determination of strength. CEN/TC 51 - Cement and building limes 33 p.;
[19] Powder Diffraction File (PDF), edited by S.Kabekkodu, copyright by International Centre for Diffraction Data (ICDD) 12 Campus Blvd., Newtown Square, Pennsylvania 19073-3272

IOP Conf. Series: Materials Science and Engineering **1323** (2024) 012006 doi:10.1088/1757-899X/1323/1/012006

Resistance of plastic anchors for fixing on ETICS in different types of base materials

Lachezar Hrischev[1] and Ivan-Aleksandar Conev[1*]

[1] Department of Construction Technology and Mechanization, University of Architecture Civil Engineering and Geodesy, Sofia, Bulgaria

E-mail: ivan_conev_fce@uacg.bg

Abstract. The durability of external thermal insulation composite systems and their resistance to wind load mainly depends on the resistance of plastic anchors by means of which the systems are fixed to the substrate. A large part of the facade walls are made of reinforced concrete and various types of masonry units. This article presents an experimental study of the resistance of plastic anchors for ETICS, with different types of base materials, typical of the facade walls of buildings implemented in Bulgaria. The resistance of anchors installed in base materials with different characteristics of hollow masonry units, solid masonry units, concrete base and autoclaved aerated concrete units was investigated. As a result of the study, it was found that the resistance under tension force of the plastic anchors was the highest for base of concrete, solid clay masonry units and high-density autoclaved aerated concrete. It was found that with the different types of masonry units, as the compressive strength of the base increases, the resistance of the plastic anchors also increases.

1. Introduction

It is characteristic of the Bulgarian construction practice that mechanically fixed external thermal insulation composite systems (ETICS) are most often applied. In this type of system, the horizontal wind load is taken up by the plastic anchors. For this reason, the bearing capacity of the plastic anchors has a significant impact on the load resistance and durability of ETICS [1].

One of the possible defects in thermal insulation systems is related to a partial or total compromise of the connection with the base material [2]. Both in the European and in the Bulgarian construction practice, there are known cases of compromised ETICS, due to poor-quality fixing to the facade [3-5]. Regarding the stability of ETICS, one may expect adverse effects due to the following: the increase in the size of the hole in which the plastic anchor is installed, the reduction of the number of plastic anchors for an area of 1 m^2, the reduction of the size of the head of the anchors, the lack of compaction in the critical (edge) areas of the building, lack of dust removal from the hole for installing the anchors, incorrectly selected installation scheme (location of anchors) and others [6,7]. The bearing capacity of the plastic anchors also depends on the type of base material, the thickness of the passing layers and the required anchoring depth [8].

As a component of ETICS, plastic anchors are used which have been assessed according to the European Assessment Document EAD 330196-00-0604 [9]. The European Technical Assessments (ETA) declare the resistance of the anchors in a base material with specific characteristics and if the base material of the structure does not meet these characteristics, it is necessary to carry out job-site pull-out tests [10, 11]. A study carried out by the authors of the article found that in the execution of the facade walls of buildings in Bulgaria, masonry units have been used with characteristics that are lower than those specified in some issued ETAs. This practically means that the data specified in the ETA cannot be directly used to determine the required number of anchors for an area of 1 m^2 and it is necessary to carry out job-site pull-out tests to determine their resistance.

The purpose of this study is to determine the resistance under tension loads of plastic anchors for fixing of ETICS, with different types of base materials, typical of the facade walls of buildings implemented in Bulgaria.

2. Materials and methods

Within the framework of the study, experimental samples of thirteen types of base materials were made, which are vertically perforated clay masonry units, horizontally perforated clay masonry units, normal weight concrete and masonry units from autoclaved aerated concrete. According to [9], the types of base materials can be classified into the following groups - group A, group B, group C and group E. Plastic anchors for fixing of ETICS were installed in the different types of base materials, and their resistance under tension loads was determined using a specialized device.

2.1 Test samples

Masonry units are of different sizes and are vertically and horizontally perforated. Five of the types are horizontally perforated, three types are vertically perforated, and one type is solid masonry units. There are three types of autoclaved aerated concrete units used. All masonry units are of different compressive strength and density, and their width, which determines the thickness of the base, is 250 mm. A fragment of concrete with a compressive strength of 41,5 N/mm^2 and dimensions of the fragment 200x200x1000 mm was also used.

The determination of the compressive strength of the masonry units and of the concrete was carried out by testing according to the applicable standards - BDS EN 12390-3, BDS EN 771-1 and BDS EN 771-4:2011+A1:2015/NA:2019 [12-14]. Examples of determining the compressive strength of masonry units and concrete specimens are presented in Figure1. The characteristics of the base material types used are summarized in Table 1.

IOP Conf. Series: Materials Science and Engineering **1323** (2024) 012006 doi:10.1088/1757-899X/1323/1/012006

a) b)

Figure 1. Determine the compressive strength of test samples
a) test sample "Type 4" – testing of horizontally perforated clay masonry units;
b) test sample "Type 10" - testing of a test sample of normal weight concrete;

Table 1. Characteristics of the types of the used base materials

Designation on the base materials (test sample)	Base materials type (test sample)	Group according to EAD 330196-00-0604	Base thickness (masonry units width), mm	Density, kg/m^3	Compressive strength, N/mm^2
„Type 1“	Vertically perforated clay masonry units	C	250	750,25	18,4
„Type 2“	Vertically perforated clay masonry units	C	250	662,86	16,6
„Type 3“	Vertically perforated clay masonry units	C	250	576,81	13,4
„Type 4“	Horizontally perforated clay masonry units	C	250	661,33	4,6
„Type 5“	Solid masonry units	B	120	1333,33	28,8
„Type 6“	Horizontally perforated clay masonry units	C	250	682,67	3,7
„Type 7“	Horizontally perforated clay masonry units	C	250	738,67	3,3
„Type 8“	Vertically perforated clay masonry units	C	250	641,31	10,3
„Type 9“	Vertically perforated clay masonry units	C	250	597,08	14,9
„Type 10“	Normal weight concrete	A	1000	2256,94	41,5
„Type 11“	Autoclaved aerated concrete masonry units	E	250	350,40	2,0
„Type 12“	Autoclaved aerated concrete masonry units	E	250	400,53	2,5
„Type 13“	Autoclaved aerated concrete masonry units	E	250	600,80	5,0

IOP Conf. Series: Materials Science and Engineering **1323** (2024) 012006 doi:10.1088/1757-899X/1323/1/012006

The investigated mechanical fasteners are a plastic anchor with a plastic push-in nail designed for fastening thermal boards, a component of ETICS. Anchors are applicable for installation in base materials of clay units for masonry, concrete, aerated concrete and others. The total length of the plastic anchor is 115 mm. The anchors have a stem diameter of 8 mm and a head diameter of 60 mm.

According to the manufacturer's instructions, the anchoring depths in the substrat are as follows: for clay masonry units ≥ 50 mm, for aerated concrete bases ≥ 65 mm, and for concrete bases ≥ 30 mm [15].

The preparation of the test samples includes drilling the holes, cleaning the holes by brushing and blowing air, and installing the plastic anchors (Figure 2). For the needs of the study, holes were drilled with rotary action and with hammer drilling of the light construction equipment (drilling machine), depending on the type of base material. In the experimental samples of horizontally and vertically perforated clay masonry units and autoclaved aerated concrete, 15 holes were drilled with rotary action, and in solid clay units for masonry and concrete, 15 holes were made with hammer drilling.

The experimental samples were pressed on a hydraulic press, in order to simulate a load characteristic of a units positioned in a facade masonry. Information about installation depth, method of drilling holes, drill diameter, etc. is summarized in Table 2.

Table 2. Plastic anchor installation information

Designation on the base materials (test sample)	Type of base materials (test sample)	Drilling method	Diameter of drill bits	Embedment depth in the base material, mm	Minimum spacing and minimum edge distance
„Type 1“ „Type 2” „Type 3” „Type 4” „Type 6” „Type 7” „Type 8” „Type 9”	Vertically and horizontally perforated clay masonry units	Rotary drilling	8 mm	≥ 50 mm	≥ 100 mm
„Type 5“	Solid masonry blocks	Hammer drilling	8 mm	≥ 50 mm	≥ 100 mm
„Type 10“	Normal weight concrete	Hammer drilling	8 mm	≥ 30 mm	≥ 100 mm
„Type 11“ „Type 12” „Type 13”	Autoclaved aerated concrete masonry units	Rotary drilling	8 mm	≥ 65 mm	≥ 100 mm

IOP Conf. Series: Materials Science and Engineering **1323** (2024) 012006 doi:10.1088/1757-899X/1323/1/012006

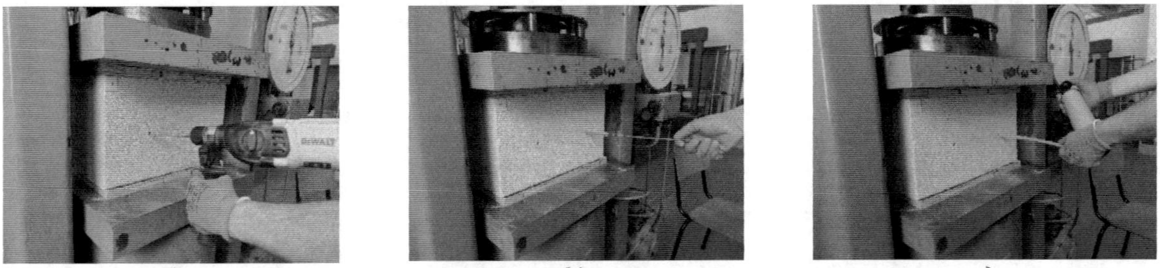

a) b) c)

Figure 2. Drilling and preparing the holes for installing the plastic anchors
a) drilling the hole; b) cleaning by brushing; c) cleaning by blowing air;

When installing the plastic anchors, additional attachments made of extruded polystyrene (XPS) of different thicknesses were used. By means of it, a layer of thermal insulation is simulated, in order to ensure the necessary depth of anchoring. The distance between the plastic anchors in the installation position, as well as between the anchors and the end of the edge of the test samples of different types of base materials, is at least 10 cm (Fig. 3).

a) b)

Figure 3. General appearance of installed plastic anchors in different types of base materials
a) installed plastic anchors in a "Type 2" base material;
b) installed plastic anchors in "Type 11" base material;

2.2 Determination of the resistance under tension loads

To test the plastic anchors, a specialized device was used to determine the resistance under tension load (Figure 4). The device with which the experiment is performed is "Hydrajaws M2000 pull-out tester" [16]. It is a specialized system for testing fasteners (plastic anchors, anchors, scaffolding anchors, etc.) under tensile load. It contains a mechanical screw and a calibrated load cell.

Its function is to measure the load applied directly to the fixing. It also contains an electronic digital manometer, on which the obtained load values are displayd in kN. There is also a built-in displacement indicator with a 50 mm scale. The maximum tensile load this device can reach is 25 kN, with an accuracy of 0,1 kN. The device's system allows tests to be automatically recorded and compiled into a digital report using the app on a mobile phone or tablet.

In the test itself, the device is placed perpendicular to the test samples after removing the attachment from the XPS and is leveled. Leveling in a horizontal position is achieved with the help of the water level built into the device. The load is increased continuously, in order to reach a final load at which destruction occurs after about 1 minute. It is loaded until the anchor pulls out of the base material or the plastic anchor breaks. The value of the force at ultimate load is recorded [17].

The testing of the plastic anchors installed in the different types of base materials is presented in Figure 4.

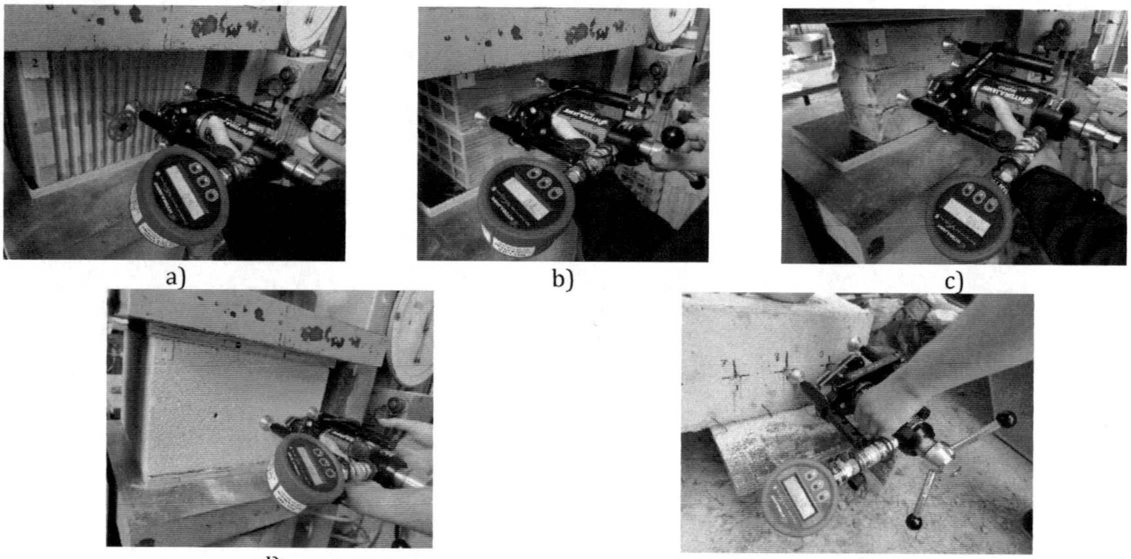

a) b) c) d) e)

Figure 4. Testing plastic anchors under tensile loading in different types of base materials
a) base of vertically perforated clay masonry units; b) base of horizontally perforated clay masonry units; c) base of solid clay units; d) base of autoclaved aerated concrete masonry units; e) base of normal weight concrete;

3. Results and discussion

In each type of base material, 15 plastic anchors for ETICS were tested. Thus, the determined values of ultimate loads make it possible to evaluate the resistance under tension loads of the plastic anchors.

The characteristic tensile resistance of the plastic anchors (N_{Rk}) is assumed to be determined by processing the results, using the equation given in BS 8539 [18]. The following dependency is used, according to the standard:

$$N_{Rk}=N_{Ru,m}.(1-K.v), \tag{1}$$

Where:

N_{Rk} – characteristic resistance under tension force;

$N_{Ru,m}$ – mean value of ultimate load in the test series;

K – coefficient that takes into account the number of tests. The values of K are taken from standard statistical tables:

- for 15 tests K =2,33;
- for 10 tests K =2,57;
- for 5 tests K =3,40.

v – the coefficient of variation of the failure loads, and is given by:

$$v=s/N_{Ru,m}\,.100\%, \tag{2}$$

Where:

s – the standard deviation of failure loads about the mean value.

The test results of the plastic anchors in the different types of base materials were processed according to formulas (1) and (2) and are presented in Table 3 and Figure 5. The failure type of the anchors at the maximum tensile load was also analyzed. Two forms of failure were found when testing the plastic anchors - failure by pulling out (slipping) the plastic anchor from the base and failure by destruction of the plastic anchor body.

Table 3. Results of the plastic anchor pull-out tests

Type of base	Method of drilling the hole	$N_{Ru,m}$, kN	*K*	*s*	*v*	N_{Rk}, kN
„Type 1“	Rotary drilling	0,66	2,33	0,05	0,08	0,54
„Type 2“	Rotary drilling	0,60	2,33	0,08	0,13	0,42
„Type 3“	Rotary drilling	0,56	2,33	0,06	0,11	0,41
„Type 4“	Rotary drilling	0,54	2,33	0,10	0,18	0,31
„Type 5“	Hammer drilling	0,89	2,33	0,10	0,11	0,66
„Type 6“	Rotary drilling	0,31	2,33	0,07	0,23	0,14
„Type 7“	Rotary drilling	0,25	2,33	0,05	0,20	0,13
„Type 8“	Rotary drilling	0,25	2,33	0,05	0,20	0,13
„Type 9“	Rotary drilling	0,52	2,33	0,08	0,15	0,34
„Type 10“	Hammer drilling	0,93	2,33	0,08	0,09	0,74
„Type 11“	Rotary drilling	0,51	2,33	0,08	0,16	0,32
„Type 12“	Rotary drilling	0,59	2,33	0,07	0,13	0,41
„Type 13“	Rotary drilling	0,83	2,33	0,07	0,10	0,66

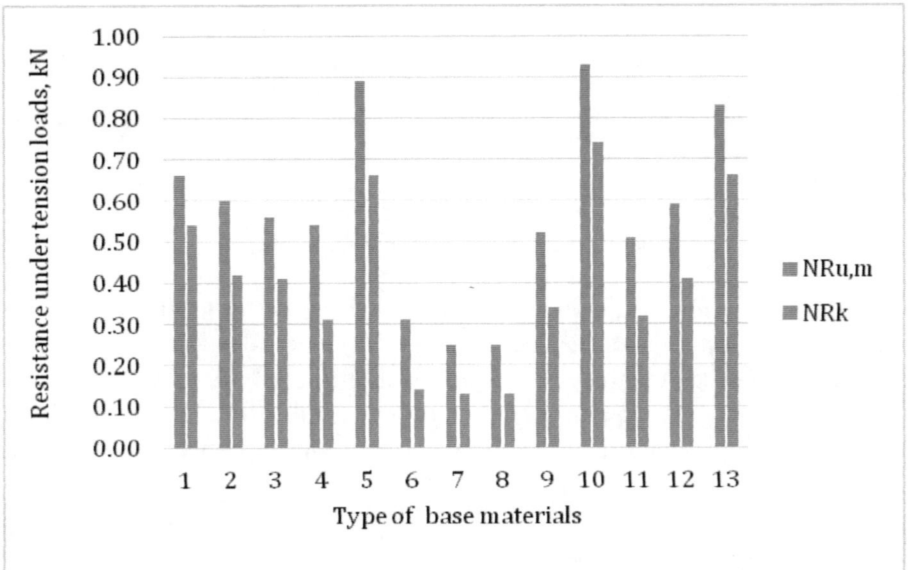

Figure 5. Mean ($N_{Ru,m}$) and characteristic (N_{Rk}) values of resistance under tension loads

From the results presented in Figure 5, it is clear that the highest mean and characteristic values of resistance under tension force have the plastic anchors installed in concrete base (base "Type 10") and in solid clay masonry units ("Type 5" base material). For perforated masonry units, the highest values of resistance were determined for base "Type 1", followed by base "Type 2", and the lowest values for base "Type 6", base "Type 7" and base "Type 8". For masonry units made of autoclaved aerated concrete, the highest values of resistance were determined for the "Type 13" base, and the lowest values for the "Type 11" base. It should be noted that the plastic anchors installed in the horizontally perforated clay masonry units are characterized by lower values of resistance under tension force compared to those with vertically perforated clay masonry units are.

During the performed visual assessment, two forms of destruction were found when testing the plastic anchors - destruction by pull-out (slipping) of the plastic anchor from the base and destruction in the body of the plastic anchor. Examples of the observed forms of failure are presented in Figure 6. Failure by pull-out (slipping) of the plastic anchors was found in all types of horizontally and vertically perforated clay masonry units, as well as in autoclaved aerated concrete "Type 11" and "Type 12" base materials. In the case of solid brick base material ("Type 5" base material), concrete base material ("Type 10" base material) and autoclaved aerated concrete base material ("Type 13" base material), both types of destruction are found. Especially with the "Type 5" base and the "Type 10" base, in most of the tested plastic anchors, the failure is in the body of the plastic anchor. Failure in the plastic anchor body occurs at a failure force in the range of 0,9 kN to 1,0 kN.

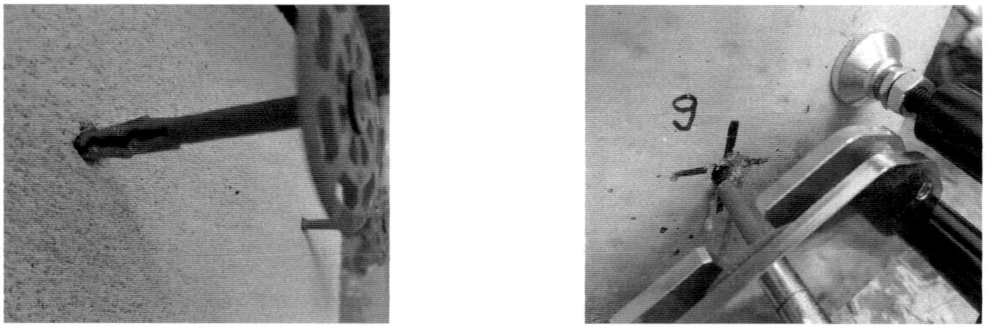

a) b)

Figure 6. Forms of damage found when testing the plastic anchors
a) destruction by pull-out (slipping) of the plastic anchor from the base material;
b) destruction of the plastic anchor body;

The analysis of the obtained results shows that the different types of bases are characterized by different compressive strength, and different values of the bearing capacity of the plastic anchors were also obtained. For example, the anchors installed in masonry units is characterized by a lower resistance compared to those installed in a concrete base. The reason being the lower strength characteristics of the masonry units themselves, the horizontal and vertical perforation, etc.

Further analysis of the results was performed using regression analysis. The relationship between the compressive strength of the clay masonry units and the load capacity of the plastic anchors, as well as between the compressive strength of the autoclaved aerated concrete units and the load capacity of the anchors installed in them, was determined. The resulting relationships, the corresponding regression equations and the determined coefficients of determination are presented in Figure 7 and Figure 8.

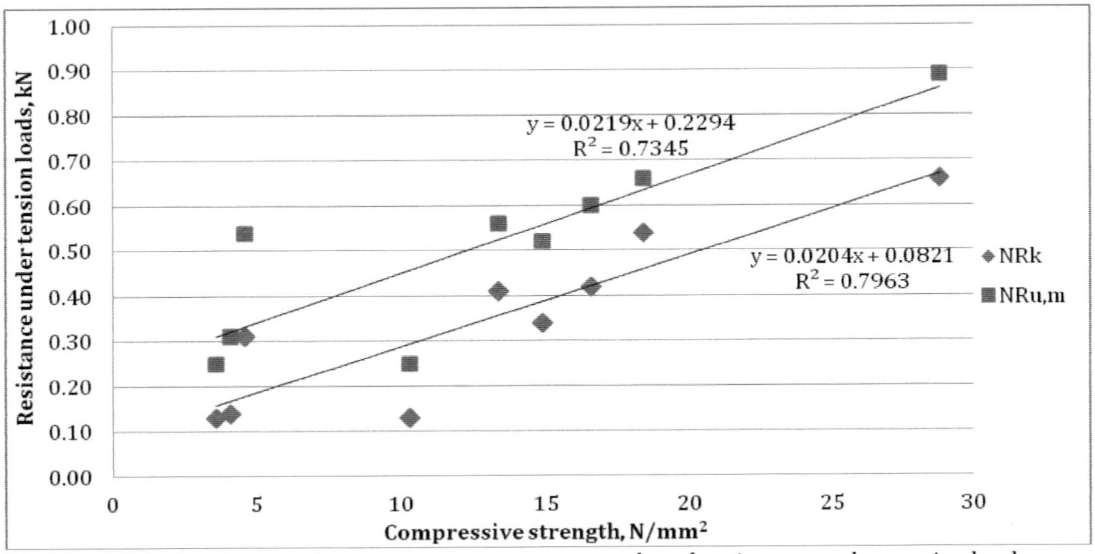

Figure 7. Relationship between compressive strength and resistance under tension loads in test samples of clay masonry units

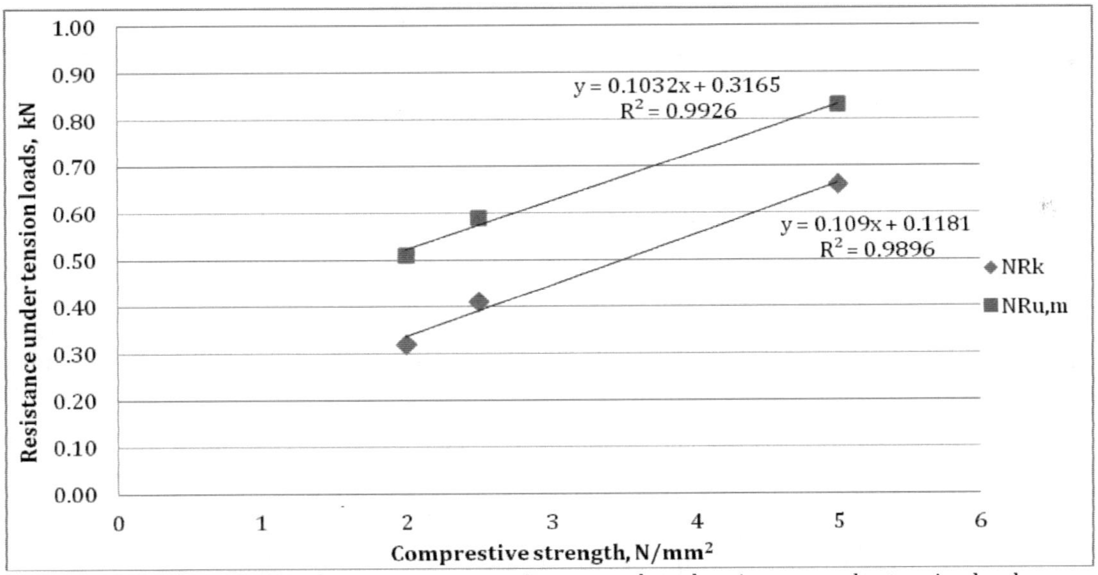

Figure 8. Relationship between compressive strength and resistance under tension loads in test samples of autoclaved aerated concrete masonry units

An analysis of Figure 7 and Figure 8 shows that there is a relationship between the compressive strength of the base and the resistance under tension loads of the plastic anchors. As the compressive strength of the base increases, the resistance of the plastic anchors also increases. For the base material of clay masonry units, the coefficients of determination are R^2=0,7963 and R^2=0,7345, which enables the correlation dependence to be defined as high. When evaluating the results, it should be taken into account that different cavity masonry units, in addition to having different compressive strength, are also characterized by different hollowness, web thickness and number of webs through which the plastic anchors pass through.

For the base material of masonry units of autoclaved aerated concrete, the determined coefficients of determination have a value higher than R^2=0,99. There is a very high correlation between the compressive strength of the base material and the bearing capacity of the anchors. It should be noted here that only three masonry units with different characteristics were studied, but nevertheless, the relationship is clearly observed.

4. Conclusions

The paper presents an experimental study of the resistance plastic anchors for fixing of ETICS installed in different types of base materials - clay masonry units, autoclaved aerated concrete masonry units and concrete.

The highest values of resistance under tension force were found for base materials of concrete, solid clay masonry units and high-density autoclaved aerated concrete, and these types of base materials are also characterized by the highest compressive strength.

A relationship was established between the compressive strength of the base materials and the resistance under tension force of the plastic anchors. For masonry unit base material types, as the compressive strength of the base material increased, the resistance of the plastic anchors also increased. A similar dependence is also observed for the types of base materials from autoclaved aerated concrete masonry units.

It can be summarized that in order to secure ETICS against wind loads, the resistance under tension load of the plastic anchors needs to be determined by site tests for each specific case.

Acknowledgments

The current research development under contract No. D-154/2023 is financially supported by the Centre for Scientific Research and Design at the University of Architecture Civil Engineering and Geodesy.

The authors express their gratitude to the company "Wienerberger Bulgaria", the company "Baumit Bulgaria" and the company "Xella Bulgaria" for the support and the provided materials.

Special thanks are also expressed to the team of the University Construction Testing Laboratory at the Department of "Construction Materials and Insulations" at the UACEG, for the comprehensive assistance during the experimental part of the research.

References

[1] EAD 040083-00-0404 xternal thermal insulation composite systems (ETICS) with renderings, January 2019.
[2] Amaro, B., Saraiva, D., Brito, J., Flores-Colen, I., October 2013. Inspection and diagnosis system of ETICS on walls. Construction and Building Materials, Volume 47, pp 1257-1267. https://doi.org/10.1016/j.conbuildmat.2013.06.024
[3] Florea, V., 2012. External thermal insulation of buildings with expanded polystyrene. Buletinul Institutului Politehnic Din Iasi, Tomul LVIII (LXII), Fasc. 3: 45-51.
[4] Angelo, Lucchini, A., Mazzucchelli, E., Stefanazzi, A., Tattoni, S., 2017. Analysis and control of façade claddings structural issues, TEMA: Technologies Engineering Materials Architecture, Vol. 3, No. 1, pp 88-100
[5] https://www.stroiteli-bg.com/novini/padna-izolatsiya-na-blok-v-studentski-grad/, (visited on 27.06.2024)

[6] Sulakatko, V., Lill I., Witt E., 2016. Methodological framework to assess the significance of External Thermal Insulation Composite System (ETICS) on-site activities. Energy Procedia, 96: 446 – 454. https://doi.org/10.1016/j.egypro.2016.09.176

[7] Sulakatko, V., Vogdt F., 2018. Construction Process Technical Impact Factors on Degradation of the External Thermal Insulation Composite System. Sustainability 2018, 10 (11), art. no. 3900. https://doi.org/10.3390/su10113900

[8] Sinica, M., Sezemanas, G., 2010. Determination of fixing loads for plastic anchors subject to structure and voidage of masonary units, 10th International Vonference Modern building materials, structurs and techniques, 19-21 may, Vilnius, Lithuania; pp 536-543.

[9] EAD 330196-00-0604 Plastic anchors for fixing of external thermal insulation composite systems with rendering, 2016.

[10] European Technical Approval ETA-07/0302, Hilti ETICS nailed-in anchor SDK-FV 8.

[11] European Technical Approval ETA-11/0192, EJOT H1 eco, EJOT H4 eco and ejotherm H1.

[12] BS EN 12390-3:2019 – Testing hardened concrete - Part 3: Compressive strength of test specimens.

[13] BDS EN 772-1:2011+A1:2015 – Methods of test for masonry units - Part 1: Determination of compressive strength (in Bulgarian).

[14] BS EN 771-4:2011+A1:2015/NA:2019 – Specification for masonry units - Part 4: Autoclaved aerated concrete masonry units - National Annex (NA).

[15] https://baumit.bg/produkti/external-wall-insulation-components/dubel/baumit-dubel-ha, (visited on 28.04.2024)

[16] https://www.hydrajaws.co.uk/product/m2000-pro-tester-kit-anchor-tester/, (visited on 28.04.2024)

[17] EOTA Technical report TR 051, Recommendations for job site tests of plastic anchors and screws, 2016.

[18] BS 8539:2012 Code of practice for the selection and installation of post-installed anchors in concrete and masonry.

IOP Conf. Series: Materials Science and Engineering **1323** (2024) 012007 doi:10.1088/1757-899X/1323/1/012007

Effect of the washers on the bearing resistance of the bolted connections

Lyubomir Zdravkov [1*], Vatyu Tanev [1] and Dobromir Dinev [2]

[1] Department "Steel, Timber and Plastic Structures", University of Architecture Civil Engineering and Geodesy, Sofia, Bulgaria
[2] Department "Structural Mechanics", University of Architecture Civil Engineering and Geodesy, Sofia, Bulgaria

*E-mail: zdravkov_fce@uacg.bg

Abstract. Bolted connections are the most common method of joining thin-walled galvanized sheets on site. Small diameter bolts (M8, M10 or M12) and the corresponding nuts are used for this purpose. Washers should be provided both under the heads of the bolts and the nuts in case of a single shear connection with one row of bolts, according to the standard BDS EN 1993-1-8. I.e. it is unnecessary to place washers under the bolt head in case of two or more rows of bolts. In laboratory tests of bolted connections with two rows of bolts without washers under the nuts, is accounted bearing capacity lower than the theoretically determined one. The initial explanation is that the lack of flat washers results in high bolt rotations and hence reduced load resistance. To verify this seemingly logical explanation, the authors has conducted series of finite element analyses (FEA) on connections with two rows of bolts. It turned out that this was not the reason for the reduced bearing resistance. The presence of washers under the bolt head and the nut had a negligible effect.

1. Introduction

For the expansion of the capacity of a warehouse for grain storage in southern Bulgaria, in 2021 five grain storage silos were ordered, delivered and mounted. They are produced of thin-walled galvanized steel sheets with a thickness of 0.8 ÷ 1.25 mm. To speed up and facilitate the mounting processes, all steel elements are connected by bolts.

On August 16, 2021, around 8:30 am, a few minutes after the beginning of first discharge of one of them, its cylindrical shell ruptured and it was completely destroyed. As it fell to the ground, the silo and the attached to it elevator tower hit and ruptured the wall of the adjacent silo.

A team from the University of Architecture, Civil Engineering and Geodesy (UACEG) was invited to the site of the accident for inspection and to establish the reasons for its occurrence. To perform this, the team's members carefully examined the construction of the ruptured facility, tested samples of the embedded material and made control calculations. In the process of work were tested several joints of overlapping sheets, realized with two bolts M10, class 8.8 each. The sheets had a thickness of t = 1.25 mm. The obtained laboratory results of the resistance of bolted connections $F_{b,test,Rk}$ are compared with the analytically determined resistance $F_{b,1,25,Rk}$, according to the methodology of BDS EN 1993-1-3:2007 [2]. It turned out that the characteristic bearing resistance $F_{b,1,25,Rk}$ is greater than the experimentally determined $F_{b,test,Rk}$. The first explanation was that this was due to the lack of flat washers under the nuts in the tested bolted connections. As a result, the rotation of the bolts is not restricted, which leads to large local deformations under the bolt's heads and nuts.

To verify this seemingly logical explanation, the authors has conducted a series of numerical tests on single-shear bolted connections. In half of them, washers are placed under the heads of the bolts and nuts. In the other half, washers do not exist. In addition, he varied the thickness of the connected sheets and the distance between the bolts.

2. Determining the bearing capacity of the bolts

The research was conducted with samples with a type of connection, shape and dimensions, shown in fig. 1. The last are selected to meet the validity limits specified in (7):

nominal thickness of the connected sheets – $t = 0.8 \div 2.5$ mm;
the sheets are hot-dip galvanized with zinc coating Z600;
joining fasteners – two bolts M10, class 8.8;
holes for the bolts – $d_o = 11$ mm;
$e_1 = 30$ mm $> 1.0d_o = 11$ mm;
$e_2 = 29$ mm $> 1.5d_o = 16.5$ mm;
$p_1 = 50$ mm or 85 mm $> 3.0d_o = 33$ mm,

where e_1 is the distance between the centre of the bolt to the adjacent edge of joined part, in the direction parallel to the direction of load transfer, see fig. 2;

e_2 – the distance between the centre of the bolt to the adjacent edge of joined part, in the direction perpendicular to the direction of load transfer;

p_1 – the distance between the centres of the bolts in the direction parallel to the direction of load transfer.

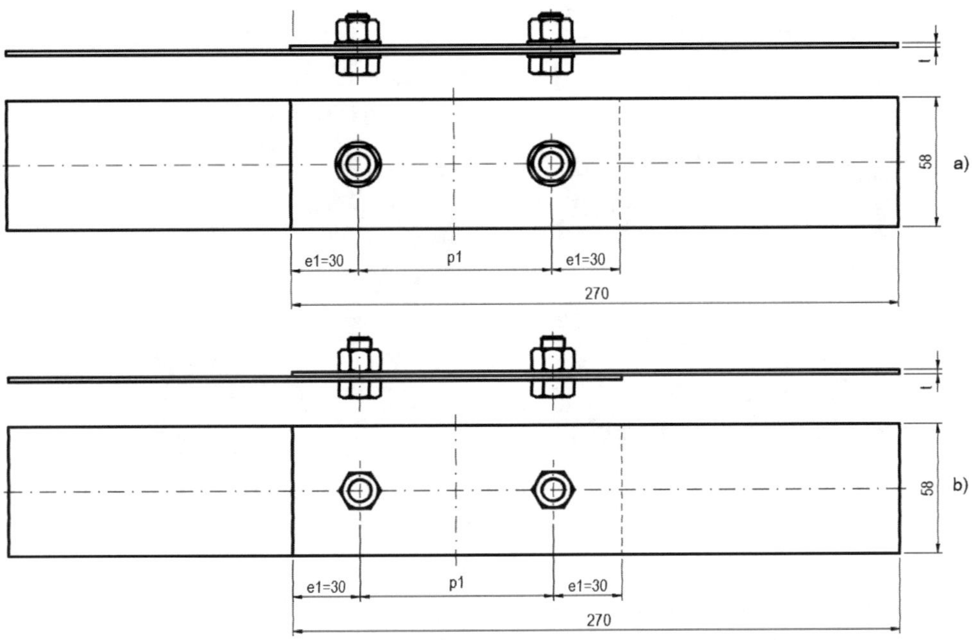

Figure 1. Shape and dimensions of tested specimens with bolts; a) with washers, b) without washers

The half of the single-shear bolted connections between the two sheets are realized by use of bolts, nuts and washers, see fig. 1*a*. The second half of the joints - only through bolts and nuts, see fig. 1*b*. The purpose is to determine the influence of washers in this type of single-shear connection.

Because the sheets are hot-dip galvanized, only their steel core $t_{core,i}$ is considered in the research. It is determined by the expression:

$$t_{core,i} = t_{nominal,i} - 2t_{coating}\,, \tag{1}$$

where $t_{nominal,i}$ are the nominal thicknesses of the sheets;
$t_{coating}$ = 42 μm – the shown in BDS EN 10346:2015 [3] average value of the thickness of zinc coating Z600 at each protected surface.

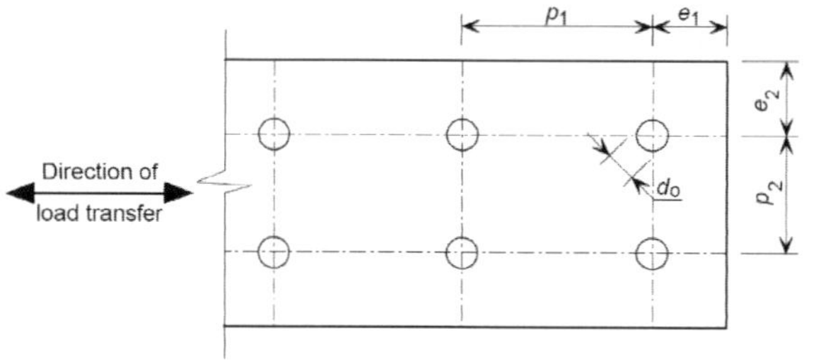

Figure 2. Symbols for end and edge distances and spacing of fasteners

The steel in the sheets is HX420LAD. Its mechanical properties were determined experimentally in the laboratory of UACEG, see fig. 3.

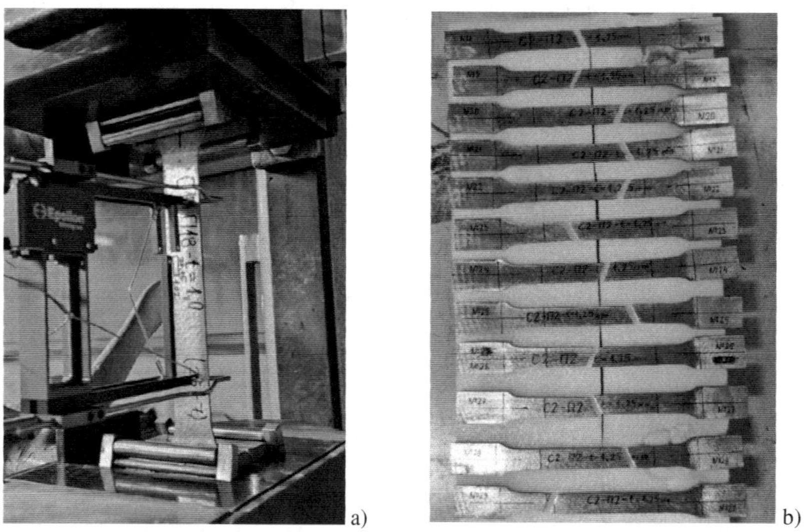

Figure 3. Laboratory tests of specimens to determine the real mechanical characteristics of steel HX420LAD, a) extensiometer, b) test specimens

For each of the tested specimens was built ratio stress–strain (σ–ε), see fig. 4.

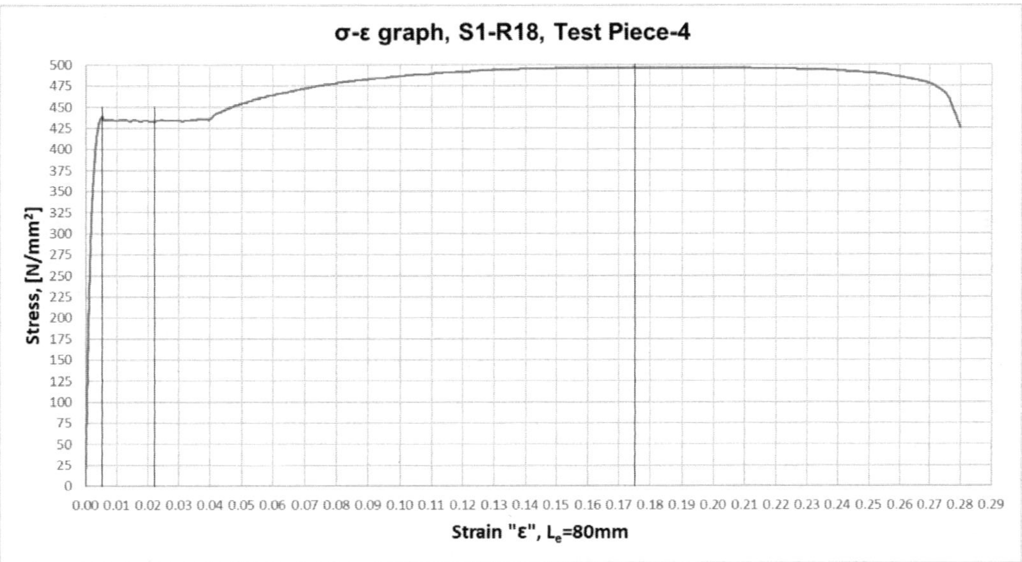

Figure 4. Real diagram σ-ε of steel HX420LAD obtained by the lab tests

The mechanical properties of the steel, determined in the laboratory of UACEG are as follows:
f_y = 412 MPa – the characteristic value of the yield strength;
f_u = 480 MPa – the characteristic value of ultimate strength;
E = 210 000 MPa – the modulus of elasticity;
$A_{80} \geq 17$ % – the relative elongation.

2.1. Capacity of the bolted connection calculated according to the standard's provisions

The analytical determination of the bearing resistance of bolts, joining thin-walled steel sheets is specified in the standard BDS EN 1993-1-3:2007 [2]. The written there formulas are:

a) bearing resistance

$$F_{b,Rd} = 2.5\alpha_b k_t d.t \frac{f_u}{\gamma_{M2}}, \tag{2}$$

where d is the nominal diameter of the bolt;
t – the core thickness of the thinner connected part of the sheets;
f_u – the ultimate tensile strength of the sheet's material;
γ_{M2} = 1.25 – the partial factor for safety;

$$a_b = \min\begin{cases} 1.0 \\ \dfrac{e_1}{3d} \end{cases} \tag{3}$$

$$k_t = \frac{0.8t + 1.5}{2.5} \text{ when } 0.75 \text{ mm} \leq t \leq 1.25 \text{ mm} \tag{4}$$

$$k_t = 1.0 \text{ when } t > 1.25 \text{ mm} \tag{5}$$

b) shear resistance

The design shear resistance of bolts class 8.8 could be calculated by the expression:

IOP Conf. Series: Materials Science and Engineering **1323** (2024) 012007 doi:10.1088/1757-899X/1323/1/012007

$$F_{v,Rd} = 0.6A_s \frac{f_{ub}}{\gamma_{M2}}, \tag{6}$$

where A_s is the cross-sectional area of the bolt in the shear plane;

f_{ub} – the ultimate tensile strength of the bolts.

Formulas (2-6) are valid when the shown in the standard BDS EN 1993-1-3:2007 [2] requirements are met:

$$\begin{array}{ll} 0.75\text{mm} \le t < 3\text{mm} & f_u \le 550\text{MPa} \\ e_1 \ge 1.0d_o & p_1 \ge 3.0d_o \\ e_2 \ge 1.5d_o & p_2 \ge 3.0d_o \end{array} \tag{7}$$

The presented above analytical methodology does not take into account the presence of washers and the distance p_1 between the centres of the bolts.

2.2. Capacity of the bolted connection obtained by a numerical analysis

Numerous numerical models of single-shear bolted connections were created using the software product ANSYS [4]. The shape and dimensions of the connections are shown in fig. 1. The bolts M10, nuts and washers have shape and dimensions according to standards EN ISO 4014:2000 [5], EN ISO 4032:2000 [6] and EN ISO 7089:2000 [7].

Translations and rotations at the left edge of the models are limited in all directions. The right edges of the models can only be translated in the longitudinal direction. The remaining two longitudinal edges are free. Tensile forces are applied to the right edges of the models as displacements parallel to their longitudinal axes. An element solid186 was used for the modelling of the thin-walled sheets. It is applied when simulating the plastic work of the material. Solid186 elements have 20 nodes - at the corners and in the middle of the edges. Each node has three degrees of freedom. The maximum size of the elements in the numerical models, see fig. 5, is 2 mm.

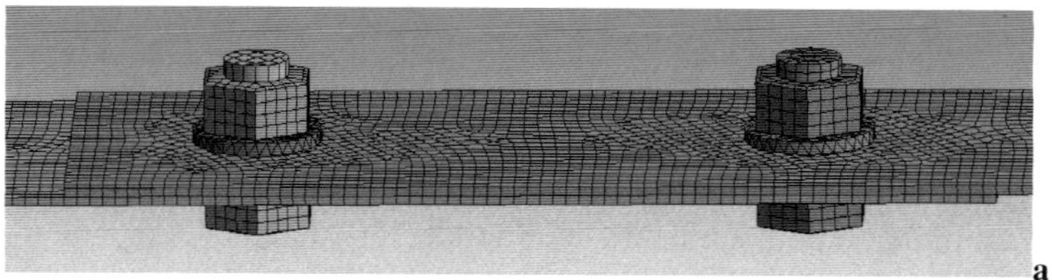

a)

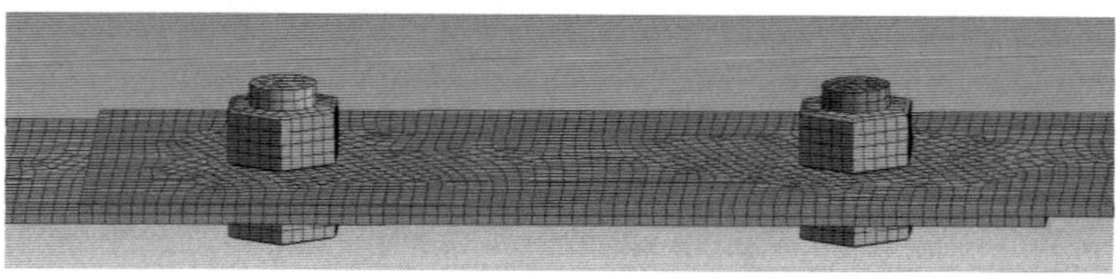

b)

Figure 5. Numerical model of bolted connection, a) bolted connection with washers, b) bolted connection without washers

IOP Conf. Series: Materials Science and Engineering **1323** (2024) 012007 doi:10.1088/1757-899X/1323/1/012007

Determined by laboratory tests in UACEG mechanical properties of steel HX420LAD are shown above.

The material nonlinearity of the steel (MNA) was taken into account in the numerical modelling. For this purpose, the relationship between stresses σ and strains ε in HX420LAD steel is represented as trilinear, see fig. 6, and is as close as possible to the determined one, see fig. 4.

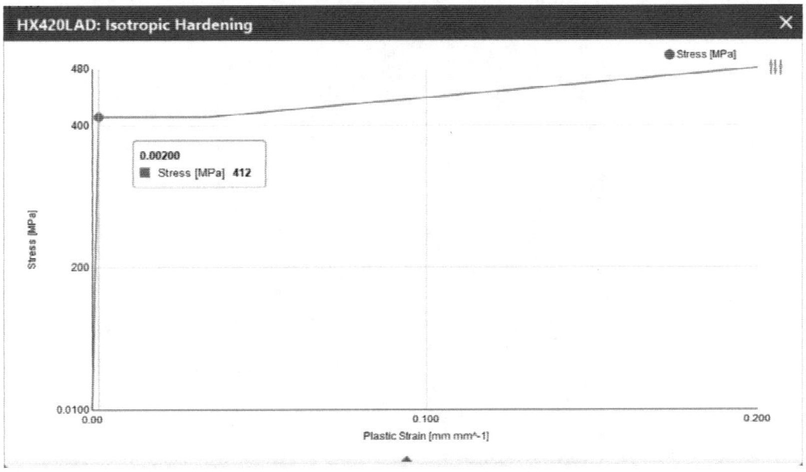

Figure 6. Multilinear ratio stress-strain (σ- ε) of steel HX420LAD

The ratio σ-ε of steel in bolts and nuts, class 8.8, is simulated as bilinear, see fig. 7.

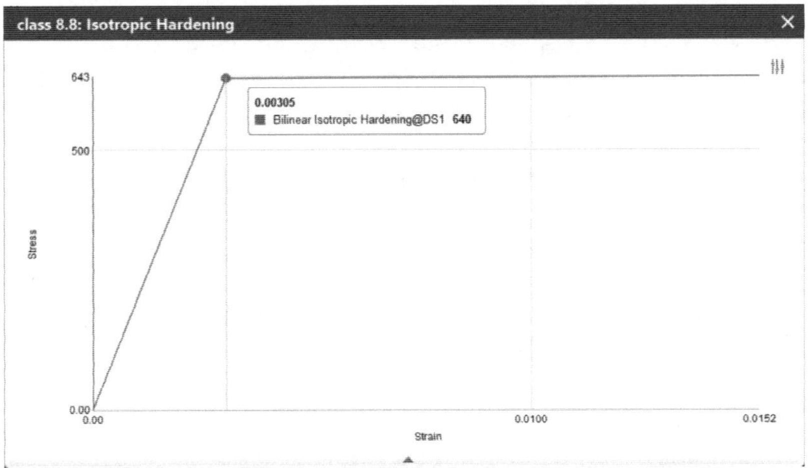

Figure 7. Bilinear ratio stress-strain (σ- ε) of the steel for bolts, class 8.8

The bearing capacity of bolted connections is determined as follows:

a) in the right, linearly movable ends of the models, were applied displacements in the longitudinal direction with a maximum value of Δ = 7 mm. The aim is that, with the chosen lengths of the models, the normal stresses in the jointed sheets reach the yield point;

b) the displacement is implemented gradually, in steps. Their minimum number is 65, and the maximum is 800. Convergence of forces, displacements and rotations is monitored in the solution;

IOP Conf. Series: Materials Science and Engineering **1323** (2024) 012007 doi:10.1088/1757-899X/1323/1/012007

c) the reactions in the left (fixed) supports give information about the forces in the bolted connections. The biggest force is the bearing capacity of the bolts.

2.3. Laboratory tests

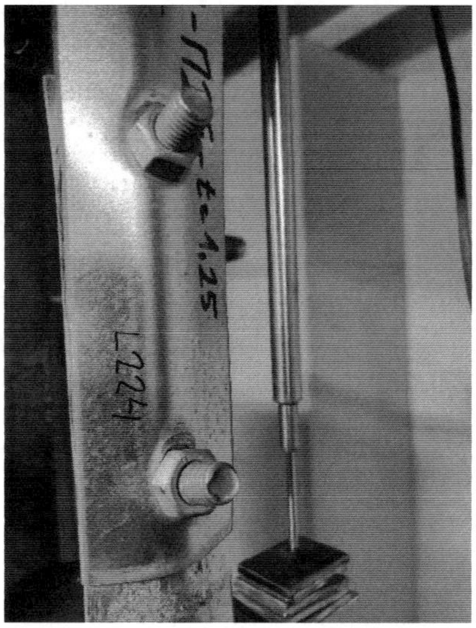

Figure 8. Testing of bolted connection

A loading machine with a constant speed of 5mm/min was used to test the bolted connections. The increase in the deformations of the test samples is 0.001 s-1, which is in accordance with the BDS EN ISO 6892-1:2020 standard [8]. Calibrated sensors for displacements, forces and strains (extensometers) were used to record the experimental data. The data from the sensors are amplified and recorded using a QuantumX universal amplifier, model 840B, of the company "HBM". The frequency for reading and recording the results in all series of experimental tests is a minimum of 20 Hz.

The extensometer has an accuracy class of 0.5 according to ISO 9513:2012 [9].

3. Results

3.1. By the analytical methodology

Determined by formulas (2-5) characteristic values of bearing resistance $F_{b,Rk}$ of one bolt M10 are shown in Table 1.

The characteristic value of the shearing resistance $F_{v,Rk}$ of one bolt M10, class 8.8 is calculated by the expression:

$$F_{v,Rk} = 0.6 A_s f_{ub} = 0.6 \cdot 0.58 \cdot 80 = 27.84\text{kN} .$$

It could be noted that for all thicknesses of the sheets, except for the biggest one, limiting is the bearing capacity $F_{b,Rk}$.

Table 1. Characteristic bearing capacity of the bolts $F_{b,Rk}$, determined analytically, according to BDS EN 1993-1-3:2007 [2]

№	$t_{nominal}$ mm	$t_{coating}$ mm	t_{core} mm	width mm	d mm	e_1 mm	p_1 mm	α_b	k_t	f_u MPa	$F_{b,Rk}$ kN
1	0.8	0.042	0.716	58	10	30	50	1.0	0.829	480	7.124
							85	1.0	0.829		7.124
2	1.02	0.042	0.936	58	10	30	50	1.0	0.900	480	10.103
							85	1.0	0.900		10.103
3	1.29	0.042	1.206	58	10	30	50	1.0	0.986	480	14.268
							85	1.0	0.986		14.268
4	1.5	0.042	1.416	58	10	30	50	1.0	1.0	480	16.992
							85	1.0	1.0		16.992
5	2	0.042	1.916	58	10	30	50	1.0	1.0	480	22.992
							85	1.0	1.0		22.992
6	2.5	0.042	2.416	58	10	30	50	1.0	1.0	480	28.992
							85	1.0	1.0		28.992

3.2. By the numerical methodology

Rotation of the bolts and related deformations in the connected steel sheets are accounted in the numerical models, see fig. 9. They are similar to the deformations in the steel specimens in the laboratory tests, see fig. 12.

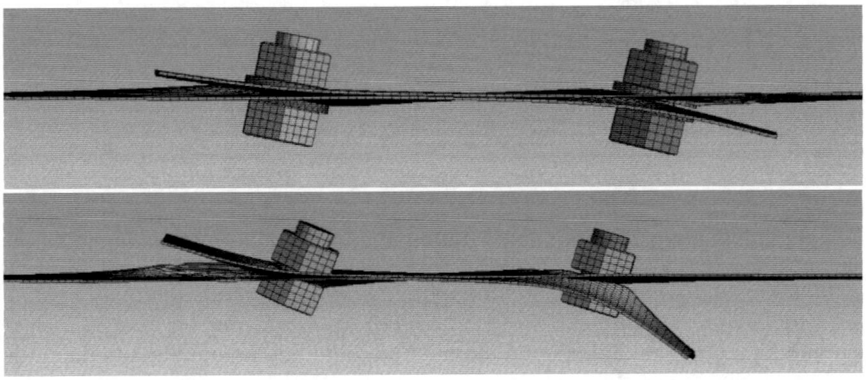

a)

b)

Figure 9. Rotation of the bolts and deformations of the sheets in the numerical models; a) in bolted connection with washers, b) in bolted connection without washers

The ovalization of the bolt holes is especially characteristic, see fig. 10. Such is observed during the tests in the laboratory of UACEG, see fig. 13, as in the studies of *Kim et al.* [11], *Kim and Cho* [12], *Lim et al.* [13], *Khalifa et al.* [14].

The relationship force-displacement in the numerical models is shown in fig. 11-12. It is characteristic here that before reaching the maximum bearing capacity $F_{b,FEA,Rk}$ of the bolted connection, an almost linear force-displacement relationship is observed. A drop in bearing capacity is accounted after reaching $F_{b,FEA,Rk}$, which is initially small, smooth. As deformations increase, the bearing capacity decreases more intensively. In the laboratory tests of *Kim et al.* [11], *Lim et al.* [13] and *Khalifa et al.* [14] also is observed a drop in the bearing capacity after reaching its limit value, but they do not have such clear almost linear dependence at the beginning.

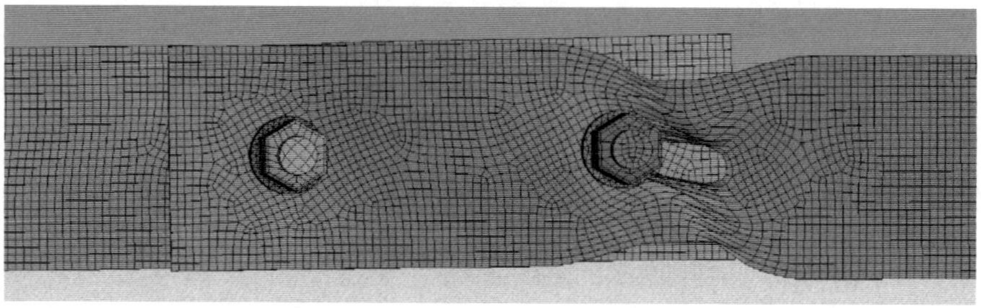

Figure 10. Ovalization in bolt's holes

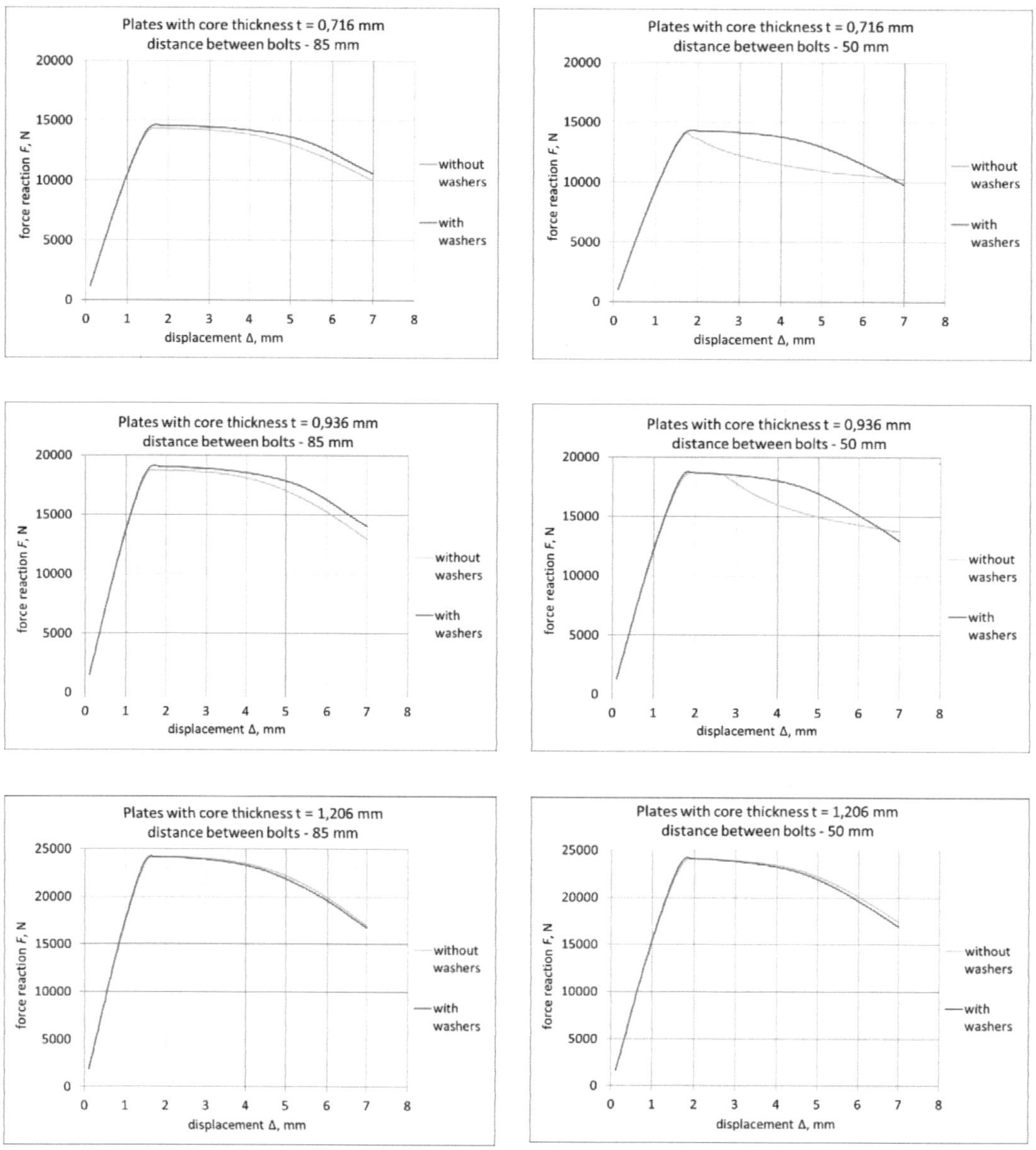

Figure 11. Force – displacement ratio in the researched single shear bolted connections, thickness 0.716- 1.206 mm

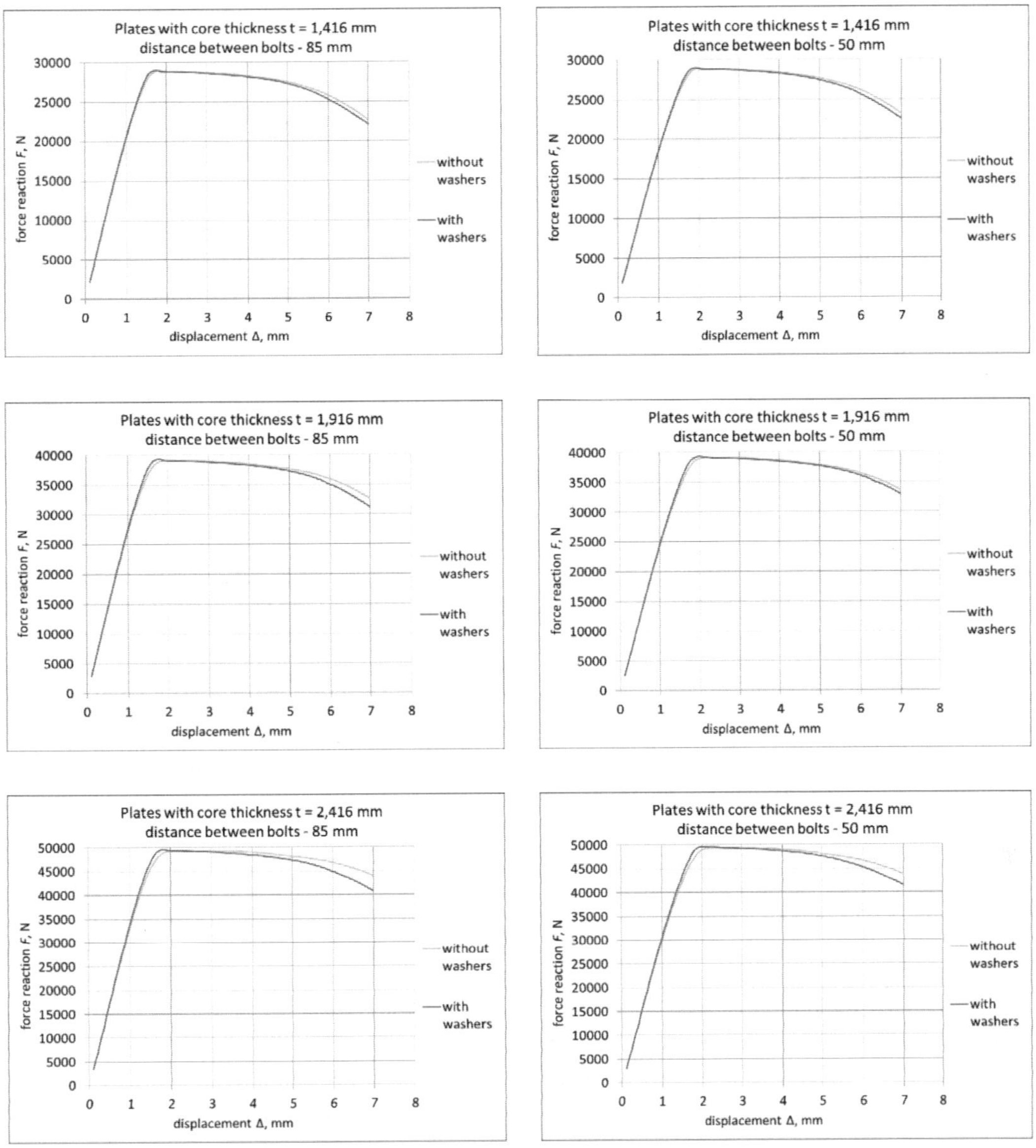

Figure 12. Force – displacement ratio in the researched single shear bolted connections, , thickness 1.416- 2.416 mm

Determined by the numerical models characteristic values of the bearing resistance $F_{b,FEA,Rk}$ of one bolt M10 are shown in Table 2.

Table 2. Characteristic bearing capacity of the bolts $F_{b,FEA,Rk}$ determined by numerical simulation

№	$t_{nominal}$ mm	$t_{coating}$ mm	t_{core} mm	width mm	d_o mm	e_1 mm	p_1 mm	washer	f_u MPa	$F_{b,FEA,Rk}$ kN
1	0.8	0.042	0.716	58	11	30	50	without	480	7.056
								with		7.166
							85	without		7.167
								with		7.308
2	1.02	0.042	0.936	58	11	30	50	without	480	9.364
								with		9.374
							85	without		9.376
								with		9.57
3	1.29	0.042	1.206	58	11	30	50	without	480	12.088
								with		12.09
							85	without		12.092
								with		12.102
4	1.5	0.042	1.416	58	11	30	50	without	480	14.462
								with		14.493
							85	without		14.458
								with		14.496
5	2	0.042	1.916	58	11	30	50	without	480	19.56
								with		19.628
							85	without		19.558
								with		19.635
6	2.5	0.042	2.416	58	11	30	50	without	480	24.681
								with		24.797
							85	without		24.748
								with		24.74

It could be seen that the presence of washers under the bolt's heads and the nuts increase, although not significantly, the bearing capacity of the bolted connection. I.e. it is confirmed what was written in BDS EN 1993-1-8 [1], that for two or more rows of bolts it is not necessary to place washers under the bolt's head. In other words, the lack of washers is not the reason for the lower load capacity of the bolted connection, determined experimentally. Moreover, increasing the distance p_1 between the centres of the bolts has practically no effect. Perhaps it is no coincidence that these parameters are unconsidered in the analytical methodology, written in the standard BDS EN 1993-1-3:2007 [2]. The disturbing thing here is that the analytically determined values of the bearing capacity $F_{b,Rk}$ are greater not only than the experimentally determined values, but also from the numerical modelling $F_{b,FEA,Rk}$. The reasons for this may be due to:

a) incorrect numerical analysis by finite elements;

b) an error in the methodology shown in the standard BDS EN 1993-1-3:2007 [2].

To prove the corrections of the numerically determined values of $F_{b,FEA,Rk}$ they are compared with those determined by laboratory tests, see p.3.3.

IOP Conf. Series: Materials Science and Engineering **1323** (2024) 012007 doi:10.1088/1757-899X/1323/1/012007

3.3. By the laboratory tests

The rotation of the bolts and the deformations in the connected steel sheets in the laboratory tests are shown in fig. 13, and the ovalization of the holes – in fig. 14.

Determined by laboratory tests characteristic values of the bearing resistance $F_{b,test,Rk}$ of one bolt M10 are shown in Table 3.

Comparing the values of the numerically and experimentally determined bearing capacities of the bolted connections, it can be seen that $F_{b,FEA,Rk}$ fall between the lowest and highest values of $F_{b,test,Rk}$. In addition, identical bolt rotation and sheet and hole deformations were observed in both test methods. I.e. it could be assumed that the finite element analysis is correctly performed. This suggests a possible error in the methodology, shown in the standard BDS EN 1993-1-3:2007 [2].

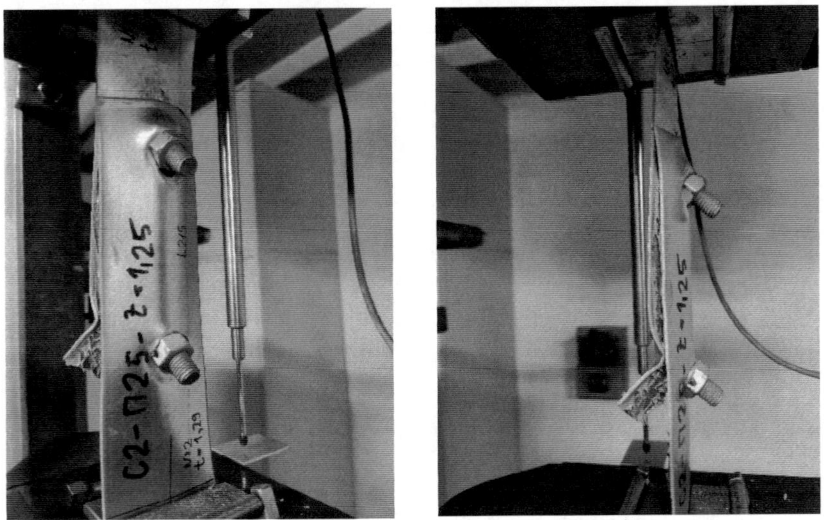

a) b)

Figure 13. Rotation of the bolts and deformations of the sheets in laboratory tests; a) front view, b) side view

Figure 14. Ovalization of the holes, observed in the laboratory tests

IOP Conf. Series: Materials Science and Engineering **1323** (2024) 012007 doi:10.1088/1757-899X/1323/1/012007

Table 3. Characteristic bearing capacity of the bolts $F_{b,test,Rk}$ determined by laboratory tests

№	$t_{nominal}$ mm	$t_{coating}$ mm	width mm	$F_{b,test,Rk}$ kN
1	1.29	0.042	58	11.9
2	1.28	0.042	58	12.3
3	1.28	0.042	49	11.5
4	1.28	0.042	58	12.55
5	1.29	0.042	58	13
6	1.29	0.042	55	12.5

The characteristic load resistance of bolted connections in steel sheets with small thickness, considering the deformations in the bolt holes, calculated by the American standard AISI S100-16 [10], should be determined by the expression:

$$F_{b,Rd} = (4.64\alpha t + 1.53) d \cdot t \cdot f_u , \tag{8}$$

where α is the coefficient for the conversation of units, as follows:

= 1.0 for U.S. customary units (with t in inches);

= 0.0394 for SI units (with t in mm).

Determined by formula (8) characteristic values of bearing capacity $F_{b,AISI,Rk}$ of one bolt M10 is shown in Table 4.

Table 4. Characteristic bearing capacity of the bolts $F_{b,AISI,Rk}$, determined analytically according to standard AISI S100-16 [10]

№	$t_{nominal}$ mm	$t_{coating}$ mm	t_{core} mm	width mm	d mm	f_u MPa	$F_{b,AISI,Rk}$ kN
1	0.8	0.042	0.716	58	10	480	5.708
2	1.02	0.042	0.936	58	10	480	7.643
3	1.29	0.042	1.206	58	10	480	10.133
4	1.5	0.042	1.416	58	10	480	12.159
5	2	0.042	1.916	58	10	480	17.293
6	2.5	0.042	2.416	58	10	480	22.865

Comparing the values of $F_{b,Rk}$ in Table 1, of $F_{b,FEA,Rk}$ in Table 2 and of $F_{b,AISI,Rk}$ in Table 4 it is obvious that $F_{b,Rk} > F_{b,FEA,Rk} > F_{b,AISI,Rk}$. I.e. the shown in BDS EN 1993-1-3:2007 [2] methodology is the most liberal, and in AISI S100-16 [10] – the most conservative. The numerically determined values $F_{b,FEA,Rk}$ are between the two extremes, but are closer to $F_{b,AISI,Rk}$. From there could be concluded that the specified in BDS EN 1993-1-3:2007 [2] methodology needs additional research and improvement.

4. Conclusions

Several joints of overlapped sheets, connected by two bolts M10, class 8.8, were tested in the laboratory of the UACEG to determine the reasons for the destruction of a steel grain silo. The obtained laboratory results for bearing capacity $F_{b,test,Rk}$ are compared with the analytically determined $F_{b,Rk}$, according to the methodology of BDS EN 1993-1-3:2007 [2]. The characteristic bearing capacity of the single-shear bolted connection $F_{b,Rk}$ is greater than the experimentally determined $F_{b,test,Rk}$. The authors conducted series of numerical tests of bolted connection to investigate the reason for this discrepancy. By them is accounted that the presence of washers under the bolt's heads and the nuts increases, although not significantly, the load resistance of the bolted connection. I.e. confirms what is written in BDS EN 1993-1-8 [1], it is not necessary to place washers under the bolt head with two or more rows of bolts. Moreover, increasing the distance p_1 between the centres of the bolts has practically no effect. The worrying thing here is that the values of the bearing capacity $F_{b,Rk}$ determined accordingly the standard BDS EN 1993-1-3:2007 [2] are bigger not only than the experimentally determined $F_{b,test,Rk}$, but also from the numerical modelling $F_{b,FEA,Rk}$. They are even bigger than those determined according to the American standard AISI S100-16 [10]. From this, it can be concluded that the European methodology needs further research and improvement.

Acknowledgements

The authors would like to express their gratitude to the Research, Consultancy and Design Centre (RCDC) of University of architecture, civil engineering and geodesy for the financial support, provided by contract BN-294/24.

References

[1] BDS EN 1993-1-8:2005, Eurocode 3: Design of steel structures - Part 1-8: Design of joints. Bulgarian Institute for Standardisation.

[2] BDS EN 1993-1-3:2007, Eurocode 3: Design of steel structures - Part 1-3: General rules. Supplementary rules for cold-formed members and sheeting. Bulgarian Institute for Standardisation.

[3] BDS EN 10346:2015, Continuously hot-dip coated steel flat products for cold forming - Technical delivery conditions. Bulgarian Institute for Standardisation.

[4] ANSYS® 2020 R2 Documentation. Ansys Inc., Canonsburg, PA, the USA.

[5] EN ISO 4014:2000, Hexagon head bolts – Product grades A and B. European Committee for Standardization, Brussels.

[6] EN ISO 4032:2000. Hexagon nuts, style 1 – Product grades A and B. European Committee for Standardization, Brussels.

[7] EN ISO 7089:2000. Plain washers – Normal series – Product Grade A. European Committee for Standardization, Brussels.

[8] BDS EN ISO 6892-1:2020. Metallic materials - Tensile testing - Part 1: Method of test at room temperature (ISO 6892-1:2019). Bulgarian Institute for Standardisation.

[9] ISO 9513:2012. Metallic materials — Calibration of extensometer systems used in uniaxial testing. ISO/TC 164/SC 1.

[10] AISI S100-16, 2016, North American Specification for the Design of Cold-Formed Steel Structural Members, 2016 Edition. American Iron and Steel Institution.

[11] T. Kim, H. Kuwamura, S. Kim, Y. Lee, T. Cho, 2009, Investigation on ultimate strength of thin-walled steel single shear bolted connections with two bolts using finite element analysis. Thin-Walled Structures, 47 pp. 1191-1202. DOI:10.1016/j.tws.2009.04.009.

[12] T. Kim, Y. Cho, 2012, Ultimate Behaviour of Single Shear Bolted Connections with Thin-Walled Aluminium Alloys (6061-T6), Advanced Materials Research, 446-449, 3441-3445. DOI:10.4028/www.scientific.net/AMR.446-449.3441.

IOP Conf. Series: Materials Science and Engineering **1323** (2024) 012007 doi:10.1088/1757-899X/1323/1/012007

[13] J. Lim, T. Kim, S. Kim, 2013, Ultimate strength of single shear bolted connections with cold-formed ferritic stainless steel. Journal of Zhejiang University-SCIENCE A (Applied Physics & Engineering), 14 (2), pp. 120-136. DOI:10.1631/jzus.A1100316.

[14] A. Khalifa, I. El-Aghoury, S. Ibrahim, 2022, Structural Performance of Bolted Thin-Walled Steel Truss Connections with C-Channel and L-Angle Sections. International Journal of Research in Engineering & Management, 5 (2), pp. 79-88. DOI: 10.13140/RG.2.2.11163.75048.

IOP Conf. Series: Materials Science and Engineering **1323** (2024) 012008 doi:10.1088/1757-899X/1323/1/012008

Numerical simulation of mechanical behavior of reinforced concrete column - short corbel assembly repaired and strengthened by bonding composite materials

Veselin Stankov[1], Ivelina Ivanova[2], Jules Assih[2] and Dimitar Dontchev[1]

[1] Department of Applied Mechanic, University of Chemical Technology and Metallurgy, Sofia, Bulgaria
[2] Department of Civil Engineering, University of Reims Champagne-Ardenne, Reims, France

*E-mail: stankov_veselin@uctm.edu

Abstract. The experimental investigation of reinforced concrete structures is a high time resource consuming process. One of the best solutions to reduce the consumed resource is the finite element simulation of those structures. The biggest challenge for the engineers modeling reinforced concrete structures is to choose the proper mathematical model who can represent with good accuracy the mechanical behavior. The study is focused on the Numerical simulation of the mechanical behavior of reinforced concrete structures column - short corbel assembly repaired and strengthened by bonding composite materials. The presented simulation is developed in two steps. The first loading step leads to complete damage to the structure. In the second step, it is assumed that the concrete section is fully repaired, but the material properties of the steel reinforcement are taken from the first step. In the second step, the structure is subjected to cyclically increasing load. The mechanical behavior is successfully simulated and the model was prepared for future parametrical study.

1. Introduction

Civil engineers are often faced with problems related to the safety and durability of building structures. According to their category of use or function, structures can obtain significant changes during their life, reducing their load-bearing capacity. The load-bearing capacity of existing building structures can be reduced due to construction or design errors, changes in category of use, changes in loads, accidental events such as earthquakes or vehicle impacts, etc. There are various methods of strengthening and repairing a structure or part. Conventional methods are limited to shotcrete (pressure spraying of cement mortar), reinforcement by gluing metal plates, reinforcement by additional prestressing, and reinforcement with composite materials.

One solution would be to strengthen these structures with a carbon fiber composite [1], [2], [3] [4]. Engineers' lack of knowledge in structural reinforcement and rehabilitation makes the industry cautious about using this type of technology. However, they recognize the need for in-depth studies of the mechanical behavior of building structures strengthened and repaired by composite materials. All these problems require a thorough knowledge of the mechanical behavior of materials. Fabricating the reinforced concrete (RC) elements, which will be then destructively examined, is a difficult, resource-consuming, and expensive process. In addition, the concrete structure could be tested after 28 days. Many researchers have worked on RC corbel or column [1], [2], [3], etc. This paper is focused on the mechanical behavior of RC column-short corbel assembly. Not much information was found in the bibliographic research information about research done on considered structure was found in the literature review [4]. The aim of this project is to develop a three-dimensional model that simulates the mechanical behavior of considered structure. These structures are used in administrative buildings, manufactures, bridges, etc. This assembly has complex behavior because it is subjected to compression, tension, shear forces, and bending. Investigating and understanding the mechanical behavior could avoid complex problems in construction using these structural elements and save lives. A 3D model was developed using two axes of symmetry. The constitutive materials were experimentally characterized and the data was introduced in FE simulation. To validate the results of the computer simulation, several test bodies were fabricated and experimentally investigated. The data from these experiments was compared with basic parameters such as load capacity, failure, cracking pattern, failure modes etc.

2. Experimental program

2.1. Materials characterisation

The constitutive materials of this research are concrete, steel reinforcement bars, adhesive, and carbon-fiber-reinforced polymer (CFRP). An experimental campaign was carried out to characterize the mechanical properties of this material. Cylindrical test specimens with 320mm height and 160mm diameter were fabricated using the same concrete as the RC column short corbel assembly. The cylinders were tested on compression. The steel reinforcement HA14, HA10, and Ø6 were tested on simple tension. The CFRP and the adhesive were also tested on simple tension. The results are presented in Table 1.

Table 1. Properties of materials

Materials	Young's Modulus (GPa)	Ultimate Strength (MPa)	Poisson's ratio
Concrete	30±2	33.2±2	0.25
Steel bars	200±1	610±10	0.3
Adhesive	4.1±1	36±1	0.41
CFRP	86±1	775±65	0.45

2.2. Structure dimensions

On the lathery sides of the RC column, two short trapezoidal corbels were cantilevering encastred symmetrically. The shear span-to-depth ratio of this structure is less than unity. In considered structure, the shear span-to-depth ratio is 0,45. The length of the column is 2660mm, with cross section 150/300mm. The cantilever projection length of the RC corbels is 250mm, with 150mm width The total depth at the column face is 360 mm and 200mm at the end. The reinforcement of the column consists of four steel bars HA14, and stirrups with 6mm diameter paced every 90mm along the column. The main reinforcement of the corbel is HA10 mm in diameter and is located at the top of the corbels with an effective concrete cover of 25 mm. To ensure the anchorage of the main reinforcements, at its ends, two crossbars with the same diameter are welded. A horizontal closed stirrup of 6mm diameter acts as additional reinforcement for the corbels, which are placed

110mm from the main tension reinforcement. The model dimensions are presented in Figure 1. The RC column short corbel assembly specimen was fabricated and tested in real dimensions.

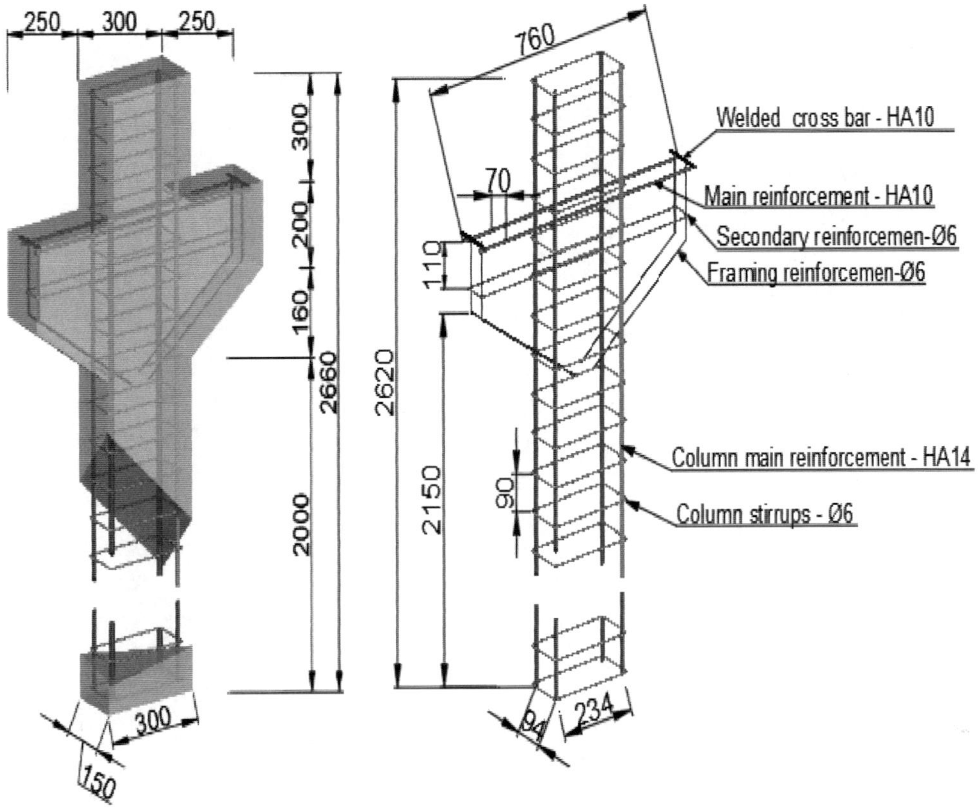

Figure 1. Dimensions and steel reinforcement of RC column short corbel assembly

2.3. Setting up the experimental study

The Department of Civil Engineering of the University of Reims Champagne-Ardenne, France possesses a test slab with a machine with a hydraulic jack with a capacity of 1000 kN. The RC column-short corbel assembly was instrumented with a strain gauge glued to the steel reinforcement and CFRP. Three-wire cable was soldered to the strain gauge to reduce interference induced during testing by other machines. All cable was connected to an acquisition system. The strain gauge was protected to prevent damage by the humidity or mechanical actions during the pouring of concrete such as vibrations. The position of the strain gauge on the steel reinforcement bars is shown in Figure 2a. The structure was subjected to static load until its total collapse in reverse position compared to reality with a loading speed of 0.02kN/s. In order to avoid the local crushing of concrete in the support zone, a 3 mm hardwood plate was installed. The load is applied to the test body through the base of the column by a jack powered by the hydraulic unit. The strain data was captured every 1,0s. Force-Strain curve of the strain gauge are given in [1].

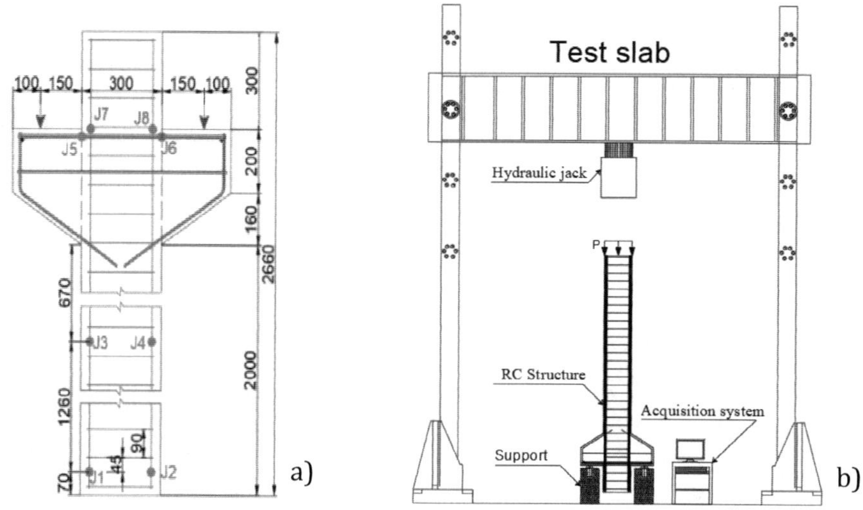

Figure 2. a) Positions of the strain gauge in the steel reinforcement bars. b) Test slab in the civil engineering laboratory of the University of Reims Champagne-Ardenne, France.

3. Finite element simulation

3.1. Properties of the constitutive materials

3.1.1. Concrete

The finite element model was developed using the software ABAQUS CAE. This software was chosen by a literature review [2], because of its ability to work with an "Explicite dynamic" type of analysis, to disperse the steel reinforcement in the concrete volume, well result on nonlinear analysis in the plastic domain, etc. In this software the model proposed by LUBLINER, et al., 1989 [3] and later modified by LEE, et al., 1998 [4] for concrete was incorporated. The model called Concrete Damaged Plasticity (CDP) uses the concept of isotropic damaged elasticity in combination with isotropic tension and compressive plasticity representing inelastic behaviour. The CDP model allowed us to reproduce the irreversible deformations. The failure mechanism assumed in this model is tensile cracking and compressive crushing of the concrete. This is represented by five parameters that have to be introduced. These parameters are as follows:

- dilation (dilatancy angle) angle ψ measured in the p–q plane at high confining pressure. Based on sensitivity analysis we took 36 as the value;
- flow potential eccentricity - ϵ, defining the rate at which the hyperbolic flow potential approaches its asymptote [5]. The taken value is ϵ=0.1;
- ratio between the biaxial compressive yield strength f_{b0} and the uniaxial compressive yield strength f_{c0}. The taken value is fb0 /fc0 = 1.16 [5];
- ratio of the second invariant of the stress on the tensile meridian, defining the form of the yield surface. The following conditions have to be met 0.5< Kc<1. For this study is Kc = 2/3 [5];
- viscosity parameter μ. This parameter is not useful in an Explicite type of analysis;

Equation (1) presents the constitutive model of CDP :

$$\sigma = (1 - d)E_0(\varepsilon - \widetilde{\varepsilon^{pl}}) \quad (1)$$

In this equation ε *is* -strain, $\widetilde{\varepsilon^{pl}}$ is the plastic equivalent strain, d is scalar damage variable, and, E_0 is initial Young`s modulus of undamaged concrete.

Defining the compression, tension behavior, and damage parameters for concrete that needed to be introduced in the FE analysis it's a big challenge and many researchers were working on that topic [6], [7], [8], [9] etc. The model proposed by F. ALFARAH [10], was chosen for our research. This model is based on the research of Z. BAZANT and B. OH [6] and D. HORDIJK [11].

3.1.2. Steel reinforcements

The steel reinforcement bars used in the model are the same as in the experimental investigation. It's defined three different sections – HA14, HA10, and Ø6. The yielding stress of the reinforcement is obtained at 557 MPa. The mechanical behavior of the steel is assumed elastic perfectly plastic.

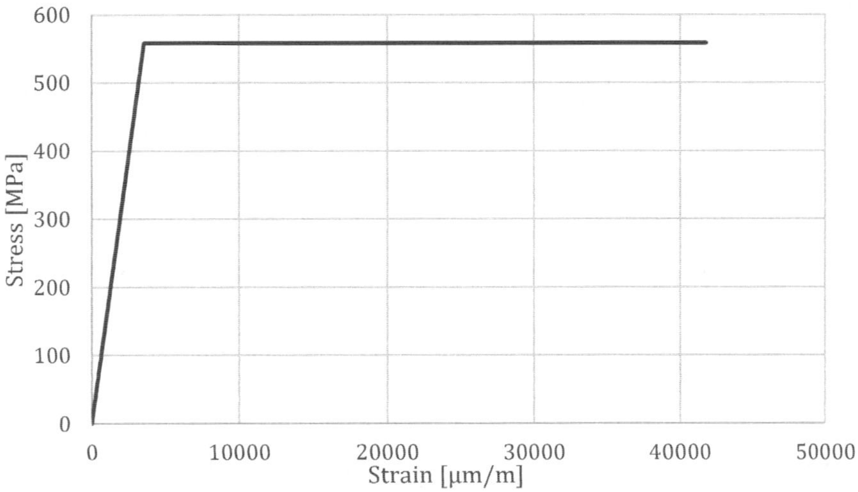

Figure 3. Elastic perfectly plastic behaviour of steel reinforcements

3.1.3. Carbon-fiber reinforced polymer (CFRP)

The RC column-short corbel assembly is strengthened using a multiple unidirectional CFRF sheet. The thicknesses of the CFRP and adhesive are considered to be constant, as is perfect contact between them. The mixing rule was used to determine Young's modulus in the fiber direction, where it is assumed that when an axial force is applied in the longitudinal direction, the load is shared between the fiber and the matrix [12]. Equations (2) and (3) describe the main condition:

$$\varepsilon_c = \varepsilon_f = \varepsilon_m, \quad (2)$$

$$P_c = P_f = P_m \quad (3)$$

where $\boldsymbol{\varepsilon_c}$, is the strain of CFRP, $\boldsymbol{\varepsilon_f}$ is the strain of fiber, $\boldsymbol{\varepsilon_m}$ is the strain of matrix, $\boldsymbol{P_c}$ is the stress of CFRP, $\boldsymbol{P_f}$ is the stress of fiber, $\boldsymbol{P_m}$ is the stress of matrix. CFRP bonded with adhesive creates an equal layer, but the adhesive layer between the concrete surface and the composite material has to be excluded [13], [14], [15]. Then the mechanical properties can be estimated as follows:

The Young's modulus along the direction of the fibers was determined by:

$$E_1 - E_f V_f + E_m V_m, \quad (4)$$

IOP Conf. Series: Materials Science and Engineering **1323** (2024) 012008 doi:10.1088/1757-899X/1323/1/012008

where E_f, is the fiber modulus, V_f is the fiber volume fraction, E_m is the matrix modulus and V_m can be determined by the equation (5).

$$V_m = 1 - V_f \tag{5}$$

The Young's modulus transverse direction of the fibers was determined by:

$$E_2 = \frac{E_f E_m}{V_m E_f + V_f E_m}, \tag{6}$$

where $\frac{1}{E_2} = \frac{V_m}{E_m} + \frac{V_f}{E_f}$. (7)

3.1.4. *Adhesive*

The adhesive used in our model is Sikadur®-330. It is a two-component structural epoxy adhesive. The main mechanical properties were taken from the datasheet given by the manufacturer. The tensile Strength is 30 N/mm², the flexural E-Modulus is 3800 N/mm², the tensile E-Modulus is 4500 N/mm² and the elongation at the break is 0.9%. The contact between the concrete surface and the composite material was modeled using the "Cohesive zone model" (CZM) according to traction-separation law. The CZM is described by three separation components: t_n – normal, t_s and t_t parallel to the interface.

The interpolation of the displacement of the integration point is done by an additional finite element named cohesive. It is assumed that those elements have only translation degrees of freedom. The applied criteria for damage initiations in simulation is quadratic traction damage (QUADS) for modes I and for mode II loading. The equation (8) describe the criteria of damage initiation:

$$\left\{\frac{\langle \varepsilon_n \rangle}{\varepsilon_n^0}\right\}^2 + \left\{\frac{\varepsilon_s}{\varepsilon_s^0}\right\} + \left\{\frac{\varepsilon_t}{\varepsilon_t^0}\right\}^2 = 1 \tag{8}$$

Based on the recommendation of [16] introduced an effective displacement. This displacement describe the evolution of damage across the interface under a combination of normal strain and shear.

$$\delta_m = \sqrt{\langle \delta_n \rangle^2 + \delta_s^2 + \delta_t^2}. \tag{9}$$

The work performed by the tensile force and the corresponding relative displacements in the normal, first, and second shear directions, are G_n, G_s, and G_t respectively. The mechanical behavior of the cohesive element is assumed to be elastic until the damage initiation. It can be presented as:

$$\begin{Bmatrix} t_n \\ t_s \\ t_t \end{Bmatrix} = \begin{bmatrix} K_{nn} & 0 & 0 \\ 0 & K_{ss} & 0 \\ 0 & 0 & K_{tt} \end{bmatrix} \begin{Bmatrix} \delta_n \\ \delta_s \\ \delta_t \end{Bmatrix}. \tag{10}$$

Where *Knn, Kss,* and *Ktt* , are the elastic stiffness of the adhesive in the normal and the two shear directions respectively – Figure 4

IOP Conf. Series: Materials Science and Engineering **1323** (2024) 012008 doi:10.1088/1757-899X/1323/1/012008

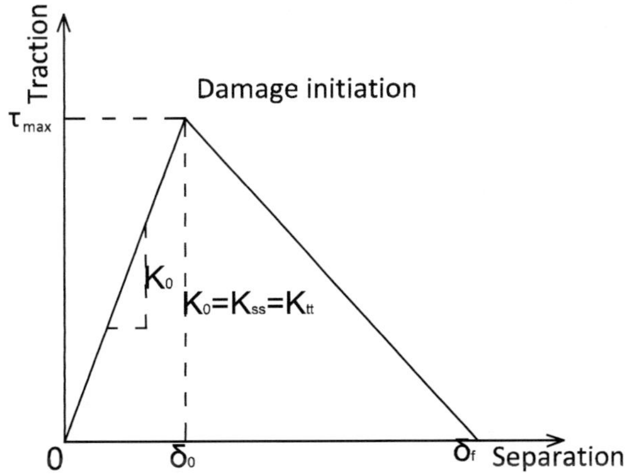

Figure 4. Traction-separation response.

Equation (11) express the loading mode.

$$K_{nn} = \frac{E_a}{T_0}. \tag{11}$$

Equation (12) present the condition for mode II - loading:

$$K_{ss} = K_{tt} = 3\left(\frac{G_a}{T_0}\right)^{0.65}, \tag{12}$$

where E_a, is the elastic modulus of the adhesive, T_0 is the original thickness of the adhesive layer and G_a is the shear modulus of the adhesive.

According to this criterion, the damage starts at the point with τ_{max} and δ_0. This mean that that the damage start when the maximum nominal stress ratio reaches one. Equation (13) describe the damage initiation:

$$max\left\{\frac{\langle t_n\rangle}{t_n^0}, \frac{t_s}{t_s^0}, \frac{t_t}{t_t^0}\right\} = 1. \tag{13}$$

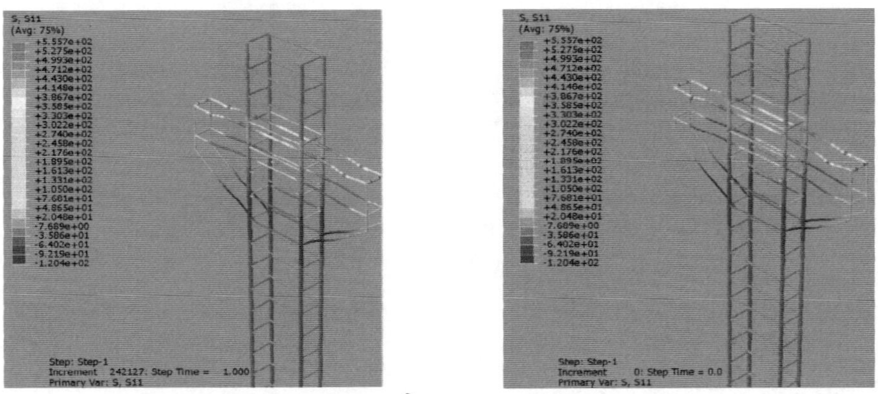

a) b)

Figure 5. Stress, increment and step time on ANR1 and ARR2 models. a) Last increment of no reinforced structure model, b) First increment of the model of strengthened and repaired structure

IOP Conf. Series: Materials Science and Engineering **1323** (2024) 012008 doi:10.1088/1757-899X/1323/1/012008

The Figure 5 shows that the stress from the last increment of no reinforced structure model, is used for the first increment of the model of strengthened and repaired structure. On the last increment of nonreinforced structure (without CFRP strengthening), the solution result is unstable because of using explicit analysis with displacement control. Nevertheless, Figure 5 shows the perfect transfer of the material properties. After this verification procedure, the material properties are transferred by force control analysis to obtain a more stable and accurate result solution.

3.2. Simulation strategy

Three models were done. The first one, named ANR1 consists the nonreinforced structure loaded until full collapse. The analysis is governed by displacement control to obtain and validate the load-bearing capacity of the structure, crack pattern. The second model, named ARR2 consists repaired structure, where it's assumed that the concrete section is fully repaired, but the material properties of the steel reinforcement are transferred to the new model. This model aims to validate the correct transmission of the steel properties from ANR1. In Figure 5 it's shown that the stress from the last increment of no reinforced structure model is transferred for the first increment of the model of strengthened and repaired structure. If we continue working with the ANR1 model, the requirement for the energy ratio will not be met [9], and the result will be influenced by "noise" and we will obtain some oscillation in the traces curves. That is why a third model is developed based on ANR1 and named ANR1-0, assigning the unloading part. After that, the mechanical properties of steel bars are transmitted from ANR1-0 to ARR2.

4. Result and discussion

Figure7 is presents the force–stress curves on nonreinforced structure. It can be noticed that the mechanical behavior of the main reinforcement is well presented. We can also compare the obtained bearing capacity obtained from the experimental study – 259 kN and the numerical simulation – 257 kN. The crack pattern is also well represented by the numerical simulation.

a)

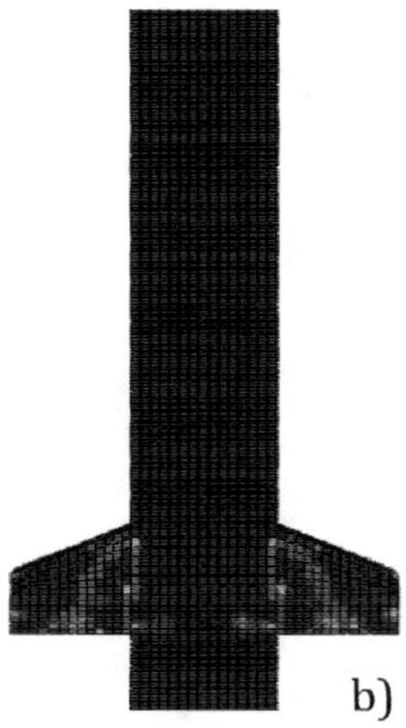
b)

Figure 6. Crack pattern on: a) experimental study b) numerical simulation – 257 kN.

IOP Conf. Series: Materials Science and Engineering **1323** (2024) 012008 doi:10.1088/1757-899X/1323/1/012008

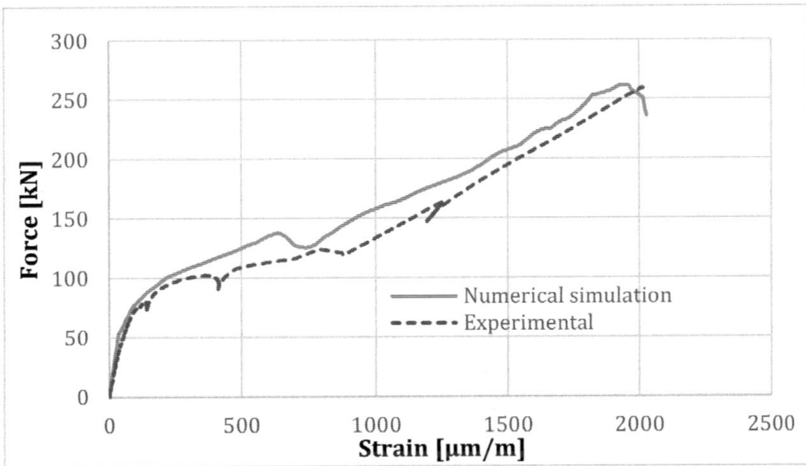

Figure 7. Force – strain curves of main reinforcement on model ANR1 compared with experimental investigation

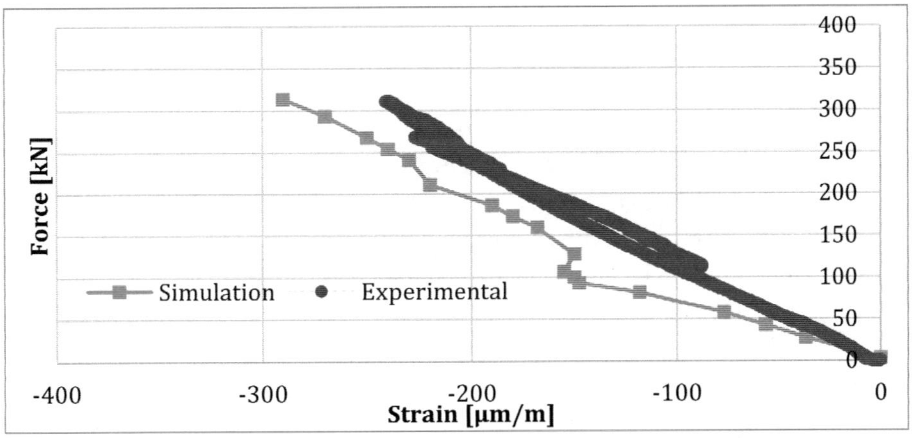

Figure 8. Force – strain curves on model ARR2 compared with experimental investigation

Figure 8 presents the force–stress curves on repaired and strengthened structure. The mechanical behavior of the structure is simulated with good agreement. The load-bearing capacity of the structure is 310 kN. obtained experimentally. By the numerical simulation, the obtained load-bearing capacity is 319 kN.

5. Conclusion

The following conclusions can be drawn as a result of the research:

- The reinforced concrete column - short corbel assembly was successfully repaired and strengthened by bonding composite materials.
- The load-bearing capacity, crack pattern, and mechanical behavior are well reproduced by the numerical simulation.
- The chosen material models successfully describe the real mechanical behavior of the structure.
- The developed simulation strategy successfully simulates the experimental study, and based on the developed model a parametrical study can be done.

Acknowledgment

This research is supported by the Bulgarian Ministry of Education and Science under the National Program "Young Scientists and Postdoctoral Students-2" at University of Chemical Technology and Metallurgy, Sofia, Bulgaria and University of Reims Champagne-Ardenne, Reims, France.

References

[1] V. Stankov, "Étude du comportement mécanique d'un assemblage poteau-console courte en béton armé renforcé ou réparé par collage de matériaux composites," 2021.

[2] M. Ben Ftima, Comparaison des trois logiciels d'éléments finis: Ansys, Adina et Abaqus, Rapport interne, École Polytechnique de Montréal, 2005.

[3] J. Lubliner, J. Oliver, S. Oller and E. Oñate, A Plastic-Damage Model for Concrete, vol. 25, International Journal of Solids and Structures, 1989, p. 299–329.

[4] J. Lee and G. L. Fenves, Plastic-Damage Model for Cyclic Loading of Concrete Structures, Vols. 124, no. 8,, Journal of Engineering Mechanics, 1998, p. 892–900.

[5] Simulia Abaqus 6.13, ABAQUS Analysis User's Guide.

[6] Z. BAZANT and B. OH, Crack band theory for fracture of concrete, vol. 16(93), Materiaux et Constructions, 1983, pp. 155-177.

[7] D. Carreira and K. Chu, Stress-Strain Relationship for plain concrete in compression, vol. 82 no. 6, Journal of the American Concrete Institute, 1985, pp. 797-804.

[8] B. Wahalathantri, D. Thambiratnam, T. Chan and S. Fawzia, A material model for flexural crack simulation, Brisbane: Proceedings of the First International Conference on Engineering, Designing and Developing the Built Environment for Sustainable Wellbeing, Queensland University of Technology, 2011, pp. 260-264.

[9] F. Aslani and R. Jowkarmeimandi, Stress-strain model for concrete under cyclic loading, vol. 64 (8), Magazine of Concrete Research, 2012, pp. 673-685.

[10] F. Alfarah, F. López-Almansa and S. Oller, New methodology for calculating damage variables evolution in Plastic Damage Model for RC structures, Engineering Structures, 2017, p. 132 (2017) 70–86.

[11] D. Hordijk, Tensile and tensile fatigue behaviour of concrete; experiments, modelling and analyses., vol. 37(1), Heron, 1992, pp. 1-37.

[12] R. M. Jones, Mechanics of composite materials., New York: Hemisphere Publishing., 1975.

[13] A. W. Al-Zand, W. H. Wan Badaruzzaman, A. Mutalib and S. J. Hilo, Modelling the delamination failure along the cfrp-cfst beam interaction surface using different finite element techniques, Vols. 12(1), , Journal of Engineering Science and Technology, 2017, pp. 214-228.

[14] H. Al-Zubaidy, R. Al-Mahaidi and X.-L. Zhao, Finite element modelling of CFRP/steel double strap joints subjected to dynamic tensile loadings, Composite Structures, 2013, p. 99:48–61.

[15] S. Fawzia, R. Al-Mahaidi and X.-L. Zhao, Experimental and finite element analysis of a double strap joint between steel plates and normal modulus, Vols. 75(1-4), Composite Structures, 2006, pp. 156-162.

[16] P. P. Camanho and C. G. Davila, Mixed-Mode Decohesion Finite Elements for the Simulation of Delamination in Composite Materials, NASA/TM-2002–211737, 2002, p. 1 –37.

Deflections in reinforced concrete slabs at an early age in the loading stage from a self-weight of the slabs and adjoining period-instant temporary load

Stanislav Bakardzhiev[1*] and Anton Gorolomov[2]

[1] Technology and Construction Management Department, University of Structural Engineering and Architecture "Lyuben Karavelov", Sofia, Bulgaria
[2] Mechanics and Mathematics Department, University of Structural Engineering and Architecture "Lyuben Karavelov", Sofia, Bulgaria

*E-mail: st.bakardziev@vsu.bg

Abstract. This study researches the behaviour of experimental reinforced concrete slabs at early age, in a loading stage for several days from the self-weight of the slabs and corresponding temporary loads for the same stage, namely parts of a formwork system used during construction with the values of the loads close to specified in the Eurocode for temporary loading during construction. The deflections for the relevant stage and period were examined as comparison of the theoretical values of deformations with experimentally measured results for the corresponding stage, as well as the relevant observations and conclusions are described.

1. Introduction

On a global scale, the building industry has always been based on development in the direction of shortening the production period of reinforced concrete elements and increasing their quality, regardless of whether we are talking about monolithic production on the construction sites as a part of a the structure or in specialized industrial units. A big part of this effort is aimed in the production of monolithic reinforced concrete floor slabs. One of the ways for faster and higher quality production is the Method of Early Striking. The different combinations of propping, casting, striking and re-propping justify the method [1], [2], [3]. For the correct, rational and safe handling of the formwork systems when using the method, the behaviour of the slabs in the different stages of their production is necessary to be known. This report is a continuation of a series of reports [1] and [2], as it examines vertical deflections for relevant stage and period were examined. A comparison of theoretical and experimental results is made, as well as the relevant observations and conclusions are described. Both this report and [2] detail the observations and research from report [1] by considering individual stages and steps of [1]. As in the previous reports, here we will cite some researches in this direction or close to the topic [4], [5], [6],[7], [8], [9], [10], [11], [12], [13], [14], [15], [16], [17], [20]. Since each numerical model of each step of the study [1] is independent, this study compares the development of elastic to elastoplastic theoretical deformations with experimentally measured results for the corresponding stage (Figure 4), and describes relevant observations and conclusions. Thus, the study aims to see the

IOP Conf. Series: Materials Science and Engineering **1323** (2024) 012009 doi:10.1088/1757-899X/1323/1/012009

differences of the theoretical deflections of the considered stage with constant stage load combination with the measured experimental ones.

2. Purpose and main tasks in the study

2.1. Purpose

The main purpose of the research is to study the behaviour of experimental reinforced concrete slabs loaded with a load combination for 3 days (Figure5) of self-weight loads of the slabs and variable actions over the second slab (level +2.32), which is close to the variable actions specified in the Eurocode for slab loading during construction. To compares the evolution of elastic to elastoplastic theoretical deformations with experimentally measured results for the corresponding stage (Figure 4), and describes relevant observations and conclusions.

2.2. Main tasks

The main task in the paper is to calculate theoretically by formulas the development of the elastic deformations calculated from the mathematical model of the structure in Step 9 to the elastoplastic ones in Step 10. Each step (Figure 4) of the experimental study has an individual numerical model. The results of the theoretical calculations are to be compared with the experimental results of deflections measured using gauges.

3. Construction

The dimensions of the slabs are given in Figure 1 [1]. The reinforcement, which is put in the slabs, is very close to the minimum reinforcement and this is done to obtain clearer results for the deflections, [1]. The loads on the slab under consideration are from self-weight, the load from the slab over it and variable actions from some formwork panels over the second slab (2.32m). The deflections are measured with gauges, which are attached to a metal box frame between the slabs which is supported independently from the construction.

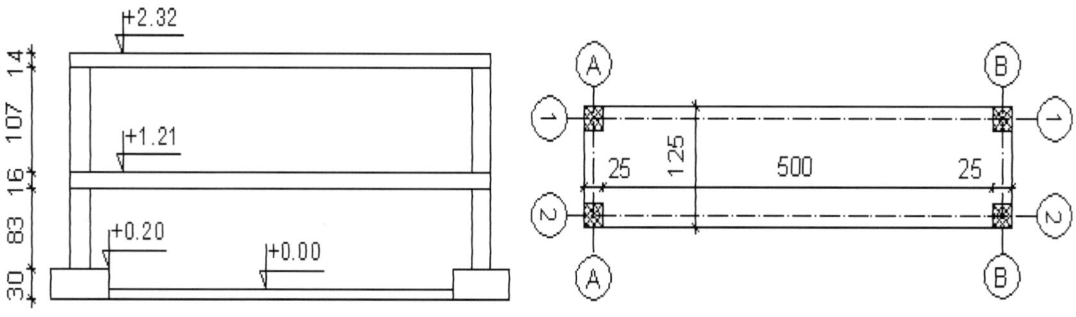

Figure 1. The experimental construction [1].

Figure 2. Stage where the formwork for the second slab is montaged.

Figure 3. Stage of loading with variable actions of formwork of the top slab.

4. Information about numerical model of main experimental methodology

In the main study [1] of this research series [1] and [2] is explained how the experimental construction was modelled with software, based on the finite element method, and it is described in detail. Each step (Figure 4) of the experimental study has an individual numerical model. More detailed information on how the experimental construction was modelled is available in [1].

5. Experimental research

In the main research [1], there are tracked in details of the deflections and the development of cracks in the two slabs, and the study [1] was done in 15 steps in Figure 4. In current study is considered in more details steps 9 and 10 of Figure 5. Each step shows the casting and striking for every level of building the experimental construction, the concrete age for both slabs and the modulus of elasticity of the slabs. One stage of the report's nine stages [1] is considered, with the steps involved in each stage listed in Table 1. In this study, we are considering stage 6, which including steps 9 and 10 (Figure 5). Different compares the development of elastic to elastoplastic theoretical deflections compared with experimentally measured results for the current stage (Table 1) are made in the graphics in Figure 7 to 12. The values of amendment of elastic to elastoplastic theoretical deformations in the stage are obtained from theoretical formulas by Eurocode and experimental deflections are measured with gauges (Table 2 and 4).

5.1. The stage of the main study [1] on which this report is developed

Table 1. Stages of the experimental methodology of loading, casting and striking [1].

Stage	Step	Period (days)	Actions
1	1-2	4	Cleaning 1st slab
2	3	4 – 7	Casting 2nd slab
3	4	7	Laying 2nd slab

4	4-5	7 - 11	-
5	5-8	11	Cleaning 2nd slab and striking 1st slab
6	9-10	11 - 14	Loading 2nd slab with temporary load
7	11	14	Unloading the temporary additional load
8	12	14th	Loading 2nd slab with temporary load
9	13-15	15	Removing the additional load and striking 2nd slab

5.2. Steps of report [1] from which the current the research considered Step 9 and 10

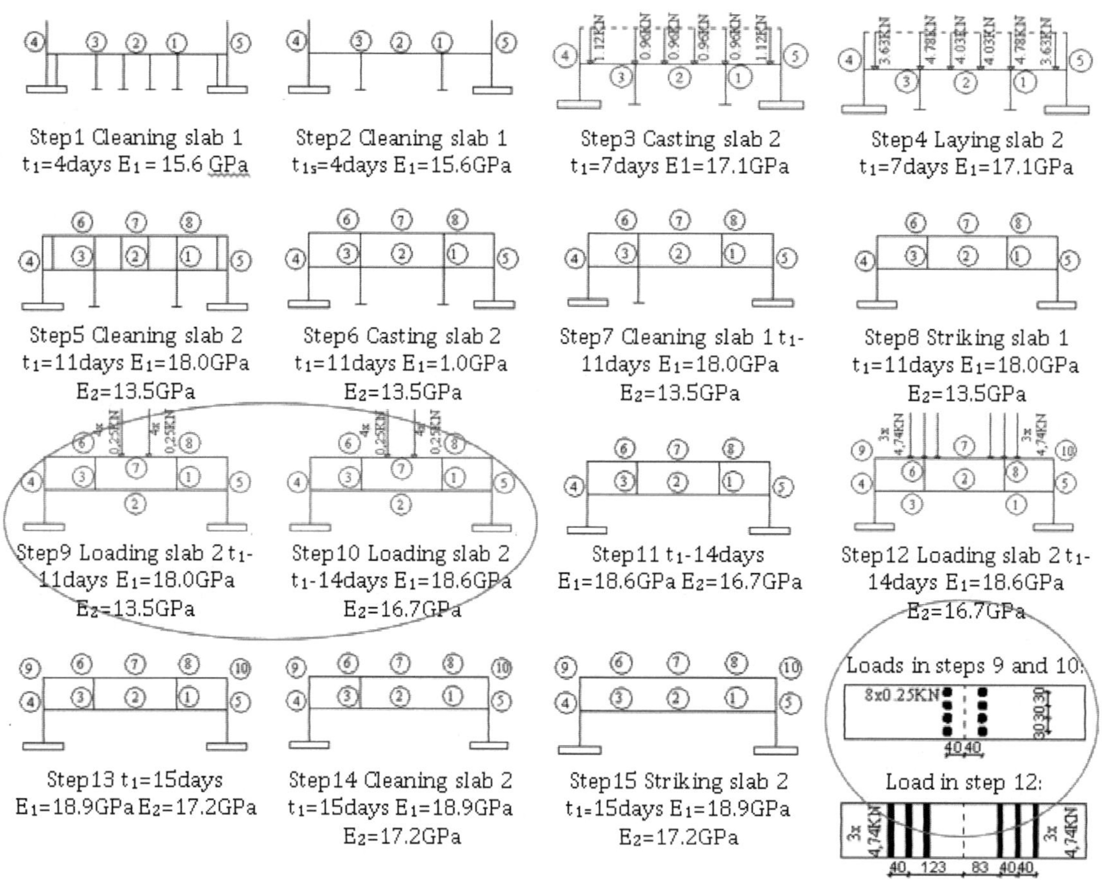

Figure 4. Experimental methodology of loading, casting and striking.

5.3. Steps in this study between step 9 and 10 of the report [1].

Figure 5 shows each of the steps of the stage of the study.

IOP Conf. Series: Materials Science and Engineering **1323** (2024) 012009 doi:10.1088/1757-899X/1323/1/012009

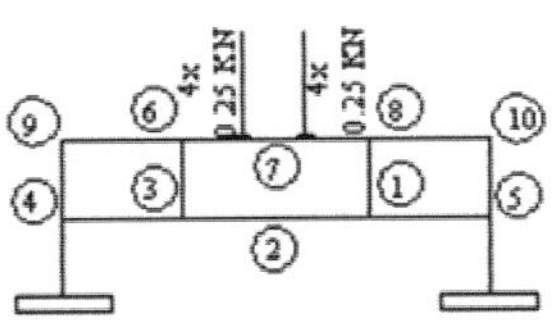

Step9 Loading slab 2 t_1-11days E_1=18.0GPa E_2=13.5GPa

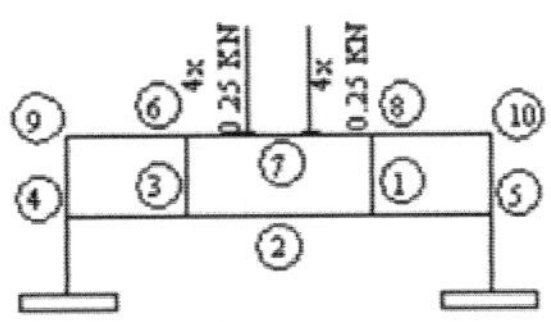

Step10 Loading slab 2 t_1-14days E_1=18.6GPa E_2=16.7GPa

Figure 5. Considered stages in this report.

5.4. Steps in photos.

Figure 6 shows each load in every step of current study

Step9 Loading slab 2 t_1-11days E_1=18.0GPa E_2=13.5GPa

Step10 Loading slab 2 t_1-14days E_1=18.6GPa E_2=16.7GPa

Figure 6. Photos of the steps.

6. Results and observations

Measured real deflections by the gauges in the slabs (level +1.21m and 2.32m) are shown in Table 2. The develop of the theoretic elastic deflections to elostoplastic deflections in the slabs derived by formulas (1), (2), (3), (4), (5), (6), (7), (8), (9) are show in Table 4 and Table 3 the cracks, as they are measured.

6.1. Deflections measured by gauges (mm).

Table 2. Real deflections from experimental measures for Stage 6.

Day	Point									
	1	2	3	4	5	6	7	8	9	10
11th	3.01	4.46	3.17	0.00	0.00	2.71	4.29	2.95	0.00	0.00
14th	4.17	6.22	4.54	0.00	0.00	3.96	5.84	4.28	0.00	0.00

6.2. Results of measuring the cracks (mm).

Table 3. Measures of cracks in the both slabs (mm).

Period	Point									
	1	2	3	4	5	6	7	8	9	10
Stage 6	0.00	0.00	0.00	0.00	0.00	0.00	0.00	0.00	0.00	0.00

6.3. Develop of the theoretical elastic deflections to elostoplastic deflections in the slabs derived by formulas (mm).

The methodology used is that specified in Annex "B" of EN 1992-1-1:2005 [19].

h=16cm – thickness of 1_{st} slab (+1.21m)

b=100cm – conditional width (strip)

t_0=11days – load start time

t=14 days – time we are searching for the creep

f cm=25+8=33MPa – average 28-day compressive strength of concrete

RH=52.5% – real measured relative air humidity

α=1 – for class R cement

A_c=b·h=100·16=1600 cm^2 – cross-section area

u=2·b=2·100=200cm – perimeter of the section where there is contact with air

Correction of time t_0 taking into account the type of cement (1):

$$t_{0,cor} = t_0 \cdot \left(\frac{9}{2+t_0^{1.2}} + 1\right)^{\alpha} = 11 \cdot \left(\frac{9}{2+11^{1.2}} + 1\right)^{1} = 16.01 \quad (1)$$

Conditional height (2):

$$h_0 = \frac{2 \cdot A_c}{u} \cdot 10 = \frac{2 \cdot 1600}{200} \cdot 10 = 160mm \quad (2)$$

$$\beta(to) = \frac{1}{0.1 + t_{0,cor}^{0.2}} = \frac{1}{0.1 + 16.01^{0.2}} = 0.543 \quad (3)$$

$$\beta(fcm) = \frac{16.8}{fcm^{0.5}} = \frac{16.8}{33^{0.5}} = 2.92 \quad (4)$$

$$\varphi_{RH} = 1 + \frac{1 - \frac{RH}{100}}{0.1 \cdot h_0^{0.333}} = 1 + \frac{1 - \frac{52.5}{100}}{0.1 \cdot 160^{0.333}} = 1.88 \quad (5)$$

Basic creep coefficient value (6), (7), (8):

$$\varphi_0 = \varphi_{RH} \cdot \beta(fcm) \cdot \beta(to) = 1.88 \cdot 2.92 \cdot 0.543 = 2.98 \quad (6)$$

$$\beta_H = min(1.5 \cdot (1 + (0.012 \cdot RH)^{18}) \cdot h_0 + 250; 1500) = 490.06 \quad (7)$$

$$\beta_c(t; t_0) = \left(\frac{t-t_0}{\beta_H+t-t_0}\right)^{0.3} = \left(\frac{14-11}{490.06+14-11}\right)^{0.3} = 0.216 \quad (8)$$

Creep coefficient (9):

$$\varphi = \varphi_0 \cdot \beta_c(t; t_0) = 2.98 \cdot 0.216 = 0.645 \quad (9)$$

Similarly, for the 14 cm thick top slab, $\varphi = 0.671$. Therefore, the elastoplastic deflections will be equal to those reported in the finite element model multiplied by 1+ φ.

Table 4. Derived elastic deflections from numeric model in 11th day (Step 9) and elastoplastic deflections from formulas in each of the points in 14th day (step 10).

Day	Point									
	1	2	3	4	5	6	7	8	9	10
11th	6.52	8.75	6.56	0.00	0.00	6.62	8.97	6.58	0.00	0.00
14th	10.73	14.39	10.79	0.00	0.00	11.06	14.99	11.00	0.00	0.00

6.4. Comparisons.

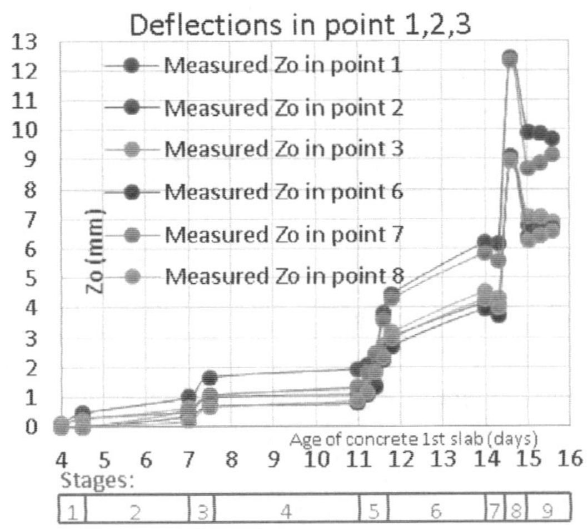

Figure 7. Deflections in point 1, 2, 3, 6, 7 and 8 [1]

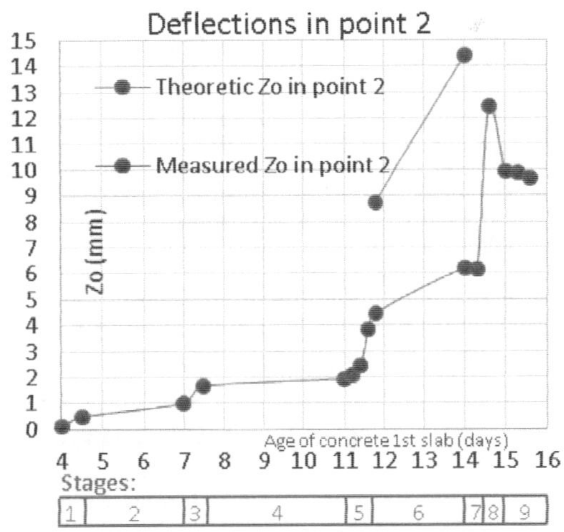

Figure 8. Deflections in point Point 2.

IOP Conf. Series: Materials Science and Engineering **1323** (2024) 012009 doi:10.1088/1757-899X/1323/1/012009

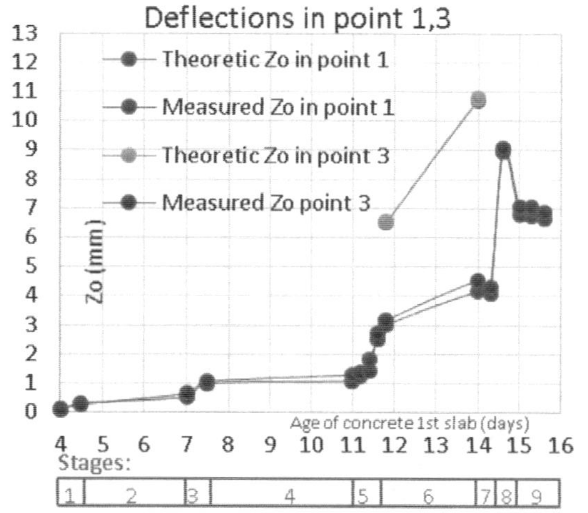

Figure 9. Deflections in point 1 and 3.

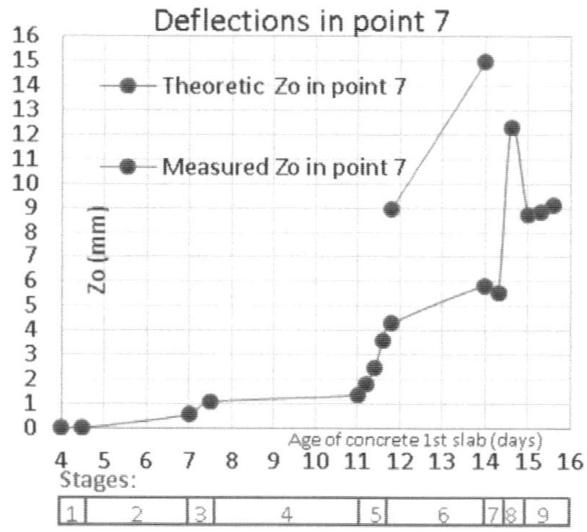

Figure 10. Deflections in point 7.

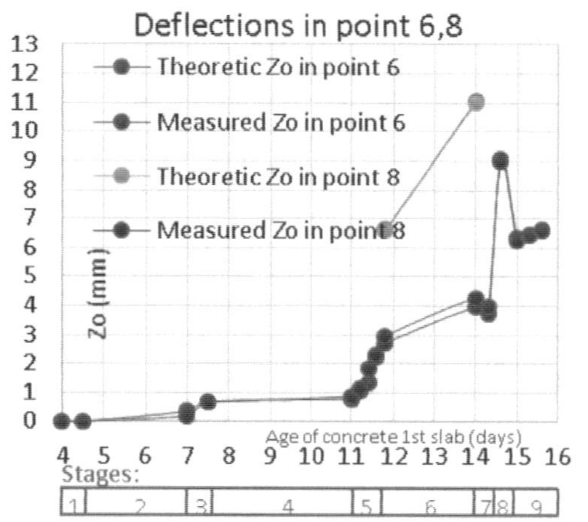

Figure 11. Deflections in point 6 and 8.

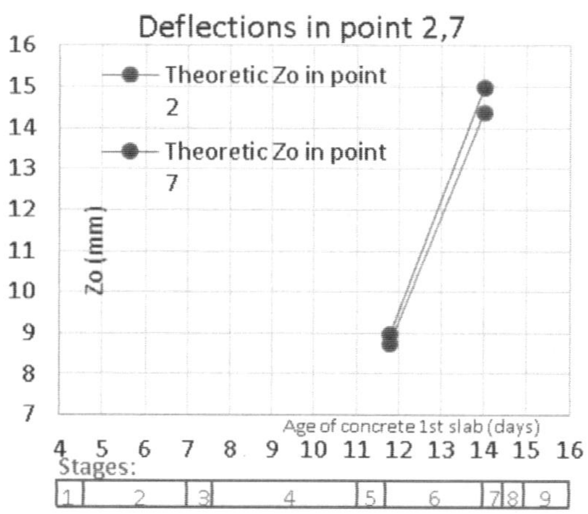

Figure 12. Deflections in point 2 and 7.

6.5. Observations on the deflections.

- Figure 7 shows the vertical displacement from experimental measures of the points in the middle of the field and above the remaining props of the two slabs. This graph is shown because it will be the basis of all comparisons in the other graphs with the deflections from the development of elastic deflections from step 9 to elastoplastic in step 10 with values derived from the theoretical formulas (1)-(9).

- In Figures 8 and 10, it can be observed that in the middle a fields for both slabs, the percentage of deflections in the development of elastic to elastoplastic for stage 6 is bigger than that measured experimentally by gauges. 64% against 39% at point 2 for the first slab (+1.21m) and 67% against 31% for point 7 in the second plate (+2.32m), respectively.

- It is also noticeable in Figures 9 and 11 that at points above the remaining telescopic struts in the fields for both slabs, the percentage increase in strain as elastic evolves into elastoplastic for Stage 6 is bigger than that measured experimentally by gauges, averaging 82% against 41%

at points 3 and 4 for slab one (+1.21m) and 67% against 46% for points 6 and 8 in slab two (+2.32m), respectively.

- We note again, as in report [1] and [2], that the experimentally measured deflections are of significantly smaller value than the theoretical elastoplastic deflections. This is due to the unsimple joint in the column-slab node in reality.

- In Figure 12, we observe close values of the theoretically obtained values of the evolution of elastic to elastoplastic strains in stage 6 for both slabs, which again points us to the trend of how, through the propping, the two plates work together.

7. Conclusions

The following conclusions can be drawn from the analysis of the results.

- After the comparisons made in graphs 8, 9, 10 and 11, it can be seen that the increase in deflections at each value of development of elastic deflections in a given loading step of the structure to elastoplastic in a subsequent one, is percentage bigger than or equal to that obtained from the experimental measurements of the deflections. Using this we can conclude that when making individual numerical models for a given stage of construction, when making checks for the development of elastoplastic deformations under constant loading for a given period of time as a percentage increase, from the side of security will be the results get in a theoretical way for the development of elastoplastic deformations in the main field, because their increase is percentage bigger than or equal to the results measured experimentally.

- Also, after the comparisons made in graphs 8, 9, 10 and 11, it is also seen that as values the results of the development of the elastic deflections taken from the numerical model for Step 9 in elastoplastic in the subsequent Step 10 are bigger than the obtained values of the deformations from the experimental measurements. It follows from here, as mentioned in [1] and [2], that again from the side of security for the main field, the results will be obtained theoretically.

Acknowledgments

Acknowledgments to “Tehstroy” Ltd., “Tirlin” JsCo, “doka” Bulgaria, eng. Radoslav Stafunski and Mr. Kostadin Simeonov.

References

[1] Bakardzhiev S, 2020 Experimental study of flat ,reinforced concrete slabs created in the method of early striking, XI ISC „Civil Engineering Design and Construction“, Sept. 10-12, 2020, Varna, Bulgaria ISSN 2603-4255 (CD-ROM); ISSN 2683-071X (online),doi:10.1088/1757-899X/951/1/012010

[2] Bakardzhiev S, 2023 Study of the behavior of a experimental reinforced concrete slab at different stages of loading from the production of the reinforced concrete slab over it, XII ISC „Civil Engineering Design and Construction“, Sept. 08-10. 2022, Varna, Bulgaria ISSN 2603-4255 (CD-ROM); ISSN 2683-071X (online),doi:10.1088/1757-899X/1276/1/012024

[3] Haidarzhynska M., Hrischev L., Traykova M., Ivanov G., 2020 Analysis of reinforced concrete slabs using formwork with dropheads during early striking, Conference Proceedings XI International Scientific Conference „Civil Engineering Design and Construction“, Sept. 10 - 12, Varna, Bulgaria ISSN 2603-4255 (CD-ROM); ISSN 2683-071X (online), pp. 312-324.

[4] Alvorado Y, Calderon P, Gasch I and Adam J, 2010 Experimental and numerical study during construction of multistory buildings using clearing or partial striking, (XXXVII IAHS World Congress on Housing)

[5] Gasch I, Alvorado Y, Calderon P and Ivorra S, 2013 Construction loads using a shoring – clearing striking process, (Structures and Buildings), Volume 167 Issue SB4, p 217-229
[6] Buitrago M, Alvorado Y, Adam J and Calderon P, 2016 Maximum loads on shores during the construction of buildings, (Article in ICE Proceedings Structures and Buildings)
[7] Buitrago M, Adam J, Moragues J and Calderon P, 2018 Load transmission between slabs and shores during the construction of RC building structures – A review, (Engineering Structures 173 (2018)) p 951–959
[8] Pallet P, 2019 Backpropping flat slabs, (Pallet Temporary Work), PFP / 136 F
[9] Zigmunt O, 2010 The analysis of slab resistance during early striking exemplified by Gridflex Peri formwork system, (Vilnius, Lithuania The 10th International Conference)
[10] Hrischev L, Tsvetkov S and Nedelchev N, 2015 Early striking calculations of formwork for reinforced concrete slabs according to Eurocode 2, (VII International scientific conference architecture, civil engineering - modernity Varna Bulgaria)
[11] Buitrago M, Adam J and Calderon P, 2018 Structural failures in cast-in-place RC building structures under construction, (ICITECH, Universitat Politècnica de València. Camino de Vera s/n, 46022 Valencia, Spain)
[12] Ivanchev I and Slavchev V, 2019 About the Possible Limitations in the Usage of the Non-Destructive Ultrasonic Pulse Velocity Method for Assessment of Cracks in Reinforced Concrete Structures, Subjected to Direct Environmental Exposure, Buildings, Volume 9, Issue 9, 202, http://dx.doi.org/10.3390/buildings9090202
[13] Bakardzhiev S and Lazova O, 2019 Research of the strenght characteristics of concretes in the quality receiving period, (XIX International scientific conference by construction and architecture VSU'2019)
[14] Ivanchev I, Slavchev V, Dobreva I and Daalov B 2018 Experimental determination of modulus of elasticity and stress-strain diagrams of concrete, (XVIII Anniversary international scientific conference by construction and architecture VSU'2018)
[15] Ivanchev I, Slavchev V, 2019 Probable compressive strength and modulus of elasticity of concrete, determined by non-destructive ultrasonic pulse velocity method (NDUPVM) in different ways of measuring transducers placement, Conference: XXIX International Scientific Symposium 'Metrology and Metrology Assurance'At: Sozopol, Bulgaria, September 6-10, DOI: 10.1109/MMA.2019.8936002
[16] Hrischev L, Rostovsky I, Conev I, Nikolov V, 2020 Establishing the strength-maturity relationship of the concrete trough laboratory tests – pioneer study in Bulgaria, Conference Proceedings XI International Scientific Conference „Civil Engineering Design and Construction", Sept. 10 - 12, Varna, Bulgaria ISSN 2603-4255 (CD-ROM); ISSN 2683-071X (online), pp. 338-349.
[17] Vona M, Paolo Del Guidice F, Manganelli B, 2024 The Key Role of Floors for the Sustainability of Retrofit Interventions in Older Existing Reinforced Concrete Buildings,Sustainability 2024, 16, 5862. https://doi.3390/su16145862
[18] Kazakov K, 2010 Finite Element Method for structural modeling, (Prof. Marin Drinov Academic Publishing House), Sofia, Bulgaria
[19] EN 1992-1-1:2005 – Eurocode 2: Design of concrete structures - Part 1-1 : General rules and rules for buildings
[20] Tsvetkov S., Helicopter service of hospital building, DCB 2022, IOP Conf. Series: Material science and engineering doi:10.1088/1757-899X/1276//1/012018

Inspection and evaluation of the technical condition of a monolithic frame reinforced concrete structure of a hotel in Borovets resort

Hristian Neshev[1]

[1] Department "Reinforced Concrete Structures", University of Architecture, Civil Engineering and Geodesy, bul „Hr. Smirnenski" № 1, 1046 Sofia, Bulgaria

*E-mail: neshev_fce@uacg.bg

Abstract. This study examines, the current research endeavours to analyse the results of structural inspection of an existing hotel building situated within the confines of Borovets resort. The structure consists of one basement level, four above-ground levels and one unusable attic level. The building was built site-cast in the period from 1983 until 1985. For the inspection, a series of activities were carried out, and the load-bearing structural elements of the building were established, the strength characteristics of the inserted materials were determined through destructive and non-destructive activities, a geometric survey of the load-bearing vertical and horizontal elements was carried out, existing defects and damages were found, an analysis of the impacts on the structure was made, the results of the structural assessment were summarized and basic recommendations were presented for bringing the building into compliance with the requirements of the current regulatory documents. The purpose of the performed in-situ, analytical, and computational activities is to establish the current technical condition of the load-bearing and non-load-bearing structural elements and to assess the possibilities for ensuring the future operation of the building, according to the construction and investment intentions of the owner.

1. Introduction

The building stock in the Republic of Bulgaria has exhibited persistent growth across various categories, including residential, public, and industrial structures. The existing edifices subject to adaptation to modern conditions of exploitation, purpose, and assurance against external influences are increasing. These existing edifices are increasingly being adapted to meet contemporary standards of load-bearing capacity, stiffness, seismic resistance, serviceability, fire resistance, execution speed, technology, durability, repairability, maintenance, adaptability to future changes in the architectural etc.

In the context of the present and future, it will be increasingly necessary to establish the structural condition of existing reinforced concrete buildings. The present inspection refers to the previous experience of many Bulgarian engineers who have observed and studied the behaviour of building structures - D. Dimov, At. Georgiev, V. Kardzhiev, D. Dimitrov, D. Nazarski, B. Petrov, D. Panichkov, P. Hristov, E. Abdulahad and many others. Some of the studies and guides that apply to the present development are [1], [2], [3], [4], [5], [6], [7], [8], [9], [10], [11], [12], [13], [14], [15] and [16].

The structure under consideration herein is typical of the cast-in-place frame constructions built according to the architectural, structural and technological requirements for the buildings of the second half of the last century. The inspection results describe the general condition of the load-bearing and non-load-bearing elements, as well as the possibilities for ensuring future operation of the building. The present article provides an opportunity to predict the expected condition of similar types of structures after 40 years of operation.

2. General information about the building

The hotel's building was designed by KIIP "Glavproekt" under KAB of the III Directorate in 1982 (Figure 1). The investor of the building is SMK "Sofia District" and the purpose is "School for training construction workers in Borovets". The construction was built from 1983 until 1985. There is no documentation available from the construction of the load-bearing structural elements, the architectural and installation works, and the permission for operation. Significant modifications were implemented during a subsequent period, approximately between 1998 and 1999. These alterations include the construction of a basement wall was built on the southern part of the hotel, a steel extension was constructed for the dining zone, and a lift in the central stairwell. The current research does not encompass any supplementary auxiliary structures beyond those previously mentioned.

The approximate dimensions of the main building are 39.2x15.5m and the steel extension - 8.75x22.75m. The hotel building has a basement level, a ground floor level, three floors with hotel rooms, and an unusable sub-roof level. The roof structure is a double-pitched cast-in-place reinforced concrete roof.

3. Inspection methodology, testing, and analysis of results

Figure 1. North and south facades of the building

3.1. Acquaintance with the available project and executive archival documentation for the building

The available project documentation for the building includes:

- Architectural plans of the basement, ground floor, first, second, and third levels. Two architectural facades, section, and site plan;
- Architectural measurements of the building from 2022.

No design or executive structural documentation was found. Figure 2 shows the original architectural plan of the second level of the building.

IOP Conf. Series: Materials Science and Engineering **1323** (2024) 012010 doi:10.1088/1757-899X/1323/1/012010

3.2. Visual and geometric inspections inspection

During the engineering team's site visits, it was observed that the hotel building was not operational.

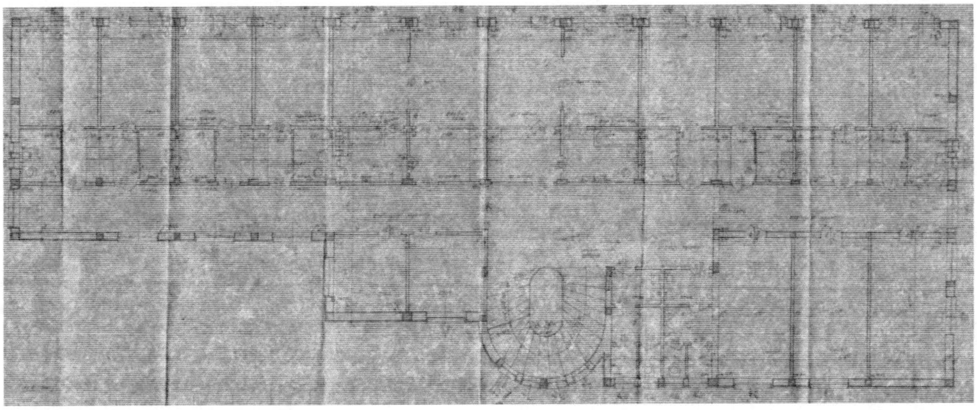

Figure 2. Architectural plan of the second level

The non-load-bearing masonry internal partitions on the levels housing the hotel rooms were dismantled to amalgamate two pre-existing hotel rooms into a singular space. Additionally, all partition walls enclosing toilets and bathrooms across the hotel rooms have been removed.

The visual and geometric inspections of the structure are performed at all levels, including the basement. The following activities were carried out:

- The supporting structural system of the building is established, which is classified as a cast-in-place reinforced concrete frame structure;
- The type of foundation of the building is identified;
- The permanent and imposed loads on the structure are established.

Figure 3 shows the formwork plan of the second level.

3.3. Concrete inspection

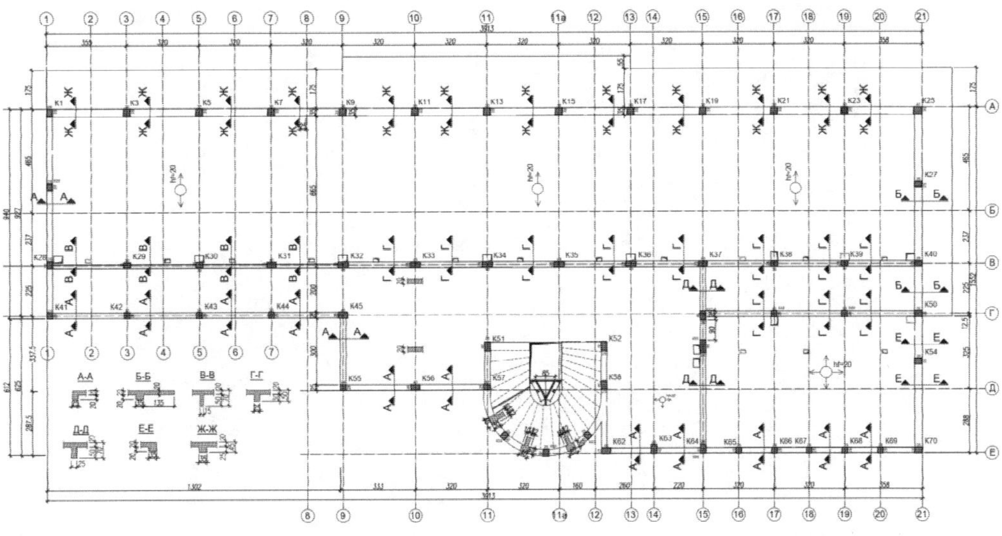

Figure 3. Formwork plan on the second level

The control and evaluation of the concrete were carried out according to standards BDS EN 12390-3 [17]; BDS EN 12504-1[18]; BDS EN 12504-2 [19], BDS EN 13791 [20] and literary sources [8], [9], [12], [13].

Nine core boreholes were executed in structural reinforced concrete elements (basement walls, columns, and slabs) with different types, locations, and construction timing (Figure 4).

Eight concrete test specimens were taken out and provided for testing in the laboratory. The compressive strength of the concrete was determined. A non-destructive method for determining the compressive strength of concrete was also applied, using Schmidt-Hammer type N. The cutting of the concrete test specimens was carried out according to the requirements of the BDS EN 12504-1 [18]. Drilling locations have been chosen in areas with minimum stresses in the reinforced concrete elements under design and seismic combinations.

Following the specimen-cutting process (Figure 5), the voids or holes created were meticulously repaired using a silicate-based construction mortar that allows for controlled drying. Significantly, it is not anticipated that the structural integrity or security of the building will be compromised as a result of these restoration activities.

Figure 4. Concrete test specimens

Figure 5. Cutting concrete specimens from a floor structure

A series of non-destructive tests were carried out using a Proceq "Schmidt N" rebound hammer on concrete surfaces prepared according to the standard. The test zones were selected in dry and accessible places, where the surface layer of the concrete is maximally preserved, undamaged, and free of cavities and contamination, as the surfaces were cleaned and smoothed before the tests.

3.4. Correlation between destructive and non-destructive test results for concrete compressive strength

Correlation of the obtained results was carried out according to the requirements of BDS EN 13791/NA [20] and BDS EN 206/NA [21]. The strengths obtained from the non-destructive method are correlated with the results of the laboratory tests of the specimens.

3.4.1. Reinforced concrete vertical elements

In this set of tests, the mechanical properties of concrete were determined for all elements by groups:

- Columns - for reinforced concrete columns, non-destructive testing covers all levels. The correlation was performed from results of the strength parameters determined by the cylindrical concrete specimens from vertical elements;
- Basement walls – three cylindrical specimens were cut and two of them were tested in structural laboratory. After drilling a borehole PT1, it was found segregation of concrete and it was impossible to obtain a test specimens for strength testing (Figure 6).

Figure 6. Concrete specimen PT1

3.4.2. Reinforced concrete horizontal elements

The mechanical properties of the reinforced concrete horizontal elements (slabs and beams) were determined by series of non-destructive tests and five test specimens.

3.4.3. Summary of concrete strength results

The subsequent conclusions are derived from the correlation and laboratory findings:

- The oven-dry density of concrete in air varies from 2160 kg/m^3 to 2300 kg/m^3;
- The density of concrete from specimens show a significant variation in the compressive strengths - from 29.9 MPa to 45.4 MPa;
- The variation in the surface strengths (non-destructive method) of the concrete is significant for all tested reinforced concrete elements – basement walls, columns, slabs, and beams;

- The significant variation of the coefficient of agreement (up to 1.28) between the determined surface strengths of the concrete and the laboratory-tested specimens is probably due to the segregation of the concrete;
- The low values of the uniformity coefficient (minimum value 0.59), for all groups of elements, show the non-constant parameters of the production process of the concrete mixture.

The standard concrete compressive strength classes determined at 95% assurance (5% fractile), as well as the average oven-dry density, are presented in Table 1:

Table 1. Indicative concrete compressive strength class and average density

Elements	Indicative concrete compressive strength class BDS EN 206-1	Average oven-dry density [kg/m^3]
Basement walls	C20/25	2160
Columns	C20/25	2190
Floor slab at ground level -0.10	C16/20	-
Floor slab at first level +2.65	C25/30	2300
Floor slab at second level +5.45	C20/25	2175
Floor slab at third level +8.25	C25/30	2280
Floor slab at fourth level +11.05	C16/20	-

Indicative concrete compressive strength classes by elements are:
- Vertical elements for all levels - C20/25;
- Horizontal elements - from C16/20 to C25/30.

3.5. Determining the structure of the concrete mixture

The concrete mixture in the load-bearing structural elements of the hotel has different physical and mechanical properties, structure, type and fraction of aggregates, degree of compaction and etc.. Concrete shrinkage and creep strains are observed in some of concrete elements. These effects were studied in a series of publications [22], [23], [24], [25], [26] and the information was used in the creation of the computer models.

The aggregate of crushed stone of marbleized limestone and rounded gravel with a maximum diameter of 30-35mm was inserted. The structure of the concrete in the individual parts of the basement walls is different, which determines non-uniform properties. In certain areas (Figure 7) the liquid phase of the mixture is absent. Due to the loose structure of the concrete, as well as compromised waterproofing, local areas with water seepage are established at the basement level.

A denser concrete mixture than in the basement level was found in the concrete mixture of the floor slabs and vertical elements of the structure. There is a segregation zones with air caverns in many places. These conditions are a prerequisite for corrosion of the reinforcement, which was established during the inspection and excavation of the reinforcing steel. The embedded aggregates are rounded gravel with a maximum grain diameter of 20-25mm.

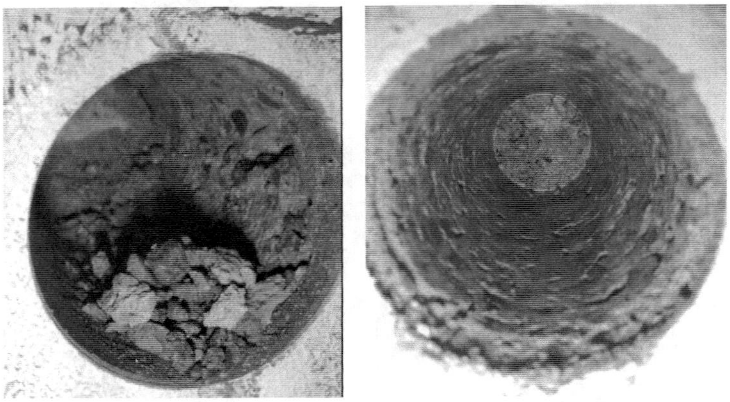

Figure 7. Drill holes in basement walls

3.6. Determination of the neutralization depth of concrete

The degree of corrosion of the concrete was determined on the basis of visual and measurement control. Visual control covers: position and type of cracks caused by corrosion of reinforcing bars, change in color of concrete, destruction of concrete cover, change of concrete structure and etc. Measurement control covers: condition of concrete cover and reinforcement, study the change of pH of concrete, width and depth of cracks, calculation criteria for changing the structure of concrete, density of concrete and change of compressive strength.

To perform its protective action against the reinforcement in the section, the concrete must have a hydrogen ion concentration index pH above 11.8. It is important to note that all vertical reinforced concrete elements, as well as the underside of the horizontal elements of all slabs, have lime plaster with a thickness of 15-20mm, which creates a passivating barrier to the concrete. Floor screeding, slope concretes, and architectural coatings (tiles, flagstones, etc.) also create a barrier and prevent the concrete. The established average neutralization depth of concrete is greater than 25mm, and in some places it is over 45-50mm (Figure 8, Figure 9). The neutralization of the concrete in the surface areas, combined with the loose structure of the concrete, is a prerequisite for corrosion of the reinforcement, which was also established during the visual inspections of the structure (Figure 10).

Figure 8. Neutralization depth of concrete in columns

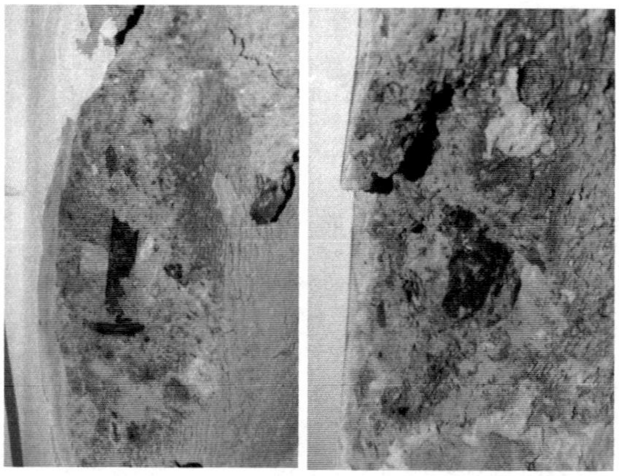

Figure 9. Neutralization depth of concrete in columns - undetermined

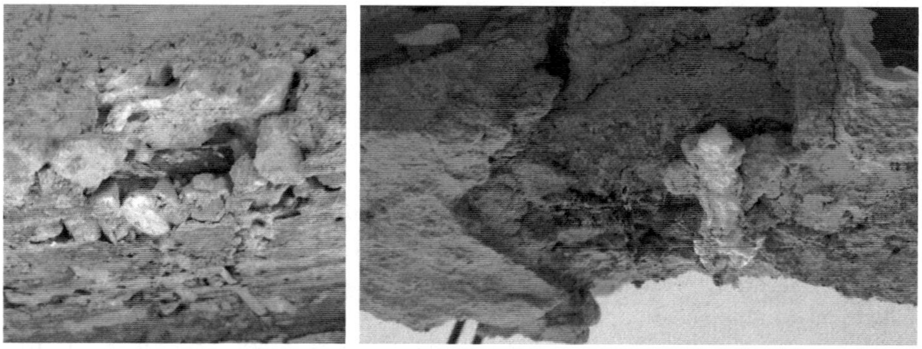

Figure 10. Corrosion in the reinforcement caused by a loose structure of the concrete and neutralization of the surface layers

3.7. Determination of the thickness of the concrete cover of the reinforcing rebars. Properties of the reinforcing steel

The concrete cover of the reinforcement was determined by direct and indirect methods.

The direct method is based on removing the cover, determining the embedded reinforcement, and measuring. Based on the conducted investigations, it has been conclusively determined that the reinforcing steel used for the transverse reinforcement in beams and columns is class A-I with a characteristic yield strength of 235 MPa. For the longitudinal reinforcement in the beams and columns, mainly steel of classes A-II and A-III was used, with a characteristic yield strength of about 310 MPa and 420 MPa. The reinforcing steel in the slabs (top and bottom reinforcement) is of classes A-I, A-II, and A-III. No embedded reinforcement was found in the basement walls detailed information for the reinforcement of the RC concrete elements is provided in Table 2.

The indirect method of determining the concrete cover, diameter, and spacing of the reinforcement specialized scanning rebar detector is used.

4. General information of the construction

The building is designed and constructed with a cast-in-place load-bearing reinforced concrete structure, consisting of strip foundations under basement walls, single footing under the interior columns, columns, beams, and slabs.

The structural elements of the building are:

- Foundation structure consisting strip foundations and single footing;

- Ground floor slab in basement level;
- Basement walls along the periphery of the entire building;
- Cast-in-place slabs with peripheral and internal reinforced concrete beams (Figure 2);
- Vertical reinforced concrete columns;
- Double pitched cast-in-place reinforced concrete roof;
- Reinforced concrete cantilever stairs.

The horizontal elements are beam and slab systems, absorbing the vertical loads from permanent and imposed loads, transmitting them to the columns and subsequently to the foundations and the ground. Reinforced concrete frames (columns and beams) are designed for seismic load. Reinforced concrete slabs also perform a diaphragmatic behaviour at storey levels. The structure is probably designed according to [27], [28], [29] and [30].

5. Structural and non-structural elements

5.1. Foundations

During the on-site visit, no geological phenomena and processes such as karst, gully, sufosis and collapses were identified. They would negatively affect the normal use of the building. There are no cracks, deformations, or displacements in the vertical and horizontal elements caused by ground collapses. It can be concluded that the supporting foundation structures of the hotel building are currently fulfilling their design functions. The water drainage system in the area around the building is functioning. There is no accumulation of water.

There is local destruction of the ground floor slab in separate rooms in the basement of the building. These deformations of the integrity of non-bearing elements do not affect the behaviour of the foundation structures and the main bearing system of the building.

5.2. Vertical elements

The vertical system is made of cast-in-place concrete elements - basement walls and columns.

The basement walls are made from the upper edge of the foundations to the ground level. They describe the outline of the building and are 35÷40cm thick. No reinforcement was found in the basement walls. All peripheral reinforced concrete columns from the ground level transmit their loads to the basement walls, and their reinforcement does not pass to the foundation structures. The columns inside the building are completed from the ground level to the foundation structures.

No deformations and cracks are observed in the concrete walls regardless of the loose structure of the concrete in some areas of the basement walls.

From the ground floor to the roof slabs, reinforced concrete columns were designed and executed, as the main step in the longitudinal direction for:

- The southern facade is 160 cm to the first level (K+2.65) and 320 cm to the roof slabs;
- The northern facade is 160 cm to the second level (K+5.45) and 320 cm to the roof slabs;
- The internal columns are320 cm for all levels.

Reinforced concrete beams in the transverse and longitudinal directions along the periphery of the building, as well as some additional continuous beams in the interior, have been implemented. Columns and beams (RC Frames) provide the structure's ability for seismic loads.

5.3. Horizontal elements

The reinforced concrete horizontal elements are slabs and beams. The thickness of RC slabs are 14-16cm on the ground floor, from 16cm to 20cm on the other floors, the unusable sub-roof slab

IOP Conf. Series: Materials Science and Engineering **1323** (2024) 012010 doi:10.1088/1757-899X/1323/1/012010

- 16cm, and the roof slabs - 20cm. At the upper levels, cantilever plates up to 300cm in length were made. No deformations or cracks have been observed, indicating that an ultimate or serviceability limit state has been reached. The main plate of the upper levels is decided as a one-way reinforced slab with a cantilever. The inserted reinforcement in the horizontal elements is systematized in Table 2.

There are many openings in the floor structures at all levels where top and bottom reinforcing bars have been interrupted (Figure 11).

In general, no mechanical failure of the concrete cover of the embedded reinforcing steel was observed. Still, it is necessary to note that in a significant number of places, a lack of concrete cover of the reinforcement (for beams and slabs) or a concrete cover of insufficient thickness and loose structure was found. Based on this and the past period of use of the building, many areas with initial corrosion in the reinforcement are observed.

Figure 11. Openings in floor structure

5.4. Reinforcement in concrete elements

The summary results for the reinforcements are presented in Table 2.

It should be noted that the reinforcement of the vertical elements does not comply with the seismic design principles of the Bulgarian codes (Regulation No. RD-02-20-2 [31]), as well as the Eurocodes.

Table 2. Reinforcing steel embedded in concrete elements

Column heading	Level/Elevation				
Reinforcement (longitudinal, transverse)	First level K+2.65 and the columns below	Second level K+5.45 and the columns below	Third level K+8.25 and the columns below it	Fourth level K+11.05 and the columns	Roof slab and the columns below it
Columns:					
- longitudinal	N18 (A-II)	N18 (A-II)	N18 (A-II)	N12 (A-III)	N16 (A-II)
- stirrups	ϕ6.5/17 (A-I)	ϕ6.5/22 (A-I)	ϕ6.5/23 (A-I)	ϕ6.5/21 (A-I)	ϕ6.5/19(A-I)
Beams:					
- longitudinal	N14 (N12) (A-III)	N14(N12) (A-III)	N12 (A-I/A-II/A-III)	N12 (A-I/A-III)	N14(A-III)
- stirrups	ϕ6.5/18 (A-I)	ϕ6.5/19 (A-I)	ϕ6.5/21 (A-I)	ϕ6.5/18 (A-I)	ϕ6.5/20 (A-I)
Slabs:					
- bottom bearing	N16/19 (A-II)	N14/15 (A-II)	N16/20(A-II/A-III)	ϕ12/15 (A-I)	ϕ10/17(A-I)
- bottom distrib.	ϕ10/25 (A-I)	ϕ10/23 (A-I)	ϕ12/22 (A-I/A-III)	ϕ8/22 (A-I)	ϕ8/21(A-I)
- top bearing	N16/24 (A-II)	N14/13 (A-III)	N14/14 (A-III)	ϕ12/17 (A-I)	ϕ10/18(A-I)
- top distrib.	ϕ10/26 (A-I)	ϕ10/23 (A-I)	ϕ10/24 (A-I)	ϕ8/19 (A-I)	ϕ8/20(A-I)

5.5. Partition walls

External and internal brick partition walls are made of hollow single or quadruple ceramic elements. Most of the external walls, as well as some internal walls with 25cm thickness, are made of single bricks with transverse cavities. Single or quadruple hollow elements were used for the internal partition walls (part of which were removed at the time of the present survey) with a thickness of 12.5 cm (Figure 12).

All completed partition walls in the building are with hollow bricks and can be classified as non-bearing for persistent or transient design situations and seismic design situations. It is necessary to take into account that masonry executed before the concrete structure (integration of the walls with the supporting structure) without an available distance with the supporting reinforced concrete structures can affect the behavior of the building under seismic situations.

Figure 12. Partition walls

6. Analytical study of construction

The analytical study of the structure was done with a spatial elastic computational model, which was developed in structural computational software [32], [33], [34] (Figure 13). In addition, simplified static models were used for individual structural elements as well as for the entire building. The research for seismic load was done by a spectral method, which was used to obtain the forces and moments in the columns and beams without considering the material and geometric nonlinearities in the structure's behavior. When determining the parameters of the models, the strength characteristics and densities of the concrete were used, according to Table 1, and literature data for the ground base in the region.

All imposed (live) loads are accepted according to BDS EN 1991-1-1 [35], and load combinations - according to BDS EN 1990 [36]. The effects of seismic loads are determined for peak ground acceleration $a_{gr}/g=0.23$, importance class – II category, regularity in plan, and elevation of the building. The combination of the seismic components by directions is performed by the square root of the sum of the squares (SRSS), and the statistical combination of the modal shape effects for the three directions is performed according to "Complete Quadratic Combination" (CQC). The behavior factor is q=1.5, according to BDS EN 1998-1 [37] and BDS EN 1998-3 [38].

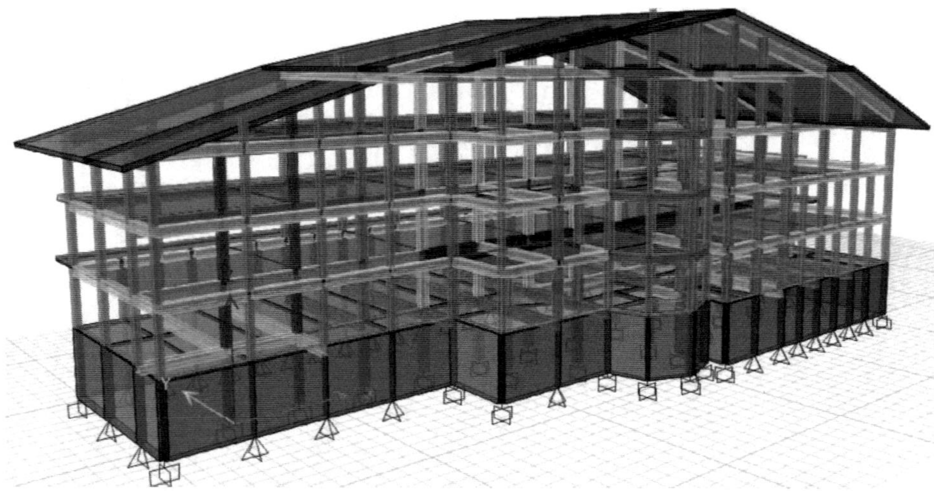

Figure 13. 3D view of the computational model

All necessary checks for ultimate and serviceability limit states have been carried out for all load-bearing horizontal and vertical elements of the building, as well as for the structure as a whole.

Based on the performed checks for all design combinations, the following conclusions were drawn, which are presented in Table 3.

IOP Conf. Series: Materials Science and Engineering **1323** (2024) 012010 doi:10.1088/1757-899X/1323/1/012010

Table 3. Reinforcing steel embedded in concrete elements

Horizontal elements / Level	Beams		Slabs
	Persistent or transient design situations	**Seismic design situations**	**Persistent or transient design situations**
Third level K+8.25	The requirements are OK	The requirements are NOT OK	The requirements are OK
Second level K+5.45	The requirements are OK	The requirements are NOT OK	The requirements are OK
First level K+2.65	The requirements are OK	The requirements are NOT OK	The requirements are OK
Ground floor K-0.10	The requirements are OK	The requirements are OK	The requirements are OK

It is important to note that during the inspection of the load-bearing reinforced concrete structure of the building, as well as the fragile non-load-bearing elements, no significant deformations and cracks in the existing elements were detected. This finding suggests that during the approximately 40-year period of operation of the building, the calculated loads for have not been reached, according to the norms according to which the structure was designed in the second half of the last century.

7. Conclusions

Based on the performed on-site, laboratory, and analytical procedures for the building, the following conclusions can be drawn:

A) Foundation structures:

- It was established that the foundation was realized on a single footing under the internal columns and strip foundations under the peripheral basement walls. The foundations were not found to be connected with beams;
- No subsidence or rotation of the foundations under the vertical elements is observed. No geological phenomena and processes (karst, gully, sufosis, and collapses) were found;
- No waterlogging is observed at the basement level with the exception of a few areas. It is probably caused by a local compromise of the waterproofing layers laid on the outer surface of the basement walls;
- There is local damage to the floor slabs in separate rooms in the basement of the building. These deformations of the integrity of non-bearing elements do not affect the behavior of the foundation structures and the elements of the building.

The foundation conditions of the hotel building can be classified as "good".

B) Materials:

- The characteristic compressive strength of the concrete is determined: for vertical elements for all levels - standard compressive strength class C20/25, and for horizontal elements (slabs and beams) - from class C16/20 to C25/30 (see Table 1). A denser concrete was found in the concrete mixture of the floor slabs and vertical elements of the superstructure than in the basement level. In all reinforced concrete elements, segregation with air cavities is noticed in many places.

IOP Conf. Series: Materials Science and Engineering **1323** (2024) 012010 doi:10.1088/1757-899X/1323/1/012010

- The reinforcing steel used in the building is class A-I (nonribbed profile) for stirrups of columns and beams, class A-II (ribbed profile) for beams, slabs and columns and class A-III (ribbed profile) – for beams, slabs and columns. The summarized results for the type (class) and amount of inserted reinforcing steel are presented in Table 2;

C) Vertical and horizontal structural system:

- In general, there are no deformations in the horizontal elements;

- No mechanical damage to the concrete cover of the reinforcement was observed. Still, it should be noted that in a significant number of places, a lack of concrete cover of the reinforcement (for beams and slabs) or a concrete cover of insufficient thickness and loose structure was found. Multiple areas of incipient corrosion are observed in the reinforcement. The reduction of the cross-sections of the reinforcement is below 1.5÷2%;

- The established average neutralization of the concrete is greater than 25mm, and in some places it is over 45-50mm. The neutralization of the concrete in the surface areas combined with the loose structure of the concrete is a prerequisite for corrosion of the reinforcement, which was also established during the visual inspections. At present, the corrosion rate can be classified as "low";

- The imposed loads in the current design codes and the codes according to which the structure was designed do not have significant differences. Based on the performed analyses and calculations, according to BDS EN 1992-1-1 [39], with established parameters of the inserted materials (concrete, reinforcing steel) and cross-sections, it was found that the structure is able to satisfy the requirements of the currently valid design codes for Persistent or transient design situations. No significant damages (deformations, cracks) were found, which would indicate that the ultimate or serviceability limit state of the reinforced concrete elements had been reached;

- It was established that the structure does not satisfy the checks for seismic impacts, according to the requirements of the currently valid Bulgarian codes [31], [37], and [38]. It was found that the structure does not satisfy the principles of capacitive design: providing the necessary ductility by constructing plastic joints in certain areas to have the required local ductility, i.e. ("strong columns - weak beams"); excluding unwanted forms of brittle failure (shearing, failure of adhesion, etc.) and preventing the formation of plastic joints in the columns. If it is necessary to satisfy the norms for seismic design, it is essential to take measures to improve the performance of the structure under seismic impacts, by increasing load-bearing capacity.

In order to ensure future regular operation for seismic situations, basic measures can be taken to improve the performance of the vertical and horizontal structural systems under seismic combinations.

The reconstructions for the vertical system can be successfully done, both by the implementation of new reinforced concrete monolithic walls, and by the implementation of steel trusses in appropriate places, taking into account the overall redistribution of vertical and horizontal forces. To ensure the normal operation of the structure, design solutions should be developed that satisfy the primary function, the corresponding requirements for strength, serviceability and resistance to fire.

The inspection provided representative information on the condition of many reinforced concrete frame structures built in the second half of the last century in Bulgaria. The accumulated experience can be helpful to similar type of constructions.

References

[1] Abdulahad E. 2023 *Instrumental monitoring of actual reinforced concrete structure of a residential building during its construction and its service life for a 10-year period*, Annual UASG, 56 (3): 1285 – 1300.
[2] Abdulahad, E., Panichkov, D., Apostolova, A. 2008 Study of the stressed and deformed state of a real reinforced concrete structure of a residential building during execution and after its entry into operation, Scientific conference with international participation - VSU'2008, Sofia.
[3] Abdulahad E. 2008 Seismic strengthening of an existing precast reinforced concrete structure with carbon FRPS, 5th International Conference on Advanced Composite Materials in Bridges and Structures (ACMBS-V).
[4] Boshnakov K., Belev B., Mollova G. 2014 Seismic performance assessment and approaches for strengthening of existing flat slab reinforced concrete building, 14th International scientific conference VSU'2014, Sofia.
[5] Boshnakov K., Hristov P. 2014 Inspection analysis and assessment of the seismic safety of existing flat slab reinforced concrete building, 14th International scientific conference VSU'2014, Sofia.
[6] Dimitrov D., Dedov K., Vassilev V. 2015 Investigation of the superstructure of the bridge over Tundja river at the town of Yambol for the carriageway widening, First scientific- applied conference with international participation reinforced concrete and masonry structures- theory and practice– Sofia.
[7] Dimov D., Georgiev A., Jelev J. 2015 Some lessons from inspection of road bridges in mountain regions in Bulgaria, First scientific- applied conference with international participation reinforced concrete and masonry structures- theory and practice– Sofia.
[8] Dimov D. 2016 Inspection and testing of building structures and bridges.
[9] Dimov D. 2011 Non-destructive testing of building structures, Sofia.
[10] Kardjiev V., Hristov P. 2015 Examination and evaluation of the technical condition of industrial buildings complex in Breznik, First scientific- applied conference with international participation reinforced concrete and masonry structures- theory and practice– Sofia.
[11] Kardjiev V., Hristov P. 2015 Reconstruction of an industrial buildings complex in the town of Breznik for the needs of a modern biotechnical manufacture, First scientific- applied conference with international participation reinforced concrete and masonry structures- theory and practice– Sofia.
[12] Panichkov, D. 2001 Inspection and testing of building structures and facilities, Sofia, UIK.
[13] Petkov V, Tsanovski S. 1990 Guide to determining the condition of concrete and the reinforcement in the load-bearing structures of buildings and facilities, Sofia, NISI.
[14] Stefanov D. 2014 Evaluation of the condition of existing buildings with masonry bearing structures, 14th International scientific conference VSU'2014, Sofia.
[15] Tsvetkov P., Chernev P, Sarafova R. 1986 Guide for the reconstruction of reinforced concrete structures of industrial buildings, NISI, Sofia.
[16] Tsvetkov P., Chernev P, Sarafova R. 1986 Guide to examination and assessment of industrial buildings and facilities subject to reconstruction, NISI, Sofia.
[17] BDS EN 12390-3:2019 - Testing hardened concrete - Part 3: Compressive strength of test specimens.
[18] BDS EN 12504-1:2019 - Testing concrete in structures - Part 1: Cored specimens - Taking, examining and testing in compression.
[19] BDS EN 12504-2:2021 - Testing concrete in structures - Part 2: Non-destructive testing - Determination of rebound number.
[20] BDS EN 13791:2019 - Assessment of in-situ compressive strength in structures and precast concrete components.
[21] BDS EN 206-1/NA Concrete - Specification, performance, production and conformity — Part 1: Performance, requirements, factory production control and assessment criteria for individual values.
[22] Dimitrov, D., Georgiev, Evg. 2012 Regarding creep coefficient according EC2. Proceedings – Volume 2 - International Jubilee Scientific Conference "70th Anniversary UACEG", 15-17 November 2012, University of architecture, civil engineering and geodesy, Sofia.
[23] Dimitrov, D., Georgiev, Evg. 2014 Determination of shrinkage deformation of buildings and bridges according to Eurocode 2". Annual of the University of Architecture, Civil Engineering and Geodesy, Fascicule V, Vol. XLVI 2013-2014, UACEG, Sofia.
[24] Dimitrov, D., Georgiev, Evg. 2014 Creep coefficient determination for high strength concrete at buildings and bridges according Eurocode 2-2. Annual of the University of Architecture, Civil Engineering and Geodesy, Fascicule V, Vol. XLVI 2013-2014, UACEG, Sofia.
[25] Dimitrov, D., Georgiev, Evg. 2016 Shrinkage strain in "thick" reinforced concrete elements with cement class N, according to Eurocode 2. Annual of the University of Architecture, Civil Engineering and Geodesy, First Scientific-

Applied Conference with International Participation "Reinforced concrete and masonry structures – theory and practice", 22-23 October 2015, Fascicule 12-I ,Part I, Vol. 48 2015, UACEG, Sofia.
[26] Georgiev, Evg. 2016 Reporting the accuracy of the creep and the shrinkage on the deflection of one-way reinforced concrete slabs according to Eurocode 2. Proceedings XVI International Scientific Conference VSU'2016, 9 – 10 June 2016, Sofia, Bulgaria, Vol. II, Sofia, 2016.
[27] Calculation of structures and earth foundation. Fundamentals of design, 1964.
[28] Codes and rules for design of concrete and reinforced concrete structures from 1969.
[29] Loads and impacts. Design standards from 1979.
[30] Regulations for construction in earthquake areas from 1964, amendments and additions 1972 and 1977.
[31] Regulation No. RD-02-20-2 from 27.01.2012 Design of structures for earthquake resistance.
[32] ETABS v.20, Computers and Structures, Inc. (CSI), https://www.csiamerica.com/products/etabs.
[33] Neshev Hr. Methodology for creating a planar computational model of a flat slab with CSI Etabs v.19 and v.20, University of architecture, civil engineering and geodesy, Sofia, 2023, ISBN 978-954-724-162-6.
[34] Neshev Hr. Methodology for creating a three dimensional computational model of a building with a wall construction system with CSI ETABS v.17, v.19 and v.20, University of architecture, civil engineering and geodesy, Sofia, 2023, ISBN 978-954-724-163-3.
[35] BDS EN 1991-1-1/NA:2004 – Eurocode 1: Actions on structures - Part 1-1 : General actions – Densities, self-weight, imposed loads for buildings.
[36] BDS EN 1990/NA:2003 – Eurocode: Basis of structural design.
[37] BDS EN 1998-1/NA:2005 – Eurocode 8: Design of structures for earthquake resistance – Part 1: General rules, seismic actions and rules for buildings.
[38] BDS EN 1998-3/NA:2005 – Eurocode 8: Design of structures for earthquake resistance – Part 3: Assessment and retrofitting of buildings.
[39] BDS EN 1992-1-1:2005/NA – Eurocode 2: Design of concrete structures – Part 1-1: General rules and rules for buildings.

IOP Conf. Series: Materials Science and Engineering **1323** (2024) 012011 doi:10.1088/1757-899X/1323/1/012011

Seismic assessment of buildings constructed under previous generations of codes

Krasimir Boshnakov[1*] and Vasil Kardjiev[2]

[1] Department of Computer-Aided Engineering, Faculty of Structural Engineering, University of Architecture Civil Engineering and Geodesy /UACEG/, Sofia, Bulgaria
[2] Department of Reinforced Concrete Structures, Faculty of Structural Engineering, University of Architecture Civil Engineering and Geodesy /UACEG/, Sofia, Bulgaria

*E-mail: krabosh_fce@uacg.bg

Abstract. The primary engineering problem of seismic assessment of existing buildings constructed under previous codes is evaluating their capacitive behavior under seismic actions defined by the codes at construction time to define the rehabilitation necessity. The paper is focused on the methodological aspects of this process. Examples from the practice present the proposed workflow.

1. Introduction

A considerable portion of the European reinforced concrete building stock was constructed under previous generations of design codes. The tolerated level of risk of structural damages due to earthquake excitations, and not only, in these historical local codes is extensively higher than in modern codes [1-6] and is rarely explicitly specified.

The buildings constructed before (1989) are considered unsecured against earthquake actions in Bulgaria nowadays. The exact numbers of these buildings are unclear but evaluations, based on different sources [16-18] reveal that more than 73% of the public and 75% of the residential buildings stock are not safe to meet strong earthquake actions. Although these numbers include poorly maintained and abandoned buildings, public buildings of primary importance for post-earthquake disaster operations, and a large part of the 5% portion of the total number of residential buildings that corresponds to app. half of their total area fall within this group.

The seismic performance assessment consists of identifying the ground shaking intensity the building can experience under certain probability of occurrence and estimating structure performance parameters, like forces and deformations and possibly damages to both structural and non-structural components, possible losses, and even casualties.

The increased uncertainty associated with the seismic performance assessment of old buildings is well recognized. The probabilistic nature of predicting the ground motion parameters and the time interval of its re-occurrence together with the relatively high uncertainty in material properties, structural geometry, adequacy of detailing, and quality of construction, among many others makes the task impossible to solve precisely in the mathematical deterministic sense. The results are naturally probabilistic whatever type of performance assessment: intensity-based, scenario-based, or time-based is considered [7].

1.1 Methodological aspects

For the cases where structural seismic upgrading is considered the first step is to evaluate the deficiencies in the structural capacity to resist the modern code-defined earthquake actions [1-6]. This involves analytical techniques that explicitly consider the expected nonlinear structure response to the seismic actions [1-6, 8, 13, 19]. The analysis is aimed at revealing the deficiencies and identifying suitable approaches to the seismic upgrade.

1.1.1 Deficiency Identification Target

The obtained results reveal the deficiencies of the existing structures and define the strengthening needs and targets to be addressed. Generally, the seismic deficiencies fall within one or more of the categories [8, 13, 21]:

- Global strength deficiency
- Global stiffness deficiency
- Configuration deficiencies – in plan, in height
- Load path deficiency
- Detailing deficiencies
- Floor diaphragm deficiencies
- Foundation deficiencies
- Other deficiencies – geologic hazards, adjacent buildings, deterioration of materials.

If necessary, the existing structure can be rehabilitated only to restore its initial capacity when the preliminary analysis does not reveal a considerable structural weakness to withstand the earthquake actions defined by the modern codes.

1.1.2 Strengthening needs, targets, and technological approaches

The seismically rehabilitated or upgraded structure shall respond without reaching the code-defined significant damages (SD) limit state [4-6] under excitation that corresponds at least to the modern code-defined response spectrum.

Classes of rehabilitation measures [8, 21]:

- Addition of elements
- Enhancing the performance of existing elements
- Improving the connections
- Seismic demand reduction

1.2 The assessed structures

For existing buildings, the preliminary assessment for identifying the need for upgrading and when necessary for defining the acceptable target performance range the upgraded structure can accept, as a starting point, the hypothesis of conformance of the existing structure to the code requirements at its design and construction time is adopted. The general process of the performance-based design and the performance-based assessment as described in [7] and suited to the current study is shown in Figure 1.

The modern code requirements define the target performance objectives and can be further elevated to satisfy owner needs.

This presented approach considers the cases of two pre-modern codes, generic reinforced concrete buildings. Both generic buildings are considered as designed in 1965 and constructed right after it, and have reinforced concrete structures. The structural type of the first one (Building 1) is taken as a framed structure with ceramic masonry infills and the second (Building 2) structure is of flat slabs with reinforced concrete shear walls around the staircase forming a structural core.

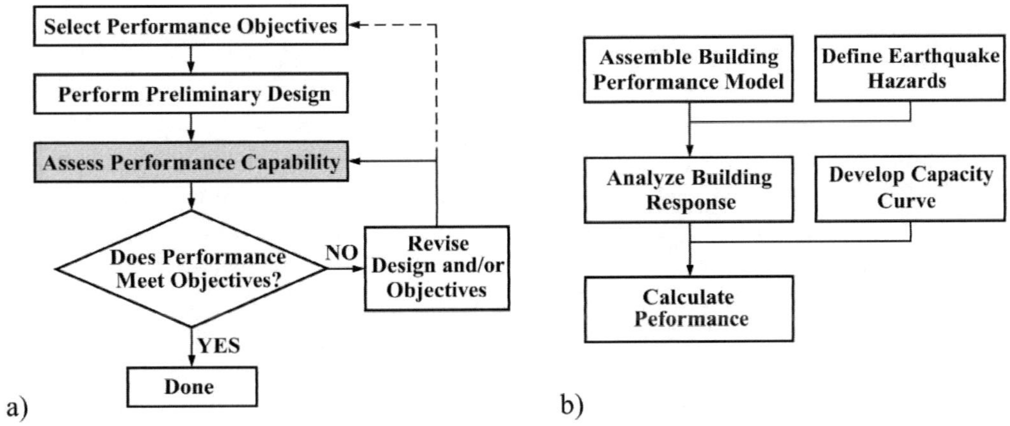

Figure 1. a) The performance-based design process as in [7] and b) The Performance-based assessment workflow adopted in the study.

The described generic buildings parametrically correspond to considerable part of the building stock for the time. The fundamental period of both Building 1 and Building 2 is equal and T = 1.0 s. This roughly corresponds to five-storey buildings. Details about the specified generic buildings are provided below. The results are indicative of the whole groups of respective buildings.

2. The earthquake hazards and the building's models

The modern formulation of the seismic hazard is based on refined probability models that reflect the almost century-long measurements of earthquake ground shakings. The reinforced concrete structures' analytical models have changed over time. The modern development of computer hardware and software provided the tools to engineers to model the real behavior of the structures without the need to apply linear analogies to the otherwise complex and essentially nonlinear behavior of the composite in their nature elements that involve different materials.

2.1 The earthquake hazards

The earthquake hazard definitions in the construction codes have substantially changed over the years. During the considered period the earthquake hazard assessment method based on observed seismicity was changed to probabilistically assessed [16, 18] seismicity. Further, its code definition was amended from earthquake intensity-based in the local codes [12] to one based purely on engineering meaningful ground shaking parameters in Eurocode 8 [1-3]. These general evolutionary changes are due to the increased knowledge and scientific development and are generally describing the earthquake construction code changes.

Both structures are taken as designed and constructed under the Bulgarian local code (1964) [12]. Both construction sites are chosen to fall within the seismic zone with expected event intensity VIII to MSK-64 (~EMS) [12] under this code. The soil on both sites is taken as class 3 of the (1964) local code and is roughly identical to soil type C from EN 1998-1 [1-3]. The seismic coefficient $K_C = 0.05$ has the meaning of reference ground acceleration in parts of g (the ground acceleration $g = 9.81$ m/s²) for the case. The design values of the induced seismic forces in this code are formulated as:

$$S_k = \psi\, \beta\, \eta_k K_C Q_k\,, \tag{1a}$$

$$0.6 \leq \beta = 0.9/T \leq 3.0, \tag{1b}$$

$$\eta_k = x_k \left(\Sigma\, Q_j x_j / \Sigma\, Q_j x_j^2 \right) \tag{1c}$$

where T is the fundamental period, $\beta(T)$ is the dynamic factor, x_k and x_j are the relevant horizontal displacements of points k and j of the structure, and ψ is the structural damping factor (accounting for the system damping or ductility in modern terms) which takes the value of 1.0 for regular structures, and 0.5 for fully reinforced concrete structures [12], but also plays as an amplification factor, when adopting the value of 1.5 for towers and masts.

The dualism in the formulation of the ψ factor is resolved by accepting that the code (1964) has defined two separate design response spectra – one for the buildings and another for the towers and masts. These response spectra cannot be considered as dependent and the possible mathematical readjustment to reorder them as in modern codes would violate the engineering terms as it would result in unsafe overestimation of the allowed, by the old code [12], non-elastic structural displacements.

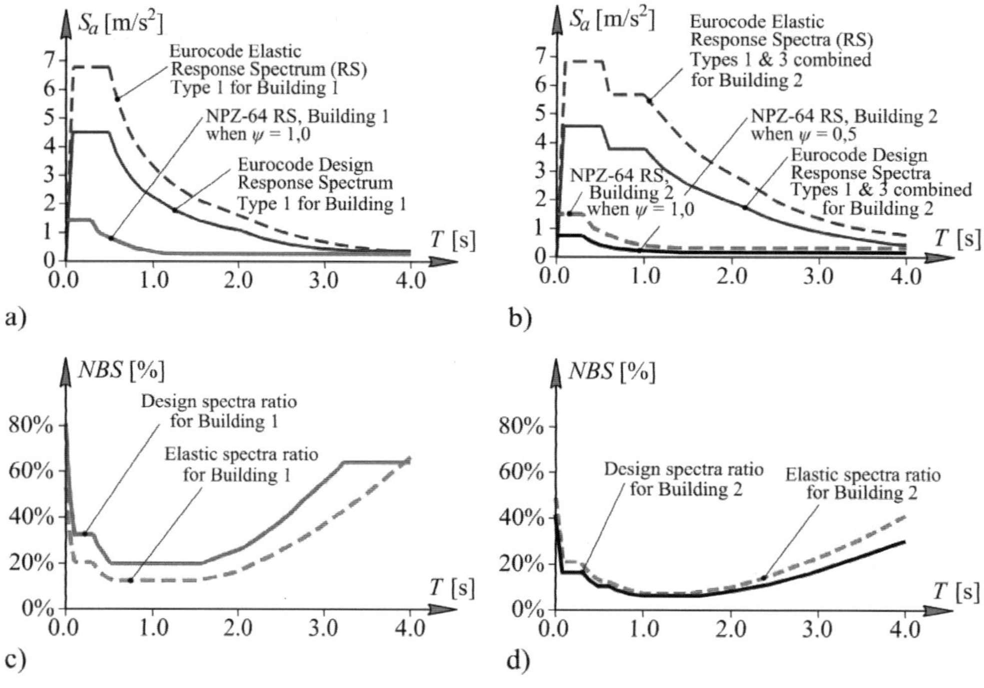

Figure 2. The response spectra by the different codes: a) Building 1, b) Building 2. The ratio of Eurocode RS to old code [12] earthquake loading: c) Building 1, d) for Building 2.

The earthquake hazard today is defined by the Eurocodes [1-3] and their national annexes. The differences between the two distinct code definitions are substantial. Eurocode explicitly defines two types (Type 1 and Type 3 in the case) [1-3] of elastic response spectra for the different soil types. The site of Building 1 is influenced by the Type 1 spectrum only, while the site of Building 2 – by both Type 1 and Type 3 response spectra which can be combined into one cummulative RS by following the formulated in [15] approches. The reference ground acceleration $a_{gR} = 0.23$ g ($g = 9.81$ m/s^2) for both construction sites. The reference probability of its exceedance $P_{NCR} = 10$ % in $T_L = 50$ years or reoccurrence period of $T_{NCR} = 475$ years. Figure 2 illustrates the different response spectra defined by codes. Eurocode 8-1 defines design response spectra for newly designed buildings to be used in linear elastic analysis.

They depend on the structural performance factors q. The design response factors $q = q_0 k_w = \mu (\alpha_U/\alpha_1) k_w$ [17], where μ is the ultimate structural ductility, considered by the code, α_U/α_1 is the structural over-strength in case of structures designed for high ductility class (DCH), and k_w is factor reflecting the prevailing failure mode in structural systems with walls.

In present cases, both structures of Building 1 and Building 2 would be assigned a value of q = 1.5 after taking into consideration their lack of seismic detailing.

2.2 The structure capacity

The previous generation of codes did consider the linear elastic analysis only. They assigned ductility to define the required structure deformation capacity in relative terms but this also fixes the absolute displacement values, when applied to a particular structure under a particular seismic action defined by the code response spectrum.

The current generation of codes and seismic performance assessment guidance documents [1-11] prefers and recommends the non-linear analysis.

2.2 The structural performance

The performance-based approach requires that existing structures do not exceed the limits of the significant damage (SD) limit state as defined in EN 1998-3 [4-6] under actions with reference mean return period T_R= 475 years.

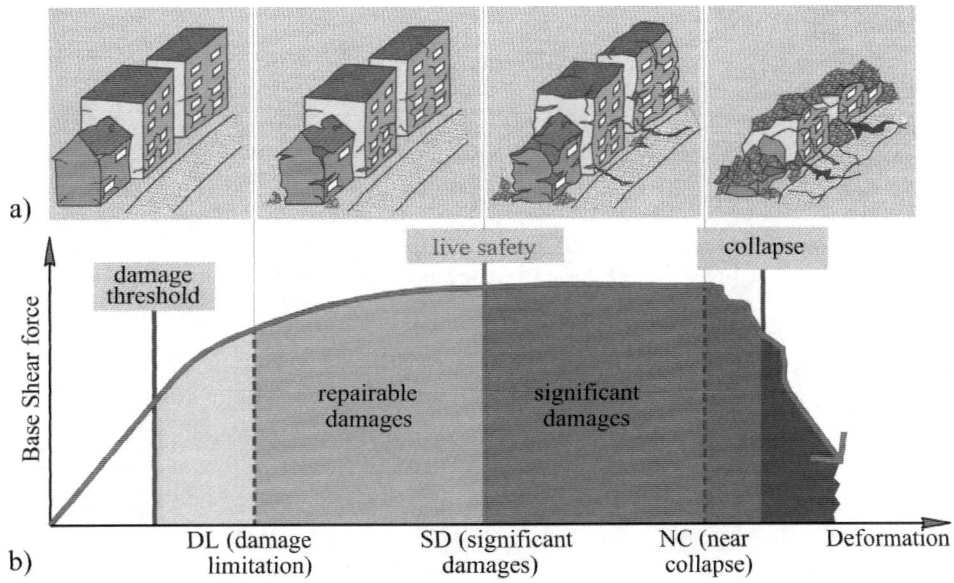

Figure 3. The Performance-based earthquake design concept as in [10] with a) visual depiction of damage, roughly correlated to b) limit states and damage levels on structure capacity curve.

Further structures shall not reach the no-collapse (NC_3) limit state, under seismic actions with return period $T_R = 2475$ years. The NC_3 requirements are relaxed compared to those for new buildings (NC_1) specified in EN 1998-1 [1-3]. The plastic displacement allowance is increased by introducing the significant damage (SD) Ultimate Limit State in EN 1998-3 as $SD = NC_1$ and extending the limits of $NC_3 = 1.33NC_1$ as shown in Figure 3. This assumption recognizes some structural over-strength, the existing differences in the material factors between different seismic code generations, and also allows increased structural damages, compared to new buildings, without allowing the structure to collapse. This plastic displacement extension is not universal and applies to EN 1998 [1-6] only.

A method to assess the risk for an existing structure is given in [11]. The assessment is based on the ratio between the ultimate seismic capacity (C_U) of the existing structure and the seismic demand ($D_{N,ULS}$) at an ultimate limit state according to the seismic standard in action, and is evaluated in percentages. The ratio of *%NBS* (new building structure) reflects the available capacity compared to the currently required one and is formulated as:

$$\%NBS = C_U \,/\, D_{N,ULS} \times 100\ \% \tag{2}$$

Seismic performance classes are then assigned according to Table 1 and are expressed in Alpha ratings. The approach applies to other codes also, when taking into consideration that the general framework of modern seismic standards is similar and that capacity components calculation and safety evaluations are subjected only to minor variations between them. In the present cases the existing structure capacity initially is adopted to satisfy the requirements of the (old) code at the time of construction [12] while the required demand ($D_{N,ULS}$) is defined by the requirements of Eurocode 8-3 [3-6]. The response spectra ratios in Figure 2 c) and d) correspond to the %NBS definition under the adopted conditions. The results for Building 1 in Figure 1 c) indicate the existence of a high to very high risk for life safety, while the results for Building 2 in Figure 2 d) show the presence of very high risk.

Table 1. Assessment outcomes (potential building status) [11, 13]

Percentage of New Building Standard (%NBS)	Alpha rating	Approx. risk relative to a new building	Life-safety risk description[a]
> 100	A+	Less than or comparable to	Low risk
80 – 100	A	1 – 2 times	Low risk
67 – 79	B	2 – 6 times	Low to Medium risk
34 – 66	C	5 – 10 times	Medium risk
20 – 33	D	10 – 25 times	High risk
< 20	E	25 times and more	Very high risk

[a] Here the context of Life safety corresponds to the Near Collapse (NC_1) Ultimate Limit State as defined in Eurocode 8-1 [1-3] and Significant Damages (SD) as in Eurocode 8-3 [4-6].

Similar conclusions from elastic analysis on detailed FEM models of existing building structures are reported in other studies [22] where historical landmarks are studied. The conclusions from these studies point to the necessity to increase the capacity of the structures to resist the seismic lateral loads practically in all of the listed cases of existing buildings. In some of them, exhaustion of the structural capacity to resist static vertical loads was identified, also.

3. The preliminarily identified approaches for seismic upgrade

3.1 The results

The results from the preliminary assessment of the seismic performance of both structures reveal considerable deficiencies in their capacities to survive the ground excitations required by modern codes. The results for Building 1 are shown in Figure 4 a) and for Building 2 – in Figure 4 b).

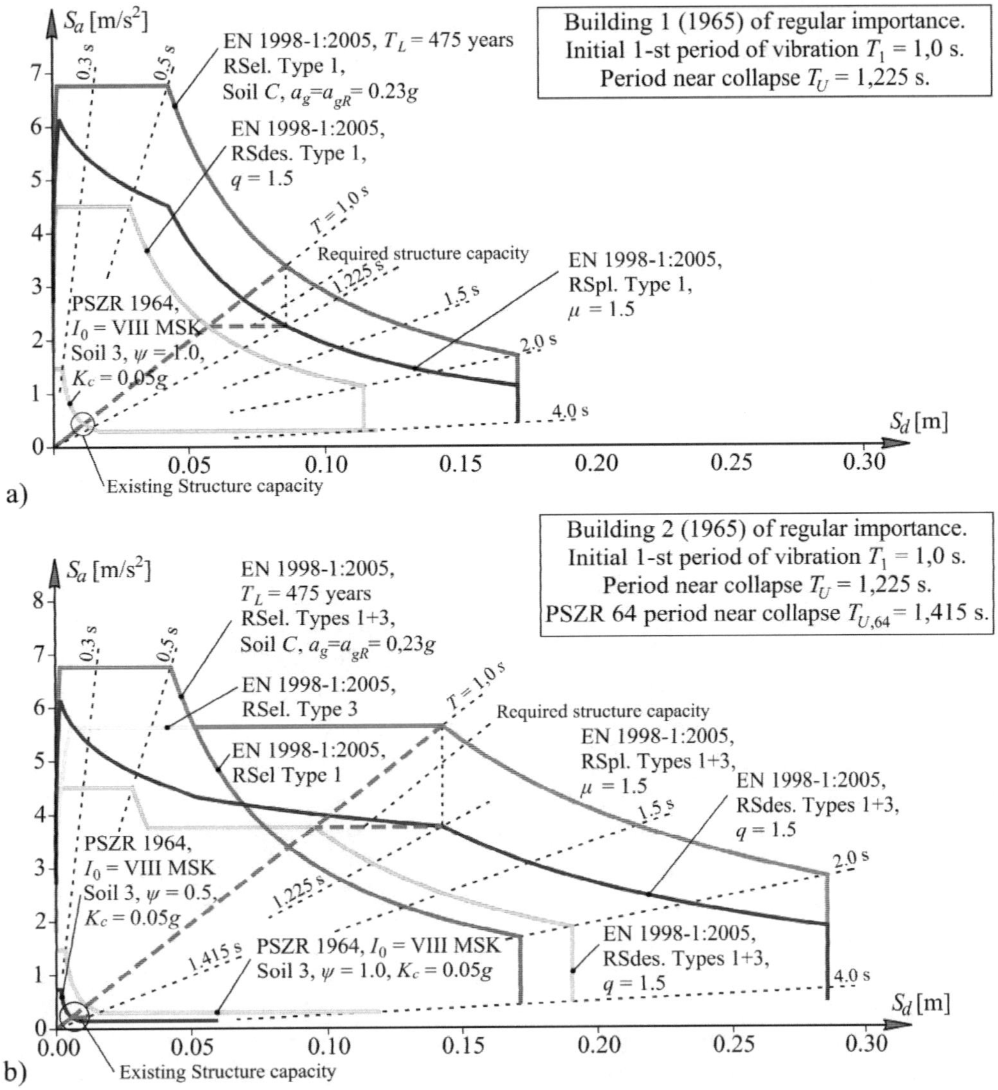

Figure 4. Evaluations for the considered structures. a) Building 1, and b) Building 2

The required capacity curves of the existing structures at design and construction time visibly fail to cross the modern seismic demand curves even in the cases of suspecting the existence of some hidden reserves like over-strength and over-ductility. The results identify the presence of strength and ductility deficiencies. Further investigation and tests on the structures can only extend the list. The fact that these buildings survived a number of seismic excitations with limited or no damages is easily explained by the relative weakness of the ground shakes in the past century in the country. The importance of the structural ability to develop non-linear deformations, especially in beam-column joints to preserve the integrity of the structure plays a vital role during strong earthquake shakings. The presence of soft-stories and other topological structural inconsistencies can dramatically influence earthquake performance. These conclusions and the recommendation to use modern codes to provide seismic resistance to buildings result not only from analysis but also from post-earthquake observations [14]. For urban buildings their collocation to adjacent buildings is important. The occurrence of earthquake-induced pounding can undermine structural integrity and lead to premature collapse[21, 23]. In such cases, seismic rehabilitation measures regarding the neighboring structures are necessary.

3.2 The feasible rehabilitation measures

In the cases where the gap between the existing supply of structural capacity and the currently required demands of the seismic actions is comparatively big, the deficiencies shall be addressed by more than one of the above listed in pt. 1.1.2 measures.

Global measures to reduce or eliminate the possible deficiencies in structural configuration, load path, floor diaphragms, foundations, and joint detailing would address the existing structure primarily.

The inclusion of new elements to address the strength deficiency would increase the stiffness and thus make the inclusion of earthquake energy dissipating devices desirable, allowing at the same time to keep the existing earthquake displacements within limits to avoid excessive damages.

Past research studies highlight the structural, technological, and economic advantages of this approach. The methodological example of an application of Concentrically Braced Frames with Modified Braces (CBF-MB) in building rehabilitation projects presented in [19] among others gives practical guidance to a case of a modern application.

The considerable demands in the case of Building 2 make almost unavoidable the inclusion of measures for seismic demand reduction in this case. Practically this can be achieved by including seismic dampers at the foundation level. Technologically this measure can be conviently combined with the required strengthening of the existing and the works on new foundations for newly added structural elements.

Implementing global measures to reduce the earthquake demand is necessary in all cases. Such measures include a reduction of the masses and the structural irregularities.

The local measures for enhancing the strength and deformation performance of all primary structural elements and their connections are necessary when the absolute seismic loads or deformations of the rehabilitated structure are increased. Excluding some of the existing elements from the list is possible only when the new structure performance point corresponds to lower displacement than the existing one and the strength requirements to them are not increased.

The most suitable mix of both global and local seismic structural rehabilitation and strengthening measures can be identified only after taking into consideration the economic, technological, and other site-specific factors. These factors play a crucial role in determining the optimal approach for enhancing a structure's seismic resilence.

4. Conclusions

The modern standards' expectations for strong earthquake demands are considerably higher than those about half a century ago. The acceptable limits of structural damages have decreased. The valuable part of the existing building stock needs seismic rehabilitation and an upgrade is expected to meet the modern social expectations and requirements.

The examples illustrate an approach that applies to the preliminary assessment of the seismic performance of existing structures, identification of the needs for seismic rehabilitation or upgrade, and initial specification of the applicable construction measures.

References

[1] BDS EN 1998-1:2005, Eurocode 8: Design of structures for earthquake resistance – Part 1: General rules, seismic actions and rules for buildings, BIS (2006)
[2] BDS EN 1998–1:2004/AC:2009, Eurocode 8: Design of structures for earthquake resistance – Part 1: General rules, seismic actions, and rules for buildings. Corrigendum, CEN (2009)
[3] BDS EN 1998–1/NA, Eurocode 8: Design of structures for earthquake resistance – Part 1: General rules, seismic actions, and rules for buildings. National annex, BIS (2011)
[4] BDS EN 1998-3:2005, Eurocode 8: Design of structures for earthquake resistance – Part 3: Assessment and retrofitting of buildings, BIS (2007)
[5] EN 1998–3:2005/AC:2010, Eurocode 8: Design of structures for earthquake resistance – Part 3: Assessment and retrofitting of buildings. Corrigendum", CEN (2010)
[6] BDS EN 1998–3/NA, Eurocode 8: Design of structures for earthquake resistance – Part 3: Assessment and retrofitting of buildings. National annex, BIS (2011)
[7] ATC. Seismic performance assessment of buildings, Volume 1 – Methodology, FEMA P-58-1, Federal Emergency Management Agency, Washington D.C. USA (2012)
[8] FEMA. Techniques for the seismic rehabilitation of existing buildings, FEMA 547:2006, Federal Emergency Management Agency, Washington D.C. USA (2006)
[9] FIB. Seismic assessment and retrofit of reinforced concrete buildings, FIB bulletin 24, International Federation for Structural Concrete (fib), Lausanne, Switzerland (2003) ISBN: 2-88394-064-9
[10] FIB. Displacement-based seismic design of reinforced concrete buildings, FIB bulletin 25, International Federation for Structural Concrete (fib), Lausanne, Switzerland (2003) ISBN: 2-88394-065-7
[11] NZSEE, The seismic assessment of existing buildings. Technical guidelines for engineering assessment, New Zealand Society of Earthquake Engineering, Wellington, NZ (2017)
[12] Regulations for anti-seismic construction, State Construction Committee (Ministry), (Bulgaria) State Gazette No 90, 1964/09/13
[13] Antoniou S. Seismic retrofit of existing reinforced concrete buildings, John Wiley & Sons Ltd. (2023) ISBN: 9781119987352, DOI: 10.11002/9781119987352
[14] Blagov D, Georgiev Tz, Kisliakov D, Petrov B, Traykova M. Rapid structural assessment of buildings – experience after 26th November 2019 earthquake in Albania, Conference on Civil Engineering Design and Construction – DCB'2020 10-12 September 2020, Varna, Bulgaria, IOP Conference Series Materials Science and Engineering, Volume 951, D Blagov et al 2020 IOP Conf. Ser.: Mater. Sci. Eng. 951 012019 DOI 10.1088/1757-899X/951/1/012019
[15] Boshnakov K, Kardjiev V, Seismic action redefinition for regions with multiple code prescribed seismic response spectra, XII International Scientific Conference "Civil Engineering Design and Construction", Varna, Bulgaria Sep. 08-10, (2022), ISSN 2683-071X
[16] Boshnakov K, Kardjiev V. Approaches for earthquake damage reduction of aged reinforced concrete structures in Bulgaria, Seismic retrofit of buildings in Bulgaria for efficient seismic risk reduction, The 12-th CMDR COE annual conference on crisis management and 3rd International conference on environmental protection and disaster risks, 04/06/2024 Sofia, Bulgaria (2024)
[17] Boshnakov K, Kardjiev V. Seismic performance assessment of RC framed system building designed to Eurocode high-class ductility. // Annuals of UACEG (2023) 56(3): 1263-1276

[18] Boshnakov K, Kouteva-Guentcheva M, Kardjev V. Seismic retrofit of buildings in Bulgaria for efficient seismic risk reduction, The 12-th CMDR COE annual conference on crisis management and 3rd International conference on environmental protection and disaster risks, 04/06/2024 Sofia, Bulgaria (2024)
[19] Georgiev Tz, Raycheva L. Chapter 5, Volume on case studies on upgrading of existing buildings, INNOSEIS – Valorization of innovative anti-seismic devices, Research Program of the Research Fund for Coal and Steel, (2017) www.innoseis.ntua.gr
[20] Kazemi F, Asgarkhani N, Manguri A, Jankowski R. Seismic probabilistic assessment of steel and reinforced concrete structures including earthquake-induced pounding, Archives in Civil and Mechanical Engineering, (24) 178 (2024) DOI: 10.1007/s43452-024-00994-7
[21] Newman A. Structural renovation of buildings – Methods, details, and design examples, 2nd edition, McGraw-Hill Education, New York (2021) ISBN: 978-1-26-045834-3
[22] Traykova MD, Traykov AV. Seismic assessment of heritage buildings in Bulgaria, 12th International Conference on Structural Analysis of Historical Constructions – SAHC 2020, Sep16-18, Barcelona (2021) DOI:10.23967/sahc.2021.222
[23] Uz ME, Jakubczyk-Gałczyńska A, Jankowski R. Numerical Analysis of Seismic Pounding between Adjacent Buildings Accounting for SSI. *Applied Sciences.* 2023; 13(5):3092. DOI: 10.3390/app13053092

IOP Conf. Series: Materials Science and Engineering **1323** (2024) 012012 doi:10.1088/1757-899X/1323/1/012012

Performance evaluation of high ductility class RC framed structure by seismic event time histories

Krasimir Boshnakov[1*]

[1] Department of Computer-Aided Engineering, Faculty of Structural Engineering, University of Architecture Civil Engineering and Geodesy /UACEG/, Sofia, Bulgaria

*E-mail: krabosh_fce@uacg.bg

Abstract. The seismic performance of new designed to high ductility class, mid-rise RC framed structure is numerically tested by non-linear dynamic time history analysis using records adjusted to represent events with different periods of recurrence. The earthquake performance assessment employs the code-defined limit states criteria. The results of the numerical non-linear studies are indicative of the seismic performance of mid-rise RC framed structures designed to Eurocode ductility classes.

1. Introduction

Destructive earthquakes cause casualties and economic losses and happen over time. Each of these events thoroughly tests the building structures for design and construction quality. Some of them cause changes in the requirements of the seismic standards.

This study is focused on the expected seismic performance of reinforced concrete building structures designed to satisfy the high ductility class (DCH) to EN 1998-1 [1-3] requirements, under records of strong earthquake actions. The performance is assessed by the code-specified limiting values for elements chord rotations EN 1998-3 [4-6] for the different limit states.

The main objectives are to get insight into the expected damage mechanisms of the structure, to estimate the expected damages under different lateral forces that are possible to act during its lifetime, and finally to improve the practical assessment workflow after comparing the results with previously obtained by different analysis methods in other studies [12-14] of the current structure.

1.1 Methodological aspects

The structural performance assessment based on modern codes [1-6] relies on non-linear models that are capable of closely reflecting in numerical form the behavior of the real structure.

The Eurocode standard implementation of the limit states approach applies two sets of performance limits – the serviceability (SLS) and the ultimate (ULS) limit states. While the seismic design of new structures is governed by EN 1998-1 [1-3], where the damage limitation (DL1) and near collapse (NC1) performance levels are defined, the seismic assessment is subject to EN 1998-3 [4-6] where the significant damage (SD) limit state is introduced as an ULS along with the near collapse (NC) and the SLS of damage limitation (DL) is preserved. Roughly the relations between the performance definitions defining the limit states in these parts of Eurocode [1-6] correspond to DL ≈ DL1, SD ≈ NC1, and NC ≈ 1.33 SD, where the NC state reflects structures still possessing the capacity to withstand the gravity loads. Still, the chances of surviving another strong earthquake or aftershock are negligible.

The considered building structure is tested for different loadings to experience the significant damages (SD) and the near collapse (NC) ultimate limit states.

1.2 The earthquake action

The code defines the earthquake action by specifying an elastic response spectrum for the construction site and soil conditions. The ground acceleration a_g is defined by the reference ground acceleration a_{gR} which corresponds to earthquake return period T_{NCR} = 475 years and is determined by the probability of exceedance P_{NCR} = 10% for structure design life $T_{L,R}$ = 50 years and the γ_I structure importance factor, used as a scaling factor ($a_g = \gamma_I\, a_{gR}$). The value of γ_I is defined by the Gutenberg and Richter dependence-based equation:

$$\gamma_I = \frac{a_g}{a_{gR}} \cong \left(\frac{T_L}{T_{L,R}}\right)^{\frac{1}{k}} \cong \left(\frac{P_{NCR}}{P_L}\right)^{\frac{1}{k}} \cong \left(\frac{T_R}{T_{NCR}}\right)^{\frac{1}{k}} \qquad (1)$$

where $k \approx 3$ according to the code. Simple recalculations suggest that $k = 3.1$ for $T_R < 475$ and $k = 3.4$ for $T_R > 475$ years give better agreement with other parameters and provide closer matching results to the explicitly specified values.

The analytically tested structure was designed to EN 1998-1 [1-3] importance class III that corresponds to $\gamma = 1.2$ reflecting the probability of exceedance P_L = 10 % during lifetime T_L = 100 years or a return period T_R = 1000 (950) years.

Table 1. The records used for evaluation.

Event ID	M	Net code	Stat. code	Epic. Dist. [km]	Component	PGA [cm/s^2]	Duration [s]
RO-1977-0001	7.5	RO	A39	160.2	N-S E-W	194.2 166.5	31.1
1986/2/21/5:39	5.1	BS	VT	17	N-S E-W	28.5 38.4	15.5
1986/12/7/14:17	5.3	BS	VT	16	N-S E-W	74.5 66.9	22.4
TK-1999-0077	7.6	TK	8101	97.5	N-S E-W	311.4 367.8	27.1
TK-1999-0415	7.2	TK	8101	5.4	N-S E-W	406.3 -516.3	85.8
INT-20230206_0000008	7.7	TK	2708	39.1	N-S E-W	1036.4 812.6	100

[a] The records for the second and third events from top originate from [25]data set. All other records are accessible in the European Strong Motion Database and are from different contributors [20-24]

For performing the time-history analysis the code requires not less than 3 records to identify the maximum response. The records shall conform to the response spectrum that defines events with the code-defined exceedance probabilities. The current study employs a variation of the traditional spectrum-based record selection workflow and criteria taking the recommendations regarding location-based event selection [15] were employed. To examine the structure performance several records with the highest PGA values among the recording stations from the most damaging earthquakes that happened in the country and its neighborhood were selected. Chosen records were grouped by 3 and scaled to spectra corresponding to the DL and SD (NC_1) limit states for the current structure and additionally, it was tested for two sets of unscaled records. The SeismoMatch software [18] was used. All sets of records are composed of 2 subsets –one for the N-S direction and one for E-W respectively. In this way, 4 sets, and 8 subsets of 3 records each were formed. The first 3 events are known to have caused structural damages in the country. The last 3 events are from strongly destructive earthquakes in the country neighborhood. The unscaled records were ordered in sequences from weakest to strongest ground shakes to resemble foreshock–mainshock events.

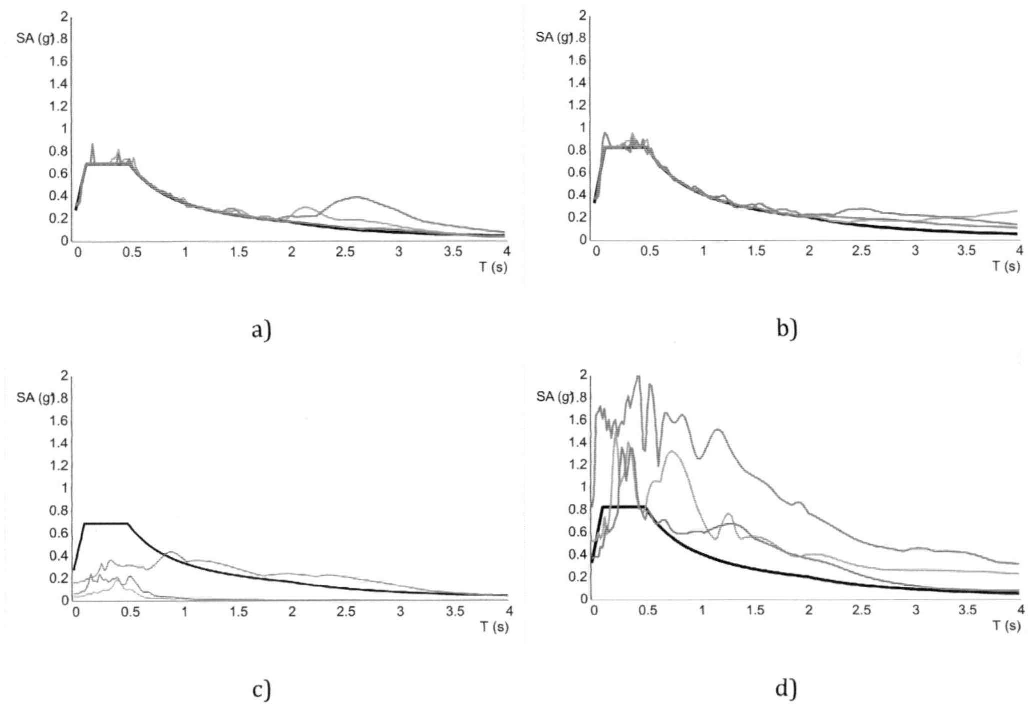

Figure 1. The response spectra of the E-W event sets: a) Set 1e, b) Set 2e, c) Set 3e, and d) Set 4e

The sets were composed and named as follows:

- Sets 1e and 1n are composed of the E-W and N-S components respectively of the first 3 events in Table 1, scaled to response spectrum (RS) Type 1 [1-3] with a_{gR} = 0.23 g and γ_I = 1.0 that represents events with return period T_{NCR} = 475.

- Set 2e and 2n are composed of the last 3 events in Table 1, scaled to RS Type 1 [1-3] with a_{gR} = 0.23 g and γ_I = 1.2 that represents events with a return period $T_{NCR} \approx$ 1000 years.
- Set 3e and 3n are the corrected and unscaled records of Set 1e and 1n.
- Set 4e and 4n are the corrected and unscaled records of Set 2e and 2n.

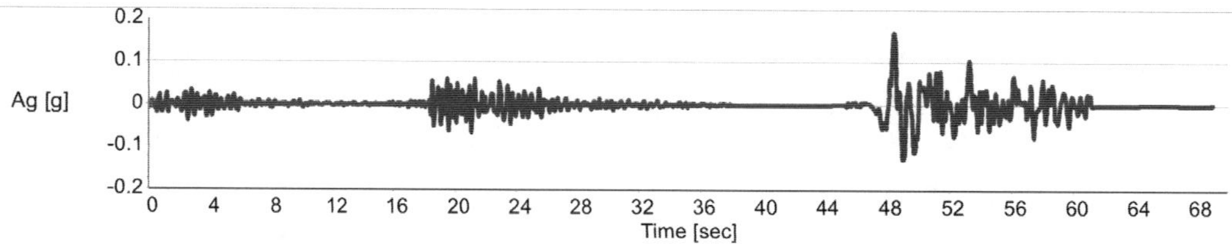

a)

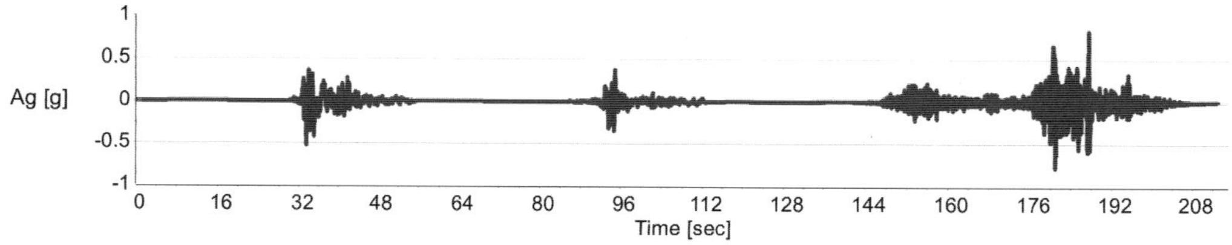

b)

Figure 2. Unscaled chains of time histories of the E-W events' components a) 1-3 (Set3e), and b) 3-4 (Set 4e).

In Figure 1 the response spectra of the E-W record of the sets are shown. The response spectrum matching is done in the period range T = 0.1 – 2.0 s. The displayed results indicate that the accelerogram scaling satisfies the requirements of the code.

1.3 The structure model

The structure was modeled by SeismoStruct [19] software. The Mander et al. [16] model is used to present the C30/37 concrete work in both confined and unconfined working conditions. The confinement factor used for scaling the stress-strain relationship is assumed constant within the entire strain range. Peak concrete compressive strength f_c' = 38 MPa, achieved at strain ε_c =0.002, and initial elastic modulus E = 33 GPa define the model parameters.

The reinforcement was modeled by the Menegotto-Pinto [17] model. The specified mean strength value is 575 MPa and the lower-bound strength value is 500 MPa.

The structure elements' hysteretic behavior is presented by the meso-scale fiber-section model of distributed plasticity. Six cross-sections with 204 control points each, along the length of all elements are used to describe the reinforcement area variation and to evaluate the results. The guidance and recommendations of [7-10] were followed to model the assessed structure.

1.4 The performance indicators

EN 1998-3 [4-6] uses chord rotation as an element performance indicator. Each standard limit state has assigned chord rotation capacities. These values are used here as a major numerical indicator regarding the performance and damage level of the structural elements. The software

[18] internally calculates the elements' demands and capacities and keeps track of the maximum demand-to-capacity ratios at each loading step. This leads to a large volume of numerical data.

2. The examined structure

2.1 Configuration, topology, and structural members

The building is a low- to mid-rise structure and consists of totally three slabs. The structure consists of 3 longitudinal 6.5 m wide and 5 transverse 6.0 m wide bays as shown in Figure 3. It is designed for high-class ductility (DCH) of EN 1998-1 as discussed in [13, 14]. The structure resulting over-strength and seismic performance evaluated by push-over analysis is discussed in [12].

The slab depths are 20 cm, the columns are 50/50 cm and the beams' dimensions are 60/40 cm. The reinforcement varies along the elements' length. The longitudinal and transverse reinforcement is unified among similar elements.

The concrete class is C 30/37 and the reinforcement steel is S 500C class.

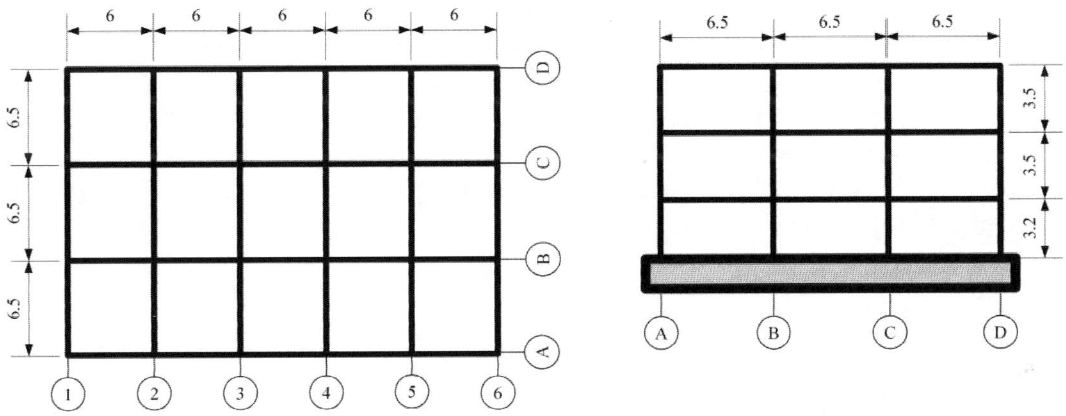

a) b)

Figure 3. The considered structure: a) Plan, b) Section

The structure is regular both in plan and in height with regularly distributed masses.

The initial design considered structure importance class III with γ_I = 1.2, the construction site soil class C, and the Type 1 earthquake response spectrum with reference ground acceleration a_{gR} = 0.23 g (g = 9.81 m/s^2).

2.2 Results from other studies

The initial design used [13, 14] the linear elastic modal analysis procedure. The obtained fundamental periods of the structure in transverse and longitudinal directions $T_1 \approx T_2 \approx 0.5$ s. The structure was adequately detailed to cover the DCH requirements.

The structure overstrength ratio is the ratio between the lateral load when sufficient numbers of plastic hinges in the structure exist (α_U) to the load when the first one emerged (α_1) as defined in EN 1998-1 [1-3]. For regular in-plan and in-height reinforced concrete framed systems in which the strict seismic detailing rules for high-class ductility (DCH) are followed the ratio values are in the range of 1.3 to 1.5. The obtained results in [12] indicate that the considered structure overstrength ratio α_U/α_1 = 1.5 in the transverse and α_U/α_1 = 1.45 in the longitudinal direction. These results predicted damage initiation on the second-floor elements and the involvement of first-floor elements in the damage development.

3. The performed analysis and results

3.1 The initial challenges and difficulties

The practical difficulty in distinguishing the performance between middle (DCM) and high (DCH) ductility classes by analytical models is widely recognized. Further, the initial design was performed by using conservatively α_U/α_1 = 1.3 while the effective α_U/α_1 = 1.5 was found.

The results from initial tests of the current non-linear model well match the results from the initial linear-elastic one within the elastic range. Anyway, the initially adopted conservative value for the overstrength factor supposes the presence of reserve in the structural capacity.

Other practical problems are connected with the big data amounts. The original records have a time step of 0.005 s. Connecting them in sequences leads to time histories reaching total durations above 3.5 min in some of the cases. Straight solution to these problems is a challenge even to modern computer hardware and software. In the current solutions, time-history stages were defined. For each set 6 stages were defined. For the 'calm' ground-shaking parts of the records with relatively negligible ground accelerations, a time step of 0.05s was assigned, while for the 'major' parts the calculations were performed with a time step of 0.01 s. The preliminary tests and comparisons show that the adopted parameters did not influence significantly the precision of the results. The amounts of data predefine the predominant application of graphical result presentation.

3.2 The results from the code-conformance-based checks

Extracts of the results are shown in Figure 4.

Figure 4 a) and b) contain the results of reaching the DL limit state element capacity criteria according to Eurocode 8-3 [4-6, 19]. The structural members that have reached the criteria of DL are marked in light green and the ovals indicate the joints where it happened. Figure 4 a) shows the results for the tested structure after the application of the whole sequence of the (Set 1e) E-W components of the first 3 events in Table 1, matched to the specified for the construction site conditions response spectrum with reoccurrence probability $T_R = T_{NCR}$ = 475 years. The obtained results indicate that practically all structural elements have reached the damage limitation (DL) limit state and no element has reached the significant damage (SD) limit state.

The results in Figure 4 b) show that even under the action of smaller ground accelerations, represented by the sequence 3e of the unscaled records of the E-W components of the first 3 events in Table 1, some elements of the structure would reach the DL [4-6, 19]criteria.

In Figure 4 c) the members marked in light red have reached the significant damage (SD ≈ NC_1) limit state and the ovals show the joints where it happened. The seismic action was represented by the Set 3e acceleration sequence, composed of the matched to the response spectrum of events with a reoccurrence period $T_R \approx$ 1000 years, and containing the E-W records of the last 3 events in Table 1.

Figure 4 d) shows the elements and the joints where the Eurocode 8-3 [4-6, 19] criteria are reached in dark red. The picture was formed as described:

- at t = 35.25 s first reaching of NC at 1st floor beam to external column joint,
- at t = 35.80 s, the external column reaches NC at column toe, more beams reach NC,

- at t = 94.15 s 3 of the columns at toe locations reach NC state, all the 1st and those on the 2nd floor to the external columns beam to column joints have reached the NC limit state,
- at t = 94.25 s the internal columns to 2nd floor joints reach the NC state.

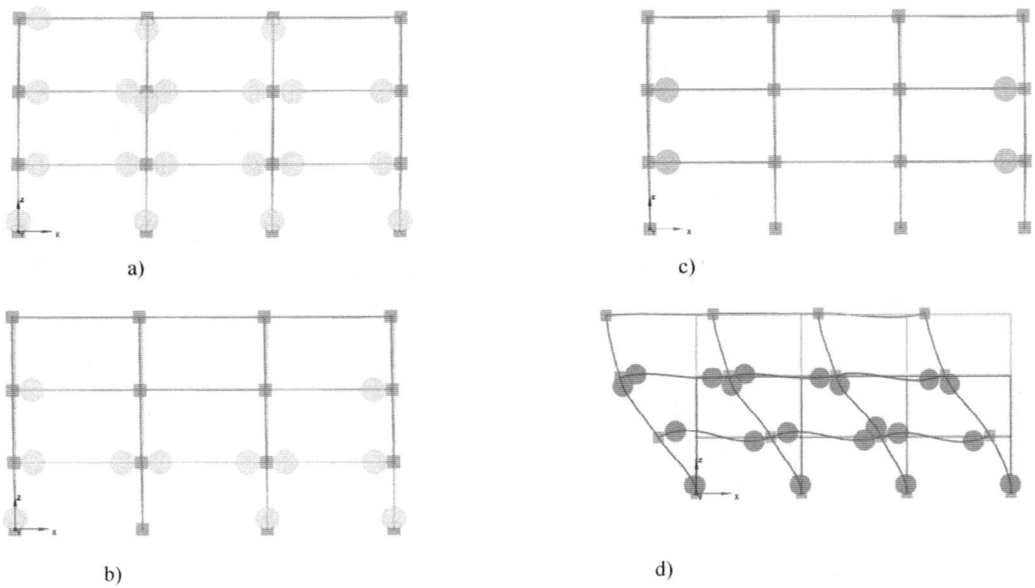

Figure 4. Limit state code checks: a) DL Set 1e, b) DL Set 3e, c) SD Set 2e, d NC Set 4e

Figure 4 d) shows the reached criteria at time t =181.20 s from Set 4e duration. The performance to the end of the set duration (212.85 s) can be numerically traced, although the results after t = 181.2 s are highly speculative.

3.3 Result assessment

The initial design of the DCH reinforced concrete structure in [13, 14] used the modal linear-elastic analysis. The static nonlinear (push-over) analysis [12] used in another seismic assessment of the same structure estimated the overstrength ratio. The current time-history based seismic assessment analysis did not reveal considerable differences in the analytical models used in the previous studies. The current assessment by nonlinear time-history analysis allows a better understanding of the dynamic behavior of the considered structure.

These results indicate a reduced damage initiation limit for the reinforced concrete frame designed for high ductility class (DCH). The significant damages (SD) limit states results suggest that the initial design satisfies the code requirements. The fact of reaching the SD criteria in two beam joints points to a precise and economical design that satisfies the weak-beam, strong-column concept.

The results from damage limitation (DL) state tests indicate an increased range to develop structural damages within the lateral load capacity curve by decreased damage initiation limit. This is in line with the applied increase in ductility performance.

The damage development trend identified in [12] was confirmed. The results reveal that the mechanism of damage development starts in the second-floor elements and spreads to the first floor. This confirms the consistency of both seismic assessment analyses.

The well-known advantages of the nonlinear time history over the push-over analysis are observed after comparing the conviction level of specifying the damage development prognosis.

4. Conclusions

The presented seismic performance assessment by non-linear time history analysis provides insights into the behavior of the considered structure when subjected to different earthquake records. The study re-confirms the need to short-list country-specific earthquake events [15] to better represent the seismic action while taking the local conditions into account.

The increased damage development range of the DCH structure through lowering the damage initiation threshold suggests of possible efficiency of the damage indices approach [11] for the evaluation and assessment of such structures in future developments.

References

[1] BDS EN 1998-1:2005, Eurocode 8: Design of structures for earthquake resistance – Part 1: General rules, seismic actions and rules for buildings, BIS (2006)

[2] BDS EN 1998–1:2004/AC:2009, Eurocode 8: Design of structures for earthquake resistance – Part 1: General rules, seismic actions, and rules for buildings. Corrigendum, CEN (2009)

[3] BDS EN 1998–1/NA, Eurocode 8: Design of structures for earthquake resistance – Part 1: General rules, seismic actions, and rules for buildings. National annex, BIS (2011)

[4] BDS EN 1998-3:2005, Eurocode 8: Design of structures for earthquake resistance – Part 3: Assessment and retrofitting of buildings, BIS (2007)

[5] EN 1998–3:2005/AC:2010, Eurocode 8: Design of structures for earthquake resistance – Part 3: Assessment and retrofitting of buildings. Corrigendum", CEN (2010)

[6] BDS EN 1998–3/NA, Eurocode 8: Design of structures for earthquake resistance – Part 3: Assessment and retrofitting of buildings. National annex, BIS (2011)

[7] ATC, Recommended modeling parameters and acceptance criteria for nonlinear analysis in support of seismic evaluation, retrofit, and design. NIST GCR 17-917-45, U.S. Departments of Commerce Engineering Laboratory, National Institute of Standards and Technology, 04.2017, DOI: 10.6028/NIST.GCR.17-917-45

[8] ATC. Guidelines for nonlinear structural analysis for design of buildings, Part I, General. NIST GCR 17-917-46v1, U.S. Department of Commerce Engineering Laboratory, National Institute of Standards and Technology, 04.2017, DOI: 10.6028/NIST.GCR.17-917-46v1

[9] ATC. Guidelines for nonlinear structural analysis for design of buildings, Part IIb – Reinforced concrete moment frames", NIST GCR 17-917-46v3, U.S. Department of Commerce Engineering Laboratory, National Institute of Standards and Technology, 04.2017, DOI: 10.6028/NIST.GCR.17-917-46v3.

[10] ATC. Recommended modeling parameters and acceptance criteria for nonlinear analysis in support of seismic evaluation, retrofit, and design. NIST GCR 17-917-45, U.S. Departments of Commerce Engineering Laboratory, National Institute of Standards and Technology, 04.2017 DOI: 10.6028/NIST.GCR.17-917-45.

[11] Bracci J.M. Reinhorn A.M. Mander J.B. Kunnath S.K. "Deterministic model for seismic damage evaluation of reinforced concrete structures", NCEER, State University of New York at Buffalo, Technical report NCEER-89-0033, 1989

[12] Boshnakov Kr. Kardjiev V. Seismic Performance assessment of RC framed system building designed to Eurocode High Class Ductility, Annual of the University of Architecture Civil Engineering and Geodesy, 56 (3) 1263-1276, Sofia (2023) ISSN 1310-814X; ISSN 2534-9759

[13] Kardjiev V. Georgiev E. Specific requirements in the design of beams of reinforced concrete frame structures for seismic action at high level of total ductility (DCH), Annual of the University of Architecture Civil Engineering and Geodesy, 56 (3) 1219-1230, Sofia (2023) (in Bulgarian) ISSN 1310-814X; ISSN 2534-9759

[14] Kardjiev V. Georgiev E. Specific requirements in the design of columns of reinforced concrete frame structures for seismic action at high level of total ductility (DCH), Annual of the University of Architecture Civil Engineering and Geodesy, 56 (3) 1231-1242, Sofia (2023) (in Bulgarian) ISSN 1310-814X; ISSN 2534-9759

[15] Kouteva-Guentcheva M. Georgiev Tz. Tsvetkova Tr. On the need of local set of accelerogram records for dynamic analysis of structures. Annual of the University of Architecture Civil Engineering and Geodesy, 56 (4) 1665-1679, Sofia (2023) (in Bulgarian) ISSN 1310-814X; ISSN 2534-9759
[16] Mander, J.B., Priestley, M.J.N., Park, R. Theoretical stress-strain model for confined concrete. // Journal of Structural Engineering, Vol. 114, No. 8, pp. 1804-1826 (1988)
[17] Menegotto, M., Pinto, P.E. Method of analysis for cyclically loaded R.C. plane frames including changes in geometry and non-elastic behavior of elements under combined normal force and bending. Symposium on the Resistance and Ultimate Deformability of Structures Acted on by Well Defined Repeated Loads, International Association for Bridge and Structural Engineering, Zurich, Switzerland, 15-22 (1973)
[18] SEISMOSOFT, SeismoMatch-Earthquake software for response spectrum matching v. 2022, https://seismosoft.com/products/seismomatch/
[19] SEISMOSOFT, SeismoStruct-Civil engineering software for structural assessment and structural retrofitting v. 2022, rel. 5, build 5, https://seismosoft.com/products/seismostruct/
[20] Engineering Strong Motion Database - https://esm-db.eu/ last accessed: 2024/06/15
[21] Luzi L., Lanzano G., Felicetta C., D'Amico M. C., Russo E., Sgobba S., Pacor, F., & ORFEUS Working Group 5 (2020). Engineering Strong Motion Database (ESM) (Version 2.0). Istituto Nazionale di Geofisica e Vulcanologia (INGV). DOI: 10.13127/ESM.2
[22] Luzi L, Puglia R, Russo E & ORFEUS WG5 (2016). Engineering Strong Motion Database. Istituto Nazionale di Geofisica e Vulcanologia, Observatories & Research Facilities for European Seismology. doi: 10.13127/ESM
[23] National Institute for Earth Physics (NIEP Romania). (1994). Romanian Seismic Network. International Federation of Digital Seismograph Networks. DOI: 10.7914/SN/RO
[24] Disaster and Emergency Management Authority. (1973). Turkish National Strong Motion Network. Department of Earthquake, Disaster and Emergency Management Authority. DOI: 10.7914/SN/TK
[25] Nenov, D. G. Georgiev, and I. Paskaleva (C.L.S.M.E.E.), and V.W. Lee and M.D. Trifunac (U.S.C.), Strong Earthquake Ground Motion Data in EQINFOS: Accelerograms recorded in Bulgaria between 1981 and 1987, Report No. CE 90-02, Bulgarian Academy of Sciences, Central Laboratory for Seismic Mechanics and Earthquake Engineering (C.L.S.M.E.E.) and University of Southern California, Department of Civil Engineering (U.S.C.), Sofia, Bulgaria and Los Angeles, California (1990)

Modelling initial conditions of free vibrations in SAP2000

Teodor Todorov[1*]

[1] Department of Technical Mechanics, University of Architecture, Civil Engineering and Geodesy, Sofia, Bulgaria

*E-mail: tstodorov_fhe@uacg.bg

Abstract. The modelling of non-zero initial conditions in SAP2000 when considering free vibrations is not a standard operation. The current paper proposes a simple method of initial condition modelling in SAP2000 when said vibrations are considered. The basic principle of impulse of a particle is used as a reference in the theoretical background. The method is illustrated by a simple single degree of freedom problem that is solved both analytically and numerically in SAP2000. Results are compared. The method is applied to a multi-degree of freedom problem as well. The multi-degree of freedom problem is undamped. Conclusions are made.

1. Introduction

Numerical analysis of structures is a standard activity for any modern-day civil engineer. Such an analysis is unthinkable without the aid that finite element method (FEM) based programs provide.

It is often the case that when analyzing a structure, civil engineers are interested in the seismic response of the later. Structures subjected to seismic loads are modelled with zero initial conditions that represent the state of equilibrium the structures are in before the earthquake takes place [1]. There are, however, cases when the initial conditions are non-zero [2]. They usually occur when a specific vibration is to be considered. Even though in civil engineering the dominant vibrations have zero initial condition, it is still useful to consider the case when non-zero initial conditions are to be modelled. To demonstrate a basic concept of such a model, the FEM program SAP 2000 is chosen for its relative simplicity and user-friendly interface.

2. Theoretical background

It is known from Mechanics that initial conditions can be defined by velocity and displacements. The simplest problem that is usually solved to demonstrate this is a single degree of freedom (SDOF) problem. It is solved analytically by using Newton's second law.

Vibrations are often divided into four categories. They can be free or forced and, furthermore, they can be damped or undamped [3]. This constitutes the four types of vibrations – free damped or undamped and forced damped or undamped.

Forced vibrations are outside the scope of the current paper for there is a simple enough built-in procedure that models them in most FEM programs. The model of forced vibrations is constructed by using a predefined function that describes the external load that excites the considered system. That is usually a model of a seismic load with zero initial conditions. In

SAP2000 it is possible to model a function with user-defined frequency of oscillation and phase using the built-in menu [4].

The non-zero initial conditions, however, require a few additional steps to be modelled in SAP 2000. The concept of an impulse of a particle is used. It is known that an impulse of a particle with average velocity $\vec{v}$, mass m and force $\vec{F}$, applied to the particle for a period Δt is:

$$\Delta t\vec{F} = m\vec{v}. \tag{1}$$

Equation (1) can be used to define velocity by a time-dependent function. Such a function is easily modelled and manipulated in SAP2000 using time-history analysis [5]. This principle is illustrated by examples.

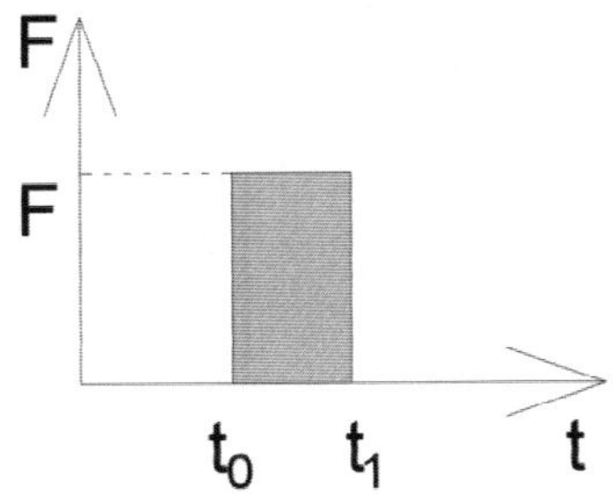

Figure 1. Basic principle of an impulse – force F applied for a period $\Delta t = t_1 - t_0$. Time is plotted on the horizontal axis, while force F – on the vertical axis.

3. Free undamped vibrations - SDOF examples

SDOF undamped problem is considered. A particle with mass $m = 5$ kg attached to a spring with stiffness $k = 100$ N/m is shown in fig. (2).

It is known that the solution of the above problem depends on the initial conditions [6].

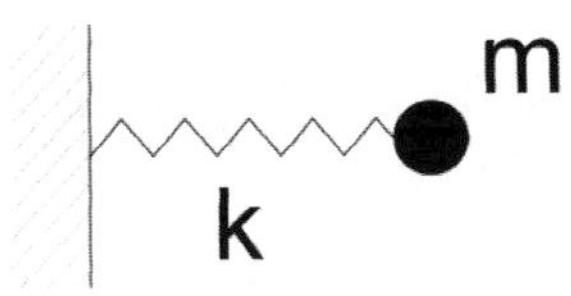

Figure 2. Free undamped SDOF vibrations – a scheme for a particle with mass m and spring stiffness k.

Hence, for initial conditions $x_0 = 0$ m and $v_0 = 1$ m/s the solution is given by:

$$\begin{gathered} x(t) = 0{,}2236sin(4{,}472t) \\ \dot{x}(t) = v(t) = cos(4{,}472t) \end{gathered} \tag{2}$$

For initial conditions $x_0 = 0{,}2$ m and $v_0 = 0$ m/s the solution is given by:

$$\begin{gathered} x(t) = 0{,}2cos(4{,}472t) \\ \dot{x}(t) = -0{,}8944sin(4{,}472t) \end{gathered} \tag{3}$$

The analytical solutions (formulas 2 and 3) are manipulated and the values for displacement $x(t)$ are obtained in a chosen time-step. For simplicity the time step is 0,5 s. The results are shown in table 1.

Table 1. Displacements calculated with the analytical solutions given in formulas (2) and (3).

Time, t [s]	Displacement x(t) for initial conditions $x_0 = 0\ m$, $v_0 = 1\ m/s$ Formula (2), [m]	Displacement x(t) for initial conditions $x_0 = 0{,}2\ m$, $v_0 = 0\ m/s$ Formula (2), [m]
0	0	0,2
0,5	0,176	-0,123
1,0	-0,217	-0,047
1,5	0,092	0,182
2	0,103	-0,177

[a] The values given in the table are calculated with time step 0,5 s for simplicity. In the SAP 2000 model the time step is smaller for more accurate results of the numerical solution. Any value can be calculated with respect to any time analytically when the governing equations are known.

The problem is modelled in SAP2000 using the definition of an impulse. The simpler initial condition to model is the one with non-zero velocity. That is because a time-history function can be defined easily. Time period of 10 seconds is considered, the value of the function is set to 1 so that more comprehensive Load Cases can be defined with a time step of 0,01 s and a load F that is multiplied by 1. The definitions can be seen in fig. 3.

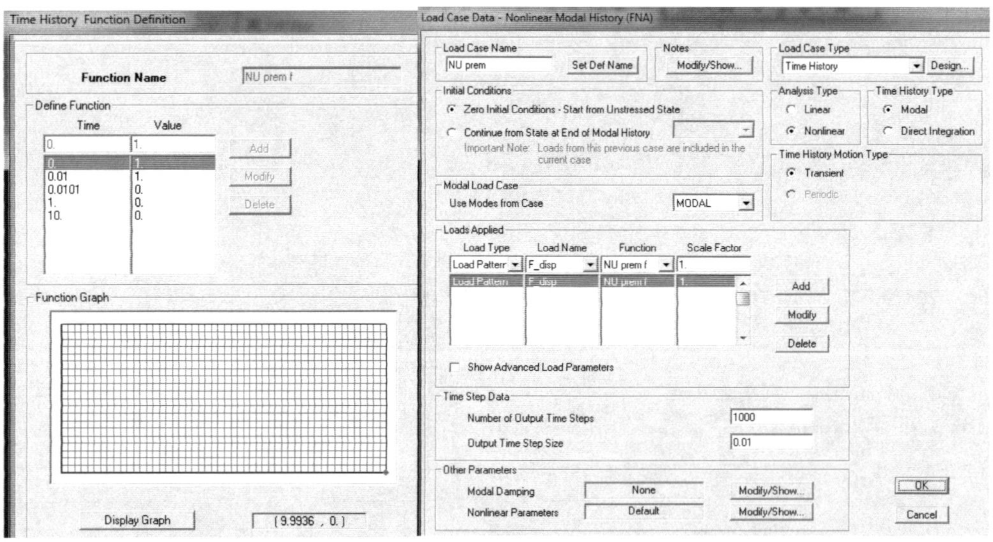

Figure 3. Definition of time-history function and load case in SAP2000.

Joint Displacements

File View Format-Filter-Sort Select Options

Units: As Noted — Joint Displacements

Joint Text	OutputCase Text	CaseType Text	StepType Text	StepNum Unitless	U1 m	U2 m	U3 m	R1 Radians
1	e-impulseNUski	NonModHist	Time	0.14	0.12755	0	0	0
1	e-impulseNUski	NonModHist	Time	0.15	0.135693	0	0	0
1	e-impulseNUski	NonModHist	Time	0.16	0.143565	0	0	0
1	e-impulseNUski	NonModHist	Time	0.17	0.151149	0	0	0
1	e-impulseNUski	NonModHist	Time	0.18	0.158432	0	0	0
1	e-impulseNUski	NonModHist	Time	0.19	0.165397	0	0	0
1	e-impulseNUski	NonModHist	Time	0.2	0.172032	0	0	0
1	e-impulseNUski	NonModHist	Time	0.21	0.178323	0	0	0
1	e-impulseNUski	NonModHist	Time	0.22	0.184257	0	0	0
1	e-impulseNUski	NonModHist	Time	0.23	0.189823	0	0	0
1	e-impulseNUski	NonModHist	Time	0.24	0.195009	0	0	0
1	e-impulseNUski	NonModHist	Time	0.25	0.199805	0	0	0
1	e-impulseNUski	NonModHist	Time	0.26	0.204202	0	0	0
1	e-impulseNUski	NonModHist	Time	0.27	0.20819	0	0	0
1	e-impulseNUski	NonModHist	Time	0.28	0.211762	0	0	0
1	e-impulseNUski	NonModHist	Time	0.29	0.214911	0	0	0
1	e-impulseNUski	NonModHist	Time	0.3	0.21763	0	0	0
1	e-impulseNUski	NonModHist	Time	0.31	0.219913	0	0	0
1	e-impulseNUski	NonModHist	Time	0.32	0.221757	0	0	0
1	e-impulseNUski	NonModHist	Time	0.33	0.223157	0	0	0

Figure 4. Table of SAP 2000 representing the calculated displacements with respect to time

The force used in the definition is calculated as follows:

$$\varDelta t = 0{,}01\ \mathrm{s}\ ;\ m = 5\ \mathrm{kg}\ ;\ v_0 = 1\ \mathrm{m/s}$$
$$\varDelta tF = mv => F = \frac{mv}{\varDelta t} = \frac{5.1}{0{,}01} = 500\ \mathrm{N} = 0{,}5\ \mathrm{kN} \tag{4}$$

Load Name *F_disp* in fig. 3 is a force of magnitude 0,5 kN that is applied along the direction of the spring.

For time $t = 0{,}3$ s the displacement x is calculated:

$$x(0{,}3) = 0{,}2236.sin(4{,}472.0{,}3) = 0{,}2236.0{,}9738 = 0{,}2177\mathrm{m} \tag{5}$$

In fig. 4 the calculation of displacements in SAP 2000 is given. At time $t = 0{,}3$ s the displacement value is 0,2176 m. The model in SAP 2000 that describes initial velocity is accurate.

A more complex procedure is required to model displacement initial conditions. It is based on the equalization of the solutions of free vibrations with different initial conditions. It is known [7] that for free vibrations with initial condition displacement x_0 (initial velocity is zero) the solution for SDOF problem is:

$$x(t) = x_0 cos(\omega t). \tag{6}$$

In the above equation ω denotes the natural frequency of the system. SDOF free vibrations with initial condition velocity v_0 (and zero initial displacement) have a trajectory:

$$x(t) = \frac{v_0}{\omega} sin(\omega t). \tag{7}$$

By equalizing (6) and (7) a system of equations is obtained:

$$\frac{v_0}{\omega} = x_0 \tag{8}$$

and

IOP Conf. Series: Materials Science and Engineering **1323** (2024) 012013 doi:10.1088/1757-899X/1323/1/012013

$$cos(\omega t) = sin(\omega t). \tag{9}$$

The principle of impulse of a particle can be used again to model displacement initial condition if equations (8) and (9) are fulfilled for the considered vibration. Equation (8) is applied to the system by defining initial velocity $v_0 = x_0\omega$. x_0 is the desired initial displacement, while the natural frequency ω is a function of the mass of the particle and the stiffness of the spring. For equation (9) to be upheld, the trigonometric relation between sine and cosine is used:

$$cos(\omega t) = sin\left(\omega t + \frac{\pi}{2}\right). \tag{10}$$

It is known that period of harmonic oscillations is related to natural frequency by the formula:

$$T = \frac{2\pi}{\omega}. \tag{11}$$

However, the phase difference between sine and cosine as shown in equation (10) is $\frac{\pi}{2}$. Then the period that is to be considered in the modelling is:

$$T = \frac{\pi}{2\omega}. \tag{12}$$

The problem of modelling initial displacement becomes a problem of modelling initial velocity. Initial displacement 0,2 m of a SDOF system with natural frequency $\omega = 4{,}472\ s^{-1}$ defines initial velocity $v_0 = 0{,}8944$ m/s. With time step $\Delta t = 0{,}01$ s and mass $m = 5$ kg the force applied to the particle is $F = 0{,}4472$ kN. Considering equation (12) the results of the vibration are read from time $t = 0{,}351$ s. At that moment the displacement is equal to 0,2 m and the results can be plotted and analysed. The result of the calculation in SAP2000 is shown in figure 5.

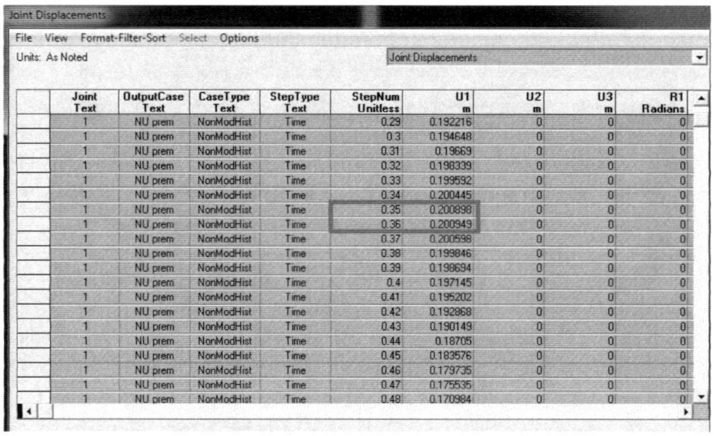

Joint Text	OutputCase Text	CaseType Text	StepType Text	StepNum Unitless	U1 m	U2 m	U3 m	R1 Radians
1	NU prem	NonModHist	Time	0.29	0.192216	0	0	0
1	NU prem	NonModHist	Time	0.3	0.194648	0	0	0
1	NU prem	NonModHist	Time	0.31	0.19669	0	0	0
1	NU prem	NonModHist	Time	0.32	0.198339	0	0	0
1	NU prem	NonModHist	Time	0.33	0.199592	0	0	0
1	NU prem	NonModHist	Time	0.34	0.200445	0	0	0
1	NU prem	NonModHist	Time	0.35	0.200898	0	0	0
1	NU prem	NonModHist	Time	0.36	0.200949	0	0	0
1	NU prem	NonModHist	Time	0.37	0.200598	0	0	0
1	NU prem	NonModHist	Time	0.38	0.199846	0	0	0
1	NU prem	NonModHist	Time	0.39	0.198694	0	0	0
1	NU prem	NonModHist	Time	0.4	0.197145	0	0	0
1	NU prem	NonModHist	Time	0.41	0.195202	0	0	0
1	NU prem	NonModHist	Time	0.42	0.192868	0	0	0
1	NU prem	NonModHist	Time	0.43	0.190149	0	0	0
1	NU prem	NonModHist	Time	0.44	0.18705	0	0	0
1	NU prem	NonModHist	Time	0.45	0.183576	0	0	0
1	NU prem	NonModHist	Time	0.46	0.179735	0	0	0
1	NU prem	NonModHist	Time	0.47	0.175535	0	0	0
1	NU prem	NonModHist	Time	0.48	0.170984	0	0	0

Figure 5. Modelling of initial displacement in SAP 2000 – starting time of analysis.

4. Modelling damping

It is known that equations of free damped vibrations of SDOF system are:

$$m\ddot{x} + b\dot{x} + kx = 0, \tag{13}$$

where b is the damping coefficient [8]. Damping is considered proportional to the first power of velocity. The solution of equation (13) is given by:

$$x(t) = C_1 e^{r_1 t} + C_2 e^{r_2 t}. \tag{14}$$

In the above equation C_1 and C_2 are constants of integration while r_1 and r_2 are the roots of the characteristic equation of (13). They are defined as:

$$r_1; r_2 = -n \pm \sqrt{n^2 - \omega^2}, \tag{15}$$

where $n = {}^{b}/_{2m}$ and $\omega^2 = {}^{k}/_{m}$. It can be shown with some simple mathematical manipulations that the solution (13) for initial conditions $v_0 = 0$ and $x_0 \neq 0$ becomes:

$$x(t) = \frac{x_0}{r_1 - r_2}\left(-r_2 e^{r_1 t} + r_1 e^{r_2 t}\right). \tag{16}$$

When the initial conditions are $v_0 \neq 0$ and $x_0 = 0$ equation (13) becomes:

$$x(t) = \frac{v_0}{r_1 - r_2}\left(e^{r_1 t} - e^{r_2 t}\right). \tag{17}$$

After some mathematical manipulations, equation (17) can be written as:

$$x(t) = \frac{v_0 e^{-nt}}{\sqrt{n^2 - \omega^2}} sinh\left(\sqrt{n^2 - \omega^2} t\right). \tag{18}$$

It was already shown in the paper that initial condition $v_0 \neq 0$ can be modelled easily in SAP 2000. Initial condition $x_0 \neq 0$ is modelled by implementing equation (18). Time t for which the maximum displacement occurs is found by calculating the derivative $\dot{x}(t)$ with respect to time of equation (18). Since the considered vibration is damped, the maximum displacement is the initial condition $x_0 \neq 0$. Thus, the starting time of the vibration is determined by solving equation:

$$\dot{x}(t) = \frac{v_0 e^{-nt}}{\sqrt{n^2 - \omega^2}}\left(-n. sinh\left(t\sqrt{n^2 - \omega^2}\right) + \sqrt{n^2 - \omega^2}. cosh\left(t\sqrt{n^2 - \omega^2}\right)\right) = 0 \tag{19}$$

The solution for t of equation (19) is:

$$t_{max} = \frac{1}{\sqrt{n^2 - \omega^2}} arctgh\left(\frac{\sqrt{n^2 - \omega^2}}{n}\right). \tag{20}$$

For the problem modelled in SAP2000 the damping coefficient is assumed $b = 50\ {}^{\mathrm{Ns}}/_{\mathrm{m}}$. The damped is modelled by using a link element as discussed and shown in [9]. With defined values:

$$m = 5\ \mathrm{kg}; k = 100\ {}^{\mathrm{N}}/_{\mathrm{m}}; b = 50\ {}^{\mathrm{Ns}}/_{\mathrm{m}}; n = 5\ {}^{\mathrm{Ns}}/_{\mathrm{m.kg}}; \omega^2 = 20\ {}^{\mathrm{N}}/_{\mathrm{m.kg}} \tag{21}$$

the time when maximum displacement occurs can be calculated by using equation (20). By substituting the values of (21) in (20) a value for t_{max} is obtained. The value is $t_{max} = 0{,}215$ s. The same value is calculated in SAP2000 for a time step when the maximum displacement of the considered problem occurs. This is shown in figure 6.

The vibration with initial displacement $x_0 = 0{,}2$ m can be modelled in SAP2000 as a vibration with initial velocity $v_0 = x_0. \omega$. The vibration is to be considered with starting time $t = 0{,}22$ s.

It is evident that the impulse of a particle can be used to model both initial velocity and initial displacement. For more complex systems (systems with multiple degrees of freedom) the process

IOP Conf. Series: Materials Science and Engineering **1323** (2024) 012013 doi:10.1088/1757-899X/1323/1/012013

of defining a proper impulse at the center of mass to model initial displacement becomes more tedious. It is easier to define an impulse at the center of mass that is enough to determine a displacement equal to the initial displacement. This can easily be done in SAP2000 as the user can always track the displacements of the center of mass. When the desired displacement is obtained, the analysis of the system can continue from the time at which the said displacement is calculated.

Joint Displacements

File View Format-Filter-Sort Select Options

Units: As Noted

Joint Displacements

Joint Text	OutputCase Text	CaseType Text	StepType Text	StepNum Unitless	U1 m	U2 m	U3 m	R1 Radians
1	TH	NonModHist	Time	0.07	0.123612	0	0	0
1	TH	NonModHist	Time	0.08	0.135844	0	0	0
1	TH	NonModHist	Time	0.09	0.146652	0	0	0
1	TH	NonModHist	Time	0.1	0.156151	0	0	0
1	TH	NonModHist	Time	0.11	0.164448	0	0	0
1	TH	NonModHist	Time	0.12	0.171641	0	0	0
1	TH	NonModHist	Time	0.13	0.177822	0	0	0
1	TH	NonModHist	Time	0.14	0.183076	0	0	0
1	TH	NonModHist	Time	0.15	0.187481	0	0	0
1	TH	NonModHist	Time	0.16	0.191109	0	0	0
1	TH	NonModHist	Time	0.17	0.194027	0	0	0
1	TH	NonModHist	Time	0.18	0.196298	0	0	0
1	TH	NonModHist	Time	0.19	0.197978	0	0	0
1	TH	NonModHist	Time	0.2	0.199122	0	0	0
1	TH	NonModHist	Time	0.21	0.199777	0	0	0
1	TH	NonModHist	Time	0.22	0.199989	0	0	0
1	TH	NonModHist	Time	0.23	0.1998	0	0	0
1	TH	NonModHist	Time	0.24	0.199249	0	0	0
1	TH	NonModHist	Time	0.25	0.19837	0	0	0
1	TH	NonModHist	Time	0.26	0.197198	0	0	0

Figure 6. Calculation of maximum displacement of free damped vibration in SAP 2000. Time of maximum displacement.

5. Modelling forced vibrations

Forced vibrations can easily be modelled in SAP2000 by using the "define function" feature of the program. Forced vibrations are usually defined with periodic sine or cosine functions [10]. The most common cases when modelling forced vibrations assume zero initial conditions and as such are not in the scope of the current paper.

6. MDOF problem

A body with a fixed point is used as an example of MDOF system. The body consists of three rods perpendicular to each other along the axes of a cartesian coordinate system, as shown in figure 7. The fixed point coincides with the origin of the coordinate system. The body has planes of material symmetry and is maintained in equilibrium by linear springs, attached to the body at four points. Point masses can also be added to the body along the length of the rods. The problem is undamped.

The analytical solution of the problem is carried through the principles of Lagrangian Mechanics. The equations of Lagrange of the second order are used to obtain the mechanical characteristics such as displacements, velocities, and accelerations. The assumed initial condition to be modelled is the displacement of the center of mass. Equations of Lagrange of the second order [11]:

$$\frac{d}{dt}\left(\frac{\partial E_k}{\partial \dot{q}_i}\right) - \frac{\partial E_k}{\partial \dot{q}_i} + \frac{\partial E_p}{\partial q_i} + \frac{\partial \Phi}{\partial \dot{q}_i} = Q_j. \tag{22}$$

In equation (22) E_k denotes the kinetic energy of the system, E_p is the potential energy of the system and Φ is the dissipative function. Q_j is the generalized force, while q_i and $\dot{q}_i$ are the generalized coordinates. For a body with a fixed point these generalized coordinates are the rotations with respect to the axes of the Cartesian coordinate system.

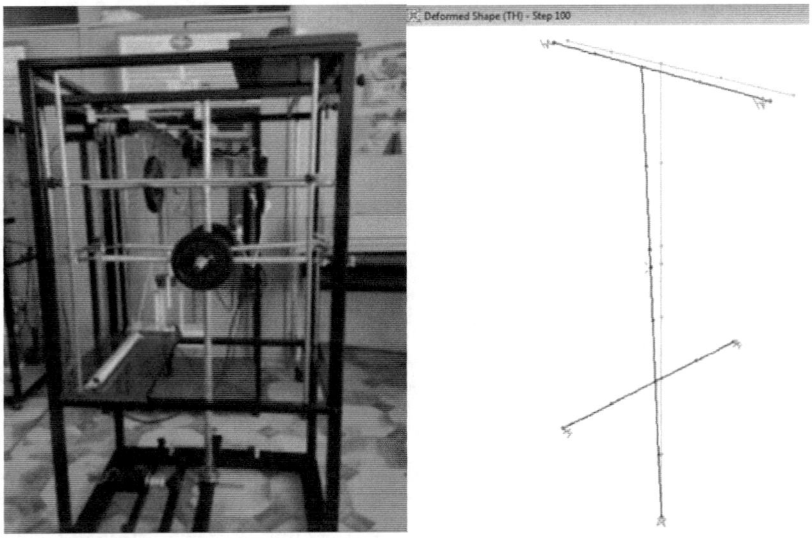

Figure 7. Model of a body with a fixed point. A system with 3 degrees of freedom (DOF). The physical body is shown in the left picture, while the model in SAP 2000 is shown in the right.

The model in SAP2000 is made with frame elements. The definition of the initial condition is achieved with a force, applied at the center of mass along an axis. The force causes a displacement at the center of mass. Such a procedure can be iterated several times to obtain the desired displacement. The process converges quickly. Not more than a few attempts are needed. The time step at which the desired displacement occurs is the beginning time of the analysis. It is worth noting that a more formal approach is possible, that allows to calculate the exact time of the start of the vibration, but the complexity of the equations in comparison with the force-approach makes it far less usable.

Figure 8. Accelerometer attached to the physical body.

IOP Conf. Series: Materials Science and Engineering **1323** (2024) 012013 doi:10.1088/1757-899X/1323/1/012013

The verification of the obtained results is achieved with the physical model of the body. An accelerometer is attached to the body (as shown in figure 8) and the real acceleration along an axis is compared with the result of the computation. The data is shown in figure 9 as a graph.

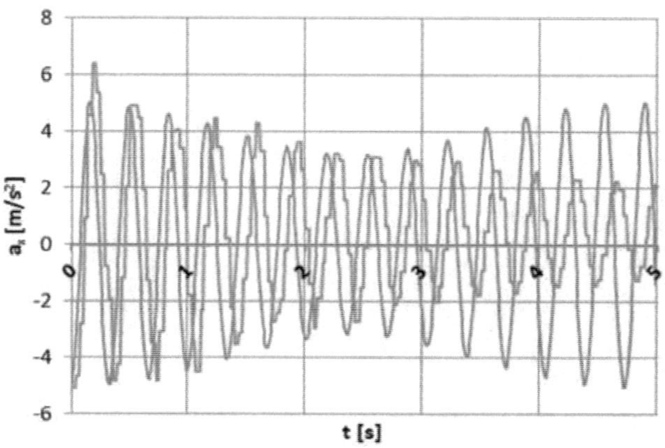

Figure 9. Accelerations of a point of the body. Blue line – calculated accelerations. Orange line – measured accelerations.

It is evident that the orange line (measured acceleration) represents the graph of a vibration with diminishing amplitude. That is to be expected because there is no undamped vibration in nature. There is always friction – in the material, between the parts of the system, because of the environment (air). On the other hand, the blue line is a sinusoidal curve. It is theoretically possible to define undamped vibration. That is why there is no damping observed in the blue line.

The first 2,5 seconds of the vibration the two curves practically overlap, while after that moment in time, the vibration of the physical model gradually fades away because of the inherent damping of a real body. The first half of the graph shows that the model in SAP2000 gives accurate results, compared with the real vibration.

7. Conclusion

The current paper offers a relatively simple method of modelling free vibrations in SAP2000. The basic idea of an impulse of a particle is used to model initial velocity. When initial velocity is obtained, initial displacement is modelled using the same principle. This is demonstrated by equalizing the solutions of SDOF problems with initial conditions velocity and displacement respectively. For MDOF problems it is simpler to iterate the desired displacement by varying the applied force than to calculate the exact value of the said force.

It is possible to model initial conditions of free vibrations where both initial velocities and initial displacements are non-zero, but in the physical world it is much simpler to model initial displacements.

The disadvantage of the method is that it requires manual operations to be carried out. The lack of programming and the simple idea behind it are its advantages. It is a tool that can be used by any engineer. No prior knowledge of coding is required.

References

[1] Chopra A 1995 *Dynamics of structures: Theory and Application to Earthquake Engineering* ed. K Scherwatzky (Englewood Cliffs, New Jersy: Prentice Hall) p 63

[2] Ding Y, Xu Y and Miao H 2024 Mathematical and physical characteristics of the phase spectrum of earthquake ground motions *Buildings* **14** 1250 https://doi.org/10.3390/buildings14051250

[3] Pavlov P, Lilkova-Markova S and Ivanova G 2016 *Modular Approach in Creating the Matrix Equations, Describing the Free Vibrations of Discrete Plane Systems* ed. ICEDM Proceedings of the 11th International Congress on Engineering, Design and Manufacturing. Sustainability and resilience approach, Barcelona, Spain, (Barcelona: IJTEE-ICEDM2016) Fullpapers E-Book pp 70-75

[4] CSI Inc. 2016 CSI Analysis Reference Manual ed. CSI Inc. (Berkeley, California: CSI Inc.) p 400

[5] Al-Kamal M and Mukai D 2017 Modeling initial velocity profiles for continuous free-vibrating beams using distributed impulse loading ARPN J. Eng. Appl. Sci. 12 pp 2888-95

[6] Tamrakar A and Harursampath D 2014 Acing the GATE Mechanical Engineering ed Wiley Editorial Office, India (New Delhi: Wiley) p 244

[7] Paz M and Leigh W 2004 *Structural Dynamics Theory and Computation* ed. Springer Science (New York: Springer Science + Business Media, LLC) p 13

[8] Grishin A and Geraschenko V 2024 Calculation of Rayleigh damping coefficients for a transient structural analysis WSEAS Trans. Appl. Theor. Mech. 19 pp 49-54

[9] CSI Inc. 2016 CSI Software Verification Example 6-005 Link-Damper Element under Harmonic Loading ed. CSI Inc. [cited 2024 June 6] Available from: https://wiki.csiamerica.com/display/kb/Damping

[10] Clough R and Penzien J 2003 Dynamics of Structures ed. CSI Inc. (Berkley, California: CSI Inc.) p 16

[11] Helrich C 2017 *Analytical Mechanics* ed. N Ashby *et al* (Goshen, USA: Springer) p 51

IOP Conf. Series: Materials Science and Engineering **1323** (2024) 012014 doi:10.1088/1757-899X/1323/1/012014

Simulating the Non-linear Bond Behaviour of Plain Rebar-Recycled Aggregate Concrete

Emad Abdulahad [1] and Ghoson Danhash[2*]

[1] Department of Reinforced Concrete Structures, Faculty of Structural Engineering, University of Architecture, Civil Engineering and Geodesy (UACEG), Sofia, Bulgaria
[2] Department of structural engineering, civil engineering college of Alba'ath University and Wadi International University, Homs, Syria

*E-mail: ghosoun.d@gmail.com

Abstract. The present research is interested in studying the plain rebar–recycled aggregate concrete bond behaviour, where the force-slip curve is obtained using the pull-out test. Five mixes were prepared, varying the replacement ratios of the natural aggregates with recycled ones: 0, 25, 50, 75, and 100%. The bond length of the rebar in the concrete was five times the rebar diameter. Fifteen pull-out specimens were cast with 14 [mm] steel plain rebar in the centre of the cubic specimen (150×150×150 mm), three specimens for each mix. The experimental procedure results showed the bond strength decreased, starting from 6% for 25% replacement ratio up to 23% for 100% replacement ratio. A numerical simulation of the pull-out test was performed using finite element method program (ABAQUS), and for describing the concrete behaviour, concrete damage plasticity was adopted in parallel with damage behaviour and cohesive behaviour for describing the interaction between rebar and recycled aggregate concrete. The cohesive behaviour of the rebar-recycled aggregate concrete interaction area was modelled using cohesive behaviour, where the parameters of the cohesive behaviour and the traction-separation behaviour were discussed for verifying the applicability of this property in predicting the rebar-recycled aggregate concrete bond behaviour. The force-slip curve obtained from the numerical simulation model exhibited good agreement with the experimental results.

1. Introduction

Recycling waste is a cost-effective and sustainable way to preserve natural resources for future generations. The recycled aggregates (RA) have an old adhered mortar, which causes the recycled aggregates to reduce durability as it is more porous, have more water absorption, and consequently have less mechanical properties [1], [2]. For verifying the structural behaviour of reinforced recycled aggregate concrete members, the steel-recycled aggregate concrete (RAC) bond behaviour must be verified [3]. While using recycled aggregates in concrete, the bond efficiency with rebar may be different compared to regular concrete and requires in-depth investigation.

Numerous studies that investigated the mechanical characteristics of RAC indicated reductions in those characteristics for recycled concretes that had an exceptionally high recycled aggregate to natural aggregate replacement ratio [4], [5]. Results of Al-Shamma research indicate that as the percentage of recycled coarse aggregate (RCA) increased, the mechanical properties of concrete like compressive, tensile, and flexural strengths decreased significantly, with up to 64%, 29%, and 38% reduction using 100% RCA. Diab and Danhash found that the compressive strength of concrete decreases with the increase of the replacement ratio, where the decrease was about 12.7% for the replacement ratio of 100%, [6], [7].

Although the mechanical properties of structural reinforced concrete are not significantly reduced when the recycled aggregate percentage is between 20 and 30 percent, recycled aggregates from concrete demolition still provide good enough features for use in structural reinforced concrete [8].

To verify the behaviour of structural members made from recycled aggregate concrete, the bond behaviour between the recycled structural concrete and rebar must be studied, as the bond between steel and RAC is responsible for transferring the axial forces between both materials and plays a major role during earthquakes, where premature bond failure may lead to catastrophic results. To define the development length and prevent early bond failure of reinforced concrete members, a comprehensive understanding of the bond behaviour is vital [3].

Certain investigations about the bond strength of recycled aggregate concrete (RAC) concluded that there is no clear association between bond strength and replacement ratio, while others discovered that bond strength declined with the increment in replacement ratio, [9], [10], [11], [12]. Prince et al. concluded that for treated recycled aggregate, bond strength increases with the increment of the replacement ratio [13].

The present research will focus on studying the effect of replacement ratio on steel-RAC bond behaviour, where a plain rebar is used, and studying the simulation of non-linear plain rebar-RAC bond behaviour using cohesive behaviour and damage behaviour in ABAQUS.

Adjusting the cohesive model parameters when damage behaviours is considered is a dilemma for researchers, where many corrections and iterations are necessary for obtaining the optimal coincidence between experimental and numerical force-slip curves.

In this research, a parametric study will be performed to conclude a relation between the stiffness parameters and the experimental results of pull-out test. The experimental results from pull-out test will be utilized to optimize the model parameters.

2. Experimental procedure

Five mixes with replacement ratios of 0%, 25%, 50%, 75%, and 100%, three specimens for each mix, and a value for the plasticizer ratio were designed in accordance with ACI 211 in order to verify the plain rebar-RAC bond behaviour experimentally. The plasticizer ratio increased as the replacement ratio increased for obtaining workability with class 4. Natural aggregate concrete, which has a replacement ratio of 0%, served as the reference mix for analysis.

Common Portland cement, type 32.5 N, with a grade of 400 kg/m3, was utilized. For the coarse aggregates, diameters ranging from 4.75 mm to 20 mm were employed. The diameters of the fine aggregates range from 1.18 mm to 4.75 mm. Natural sand was utilized in all mixes, and no recycled sand was used. In the laboratory, recycled aggregates were obtained by manually crushing and sorting, and the original waste resource was construction parts, like columns and solid slabs. Coarse aggregates, varying in size from 4.75 mm to 20 mm, were also employed. Depending on the mass replacement, recycled coarse aggregates were used in place of natural coarse aggregates at ratios of 25%, 50%, 75%, and 100%.

A high-performance plasticizer, RSP was employed at a ratio of 0%, 0.5%, 0.75%, 1%, and 1.5% of the cement weight for replacement ratios of 0%, 25%, 50%, 75%, and 100%, respectively.

IOP Conf. Series: Materials Science and Engineering **1323** (2024) 012014 doi:10.1088/1757-899X/1323/1/012014

The water-to-cement ratio was maintained at (w/c = 0.5) in all mixes using drinking water. Table 1 represents the constituents of the reference mix.

Table 1. The constituents of the reference mix.

Constituents	Coarse gravel	Fine gravel	Coarse sand	Fine sand	Water	Cement
Dosage kg/m³	570	270	590	355	200	400

A total of fifteen cylinders comprise the three cylindrical samples (diameter: 150 mm, height: 300 mm) that were cast for each mix. The samples were taken out of the mould after roughly 24 hours, and they were allowed to cure for 28 days in water. The other 15 cubic test specimens were manufactured by following the same procedure as this for the cylindrical ones; however, the pullout specimens were cured for 56 days in water. Figure 1 represents the current study's tested specimens and experimental approach.

2.1 Experimental results

After four weeks of water curing, the specimens were removed from the water, and after 12 hours, they were tested under axial compression with a Technotest (5000 KN) with a manually regulated loading speed. The average values of the compression test results are displayed in Table 2. A pull-out test was performed, where a special metal holder was manufactured especially for holding the pull-out specimen on the steel tension machine, and the pull-out test results are shown in Table 2.

Table 2 shows how the strength dropped from 6% for the 25% ratio up to 13% at the replacement ratio of 100%. The outcomes attained matched those of other research [14], [15], and [16].

Table 2. Experimental compressive strength values and pullout test results.

Replacement ratio R%	0%	25%	50%	75%	100%
f_c $[MPa]$	25.95	24.24	23.92	23.5	22.66
Pullout force Fmax [KN]	18.65	16.89	15.78	15.74	13.99
Slip under Fmax [mm]	0.09	0.11	0.12	0.08	0.11
Bond stress [MPa]	6.06	5.49	5.12	5.11	4.54

The bond strength dropped as the amount of recycled aggregate grew; for the ratios of 25%, 50%, 75%, and 100%, the drops were 9.4%, 15.5%, 15.5%, 15.67%, and 25.08%. The RAC undergoes the same failure process as regular concrete, see Figure 2, with the force curve moving through three distinct phases: the elastic ascending line, the yielding stage, and the softening descending stage. The force-slip curves of the mixes with a replacement ratio of 50% and 75% were close together. The reference mix containing natural aggregates has a slip of 0.09 [mm] at the maximum load; nevertheless, there is no obvious correlation between the replacement ratio and the slip at the maximum load.

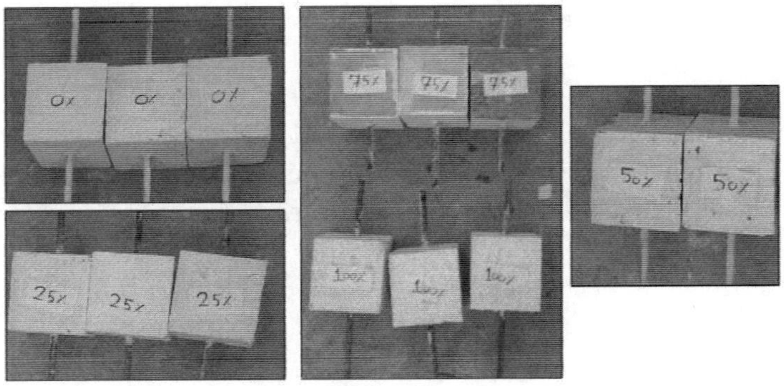

(a)

(b)

Figure 1: The tested specimens (a) and the experimental procedure of the present study (b).

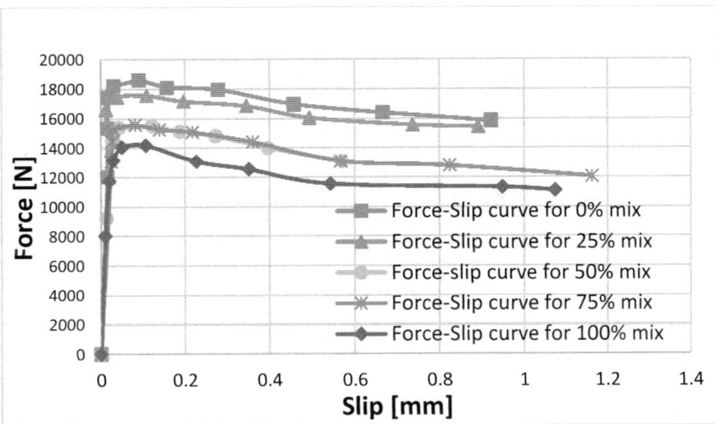

Figure 2. The average experimental Force-Slip curves for the present study mixes.

3. Finite element method (FEM) simulation

By applying a fundamental computational method, Finite Element Analysis (FEA) can be used to investigate and predict material behaviour under a range of stress conditions [15]. One of

ABAQUS's distinctive features is its capacity to handle complex simulation cases, such as dynamic analysis and nonlinear modelling, which are required for analysing materials such as concrete. Precise modelling is critical to the computer program's capacity to accurately represent material properties, so precise and comprehensive calibration is required to forecast the behaviour of the steel-RAC bond. Cohesive behaviour in ABAQUS can characterize the interaction between steel and RAC; nevertheless, accurate bond behaviour necessitates accurate calibration of the model parameters, describing the steel-RAC interaction.

To acquire a thorough knowledge of the impact of stiffness parameters on bond behaviour in the pull-out test, a parametric study of the stiffness parameters of the steel-concrete interaction using cohesive and damage behaviour will be carried out in this study. An example of disregarding this property is presented and compared with the model adjusted with the experimental results, highlighting the significance of taking the steel-concrete bond behaviour in reinforced concrete members into account.

3.1 Model building

ABAQUS 6.14 was used to numerically simulate the steel-concrete pull-out test, using damage and cohesive models for describing the steel-concrete interaction zone, adding to hard contact and normal behaviour. The concrete damage plasticity property was used to define the mechanical characteristics of the concrete part, while the experimental tensile stress-strain curve was used to describe the plastic behaviour of the steel bars' material. The steel bar was positioned inside the concrete specimen using the assembly, and a cohesive interaction with adjusted damage initiation and evolution criteria was defined at the interaction. The mesh was fine-tuned close to the contact to guarantee precision, with 1 mm element size near the interaction and 2.5 mm outside the interaction, and the concrete block was fixed using boundary conditions. Figure 3 presents the pull-out test numerical modelling.

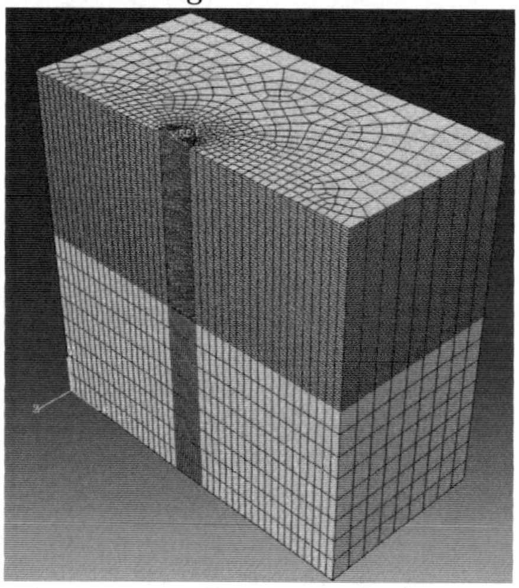

Figure 3. The numerical model in ABAQUS.

In the assembly process, two parts were created, the C3D8R element was used for concrete and steel. A displacement was applied to the rebar with a smooth function, and the force – slip curve was obtained, indicating that the force was multiplied by two as half the specimen was modelled, while the slip was the same. Table 3 shows the adopted parameters in the present study.

When a steel-concrete interaction's tensile, shear, or tangential stresses reach their peak values, or when a steel-concrete interaction's normal, shear, or tangential deformations reach

IOP Conf. Series: Materials Science and Engineering **1323** (2024) 012014 doi:10.1088/1757-899X/1323/1/012014

their peak deformations, that is when the failure initiation begins. However, for steel-concrete bond behaviour, the critical point is the small tensile strength of concrete, and the small deformation under peak concrete tensile stress, which makes the tensile ascending curve very important in describing the initiation of bond failure in steel-concrete interaction, see Figure 4. As the force is applied to the steel, the steel deforms, and this deformation is transferred to the concrete. When this deformation reaches the deformation value under peak stress in tensile or shear, the damage process starts.

Table 3. The parameters of cohesive and damage behaviour.

The interaction property	The parameters	Value	Friction coefficient in tangential behaviour
	Knn [MPa/mm]	1125	
Cohesive	Kss [MPa/mm]	400	
	Ktt [MPa/mm]	400	0.45
	Normal nominal stress [MPa]	110	
Damage	Shear_1 nominal stress [MPa]	7.885	
	Shear_2 nominal stress [MPa]	7.885	

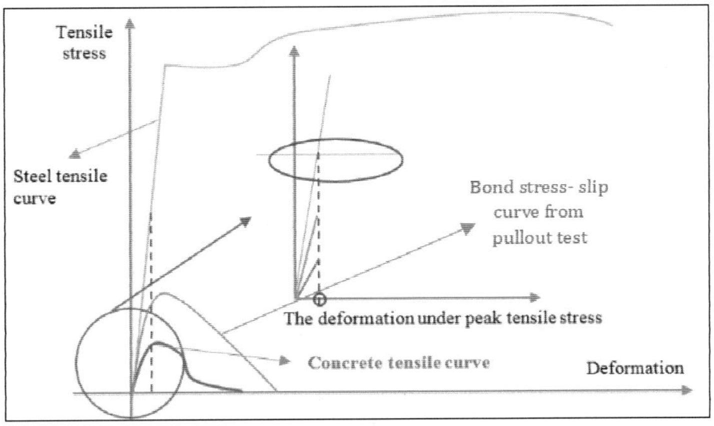

Figure 4. Steel-concrete bond failure initiation depending on the experimental behaviour.

When comparing numerical and experimental findings for the reference mix 0%, the two curves were comparable with the relative error of 2% for force and 4.2% for slip at the maximum force. With the same parameters shown in Table 3, the relative errors for the remaining mixes; 25%, 50%, 75%, and 100%; reached 14% between the numerical and experimental forces for the 100% mix and 11% between the numerical slip and the experimental slip for the 50% mix.

It can be noticed from Figure 5 and Figure 6 that the stiffness of the ABAQUS model is fairly comparable to the experimental force-slip, as seen by the beginning slopes of the experimental and numerical curves. This demonstrates that simulation precisely captures the behaviour of the bond during the first elastic phase. However, the descending part of the numerical curve is not similar to the experimental one. For the other mixes with different replacement ratios, the parameters of cohesive and damage parameters are the same; however, the difference in tensile and compressive stresses is responsible for describing the difference in the steel-concrete bond behaviour between the five studied mixes, with a maximum error of 14% between the numerical and experimental results.

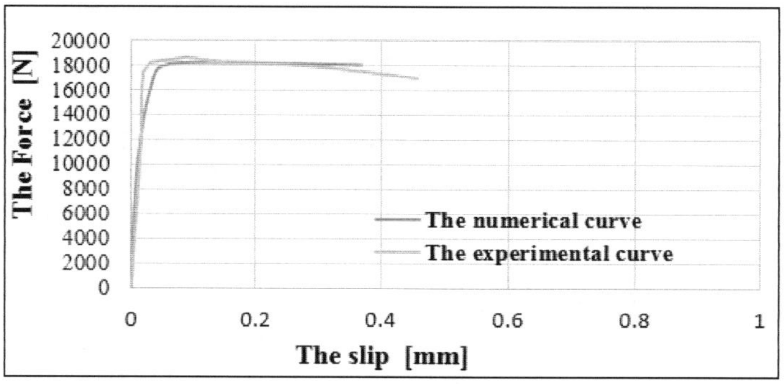

Figure 5. The comparison between the experimental and numerical force-slip curves for the (0%) mix.

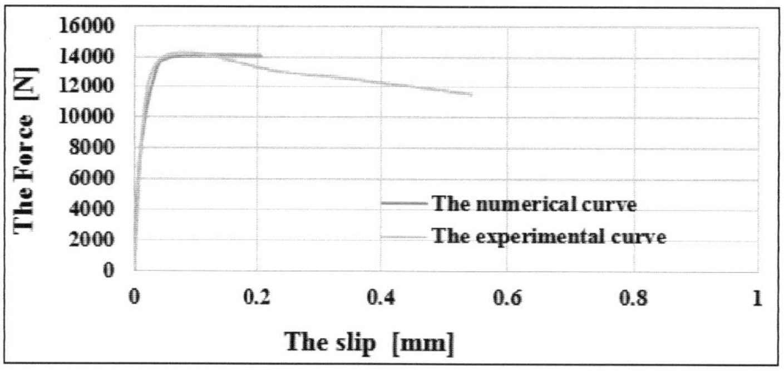

Figure 6. The comparison between the experimental and numerical force-slip curves for the (100%) mix.

3-2 Parametric study

After adjusting the model, and for a deep understanding of how the values of Kss, Ktt and Knn change the slip and force, a parametric study was performed by changing Kss, Ktt, and Knn. The resultant maximum slip and the force obtained from the ABAQUS model were registered, see Table 4. To indicate the importance of the exact modelling of the steel-concrete interaction zone on the steel and concrete mutual work, a new model was simulated without taking into account the cohesive and damage behaviour, and the force – slip curve was obtained, adding to the tensile cracking (DamageT) and compressive cracking (DamageC).

Table 4. The parametric study of the stiffness interface Kss, Ktt.

Kss, Ktt [MPa/mm]	100	150	200	300	350	400	450	500
The maximum force [N]	17355	16183	16341	17461	18279	18196	19504	20192
The slip under maximum force [mm]	0.091	0.069	0.065	0.073	0.087	0.087	0.091	0.092

The other parameters were: Knn=1125 [MPa/mm], the friction coefficient=0.45, $t_n^0 = 110\ [MPa]$, $t_t^0 = t_s^0 = 7.885\ [MPa]$.

Figure 7 demonstrates the force-slip curves for pull-out test for various tangential and shear stiffness factors Ktt=Kss. The figure shows that changing the shear and tangential stiffness factors changes the slope of the first ascending part of the curve. The maximum force increases with increasing shear and tangential stiffness factors. Figure 8 shows the maximum pull-out force for various Ktt = Kss values, where a nonlinear relationship was found with a correlation factor equal to 0.9238. Figure 9 demonstrates how Ktt, Kss, and the experimental slip under maximum force are related to one another. A linear relationship was found with a correlation factor equal to 0.7291. Table 5 presents the parametric study of the normal stiffness parameter Knn. Figure 10

IOP Conf. Series: Materials Science and Engineering **1323** (2024) 012014 doi:10.1088/1757-899X/1323/1/012014

shows the force-slip curves for pull-out for various Knn values of normal stiffness. The figure shows that changing the normal stiffness does not change the first ascending part of the curve, but it controls the yield and descending part.

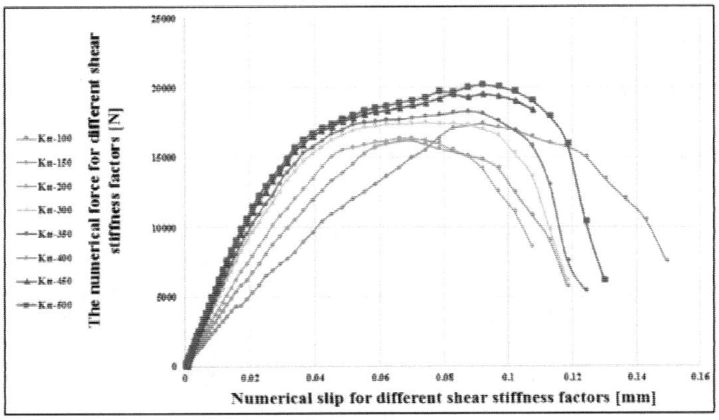

Figure 7. The pull-out force-slip curves for different shear and tangential stiffness factors Kss=Ktt.

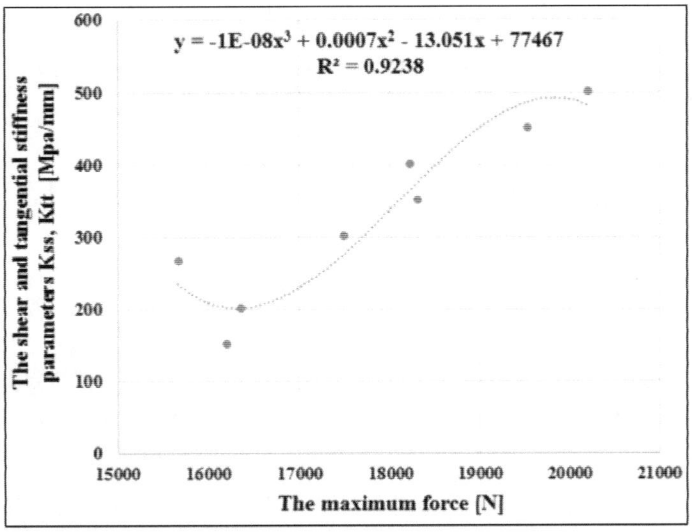

Figure 8. The maximum pull-out force for different values of Ktt = Kss

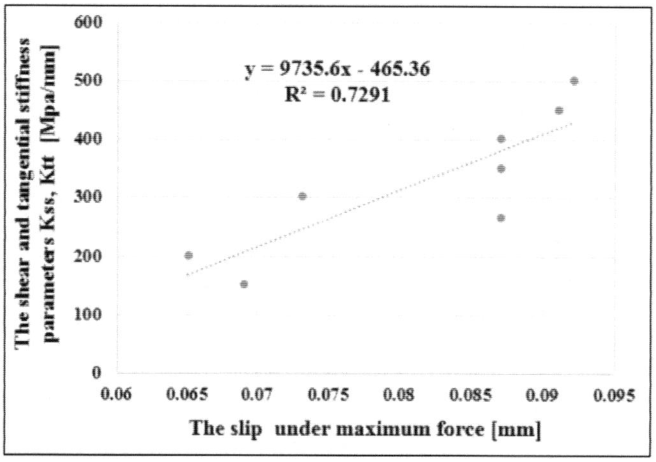

Figure 9. The relation between the shear and tangential stiffness parameters Ktt, Kss, and the experimental slip under the maximum force.

Table 5. The parametric study of the normal stiffness parameter Knn.

Knn [MPa/mm]	500	1000	2000	3000	4000	5000
The maximum force [N]	18865	18938	18871	19683	20002	20074
The slip under maximum force [mm]	0.087	0.087	0.092	0.092	0.087	0.084

The other parameters were: Kss=Ktt=400 [MPa/mm], the friction coefficient=0.45, $t_n^0 = 110\ [MPa]$, $t_t^0 = t_s^0 = 7.885\ [MPa]$.

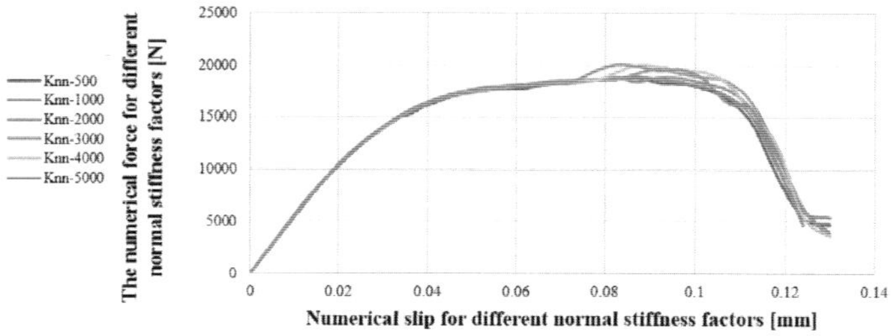

Figure 10. The pull-out force-slip curves for different normal stiffness factors Knn

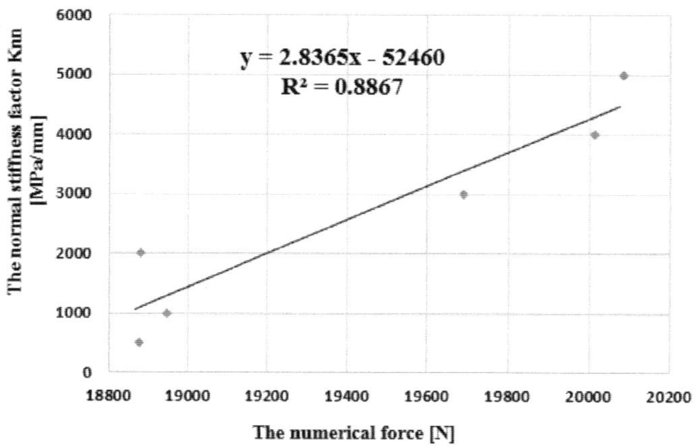

Figure 11. The relation between the numerical force and the normal stiffness factor

IOP Conf. Series: Materials Science and Engineering **1323** (2024) 012014 doi:10.1088/1757-899X/1323/1/012014

Most researchers ignore the bond relationship between steel in FEM environment and concrete and define it as a perfect bond with no slip; however, this does not represent the real behaviour of reinforced concrete members. Figure 12 illustrates the evolution of tensile damage while taking into account the cohesion and damage behaviour of steel-concrete for (a) and without it for (b), and Figure 13 shows the progression of compression damage for (a) and (b) with and without the steel-concrete cohesive and damage behaviour.

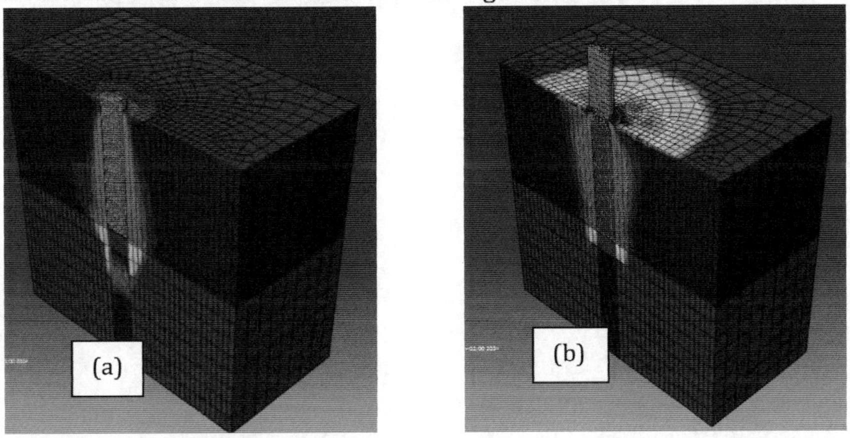

Figure 12. The tensile damage evolution with the considering steel-concrete cohesive and damage behaviour for (a), and without it for (b)

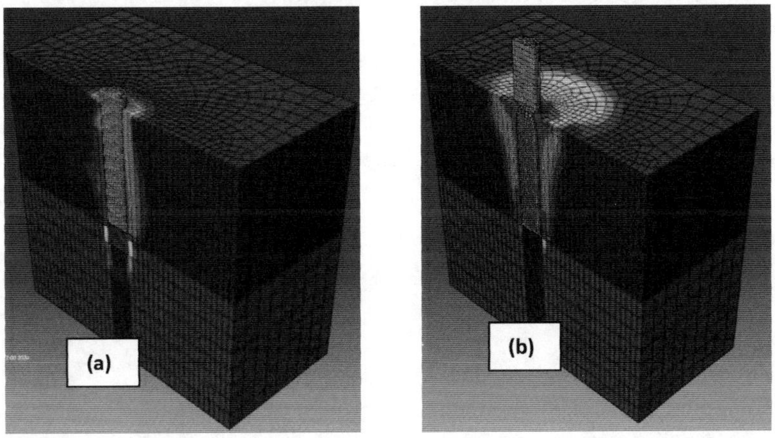

Figure 13. The compression damage evolution with the steel-concrete cohesive and damage behaviour for (a), and without it for (b).

When the perfect bond behaviour was adopted, the maximum force was larger than the experimental one, as shown in Figure 14. The perfect behaviour does not represent the real steel-concrete bond behaviour, however, it is easier in the simulation process and model calibration.

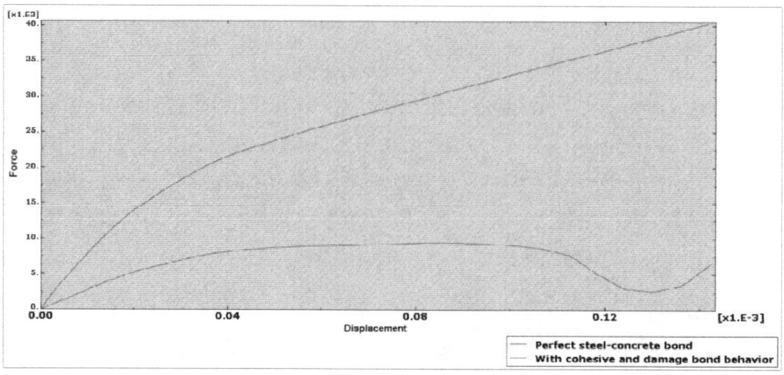

Figure 14. Force (N)-slip (m) curves for cohesive and damage behaviour (green), and without it (red).

4. Results

The experimental procedure results showed that for the mixes with replacement ratios of 25%, 50%, 75%, and 100%, respectively, the compressive strength decreased from 6% for the 25% replacement ratio up to 23% for the 100% replacement ratio, see Table 2.

The ideal replacement ratio was 25%, as the mechanical properties were almost comparable to those of the mix with natural aggregates; see Table 2.

As the replacement ratio increased, the bond strength declined as well; for the 25%, 50%, 75%, and 100% ratios, the drops were 9.4%, 15.5%, 15.5%, 15.67%, and 25.08%, see Table 2.

The cohesive and damage behaviour can describe the bond-slip behaviour of plain rebar-RAC in the elastic and yielding stages, see Figure 5, Figure 6.

Due to its impact, the stiffness parameters of the steel-concrete interaction should be chosen correctly in the modeling.

Changing the shear and tangential stiffness factors changes the slope of the first ascending part of the curve, see Figure 7. The slip under the maximum force increases with increasing shear and tangential stiffness factors, see Figure 9.

The normal stiffness controls the yield and descending part of the force-slip curve, see Figure 10. When the normal stiffness increases, the force increases, see Figure 11.

Ignoring the cohesive and damage behaviour by defining the perfect steel-concrete bond connection does not represent the bond behaviour, see Figure 12, Figure 13, and gives larger forces, see Figure 14.

References

[1] Sankalp, Kaushik., Pravindra, Singh, Bhan. (2024). Chemical Modifications of Recycled Concrete Aggregate. International journal of emerging science and engineering, doi: 10.35940/ijese.g9900.12060724

[2] Zaharieva, R., & Dimitrov, G. (2023a). Impact of recycled aggregate properties on concrete performance. IOP Conference Series: Materials Science and Engineering, 1276(1), 012003. https://doi.org/10.1088/1757-899X/1276/1/012003

[3] Georgiev, Evg. Improvement of the deformation characteristics of reinforced concrete slabs by varying the quantity of embedded reinforcement. Proceedings – Volume 2 - International Jubilee Scientific Conference "70th Anniversary UACEG", 15-17 November 2012, University of Architecture, Civil Engineering and Geodesy, Sofia, 2012, ISBN 978-954-724-049-0, p. 261÷266, 506 p.

[4] L., Dadd., T., Xie., B., Bennett., Phillip, Visintin. (2024). Exploring the physical and mechanical characteristics of multi-generation recycled aggregate concrete at equivalent compressive strengths. Journal of Cleaner Production, doi: 10.1016/j.jclepro.2024.141965

[5] M., F., Al-Shamaa., Ammar, A., Ali., Ikram, F., Ahmed, Al-Mulla. (2024). Mechanical characteristics of structural concrete using building rubbles as recycled coarse aggregate. Journal of the mechanical behaviour of materials, doi: 10.1515/jmbm-2024-0001
[6] Diab, W. Danhash, G. (2023). " Predicting Compressive Strength of Recycled Aggregate Concrete Using Natural Aggregate Concrete Compressive Strength and Material Proportions" International Journal of Masonry Research and Innovation, 1(1). https://doi.org/10.1504/IJMRI.2024.10064603
[7] Diab, W. Danhash, G. (2023). The effect of replacement ratio on the compression strength of recycled aggregate concrete". Albaath university journal, issue16, volume 45, 95-124.
[8] Lynda, Amel, Chaabane., Hamza, Soualhi., Ilies, Fellah., Yassine, Khalfi., Nadia, Sirine, Bouayed. (2024). Physical and mechanical behaviour of recycled concrete under destructive and non-destructive testing. Budownictwo i Architektura, doi: 10.35784/bud-arch.5840
[9] Oudeh, Y. Diab, W. Danhash, G. (2023). Using Digital Image Correlation for bond - slip curve registration in pull-out test for steel -recycled aggregate concrete. ICEE November 2022, Damascus, Syria.
[10] SADEK, E. DANHASH, G. (2024) "Suggesting a New Empirical Relationship for Steel-Recycled Aggregate Concrete Bond Strength" Albaath University Journal. Volume 46, issue 1.
[11] Wardeh G, Ghorbel E, Gomart H, Fiorio B. Experimental and analytical study of bond behaviour between recycled aggregate concrete and steel bars using a pullout test. Struct Concrete. 2017:0:1-15.
[12] Danhash, G. Oudeh, Y. Diab, W. Wardeh, G. (2023). Studying Steel-Recycled Aggregate ConcreteBond–Slip Behaviour Using Digital Image Correlation. Journal of Duhok university, 2023.
[13] Prince, M. J. R. and B. Singh. Bond behaviour of deformed steel bars embedded in re-cycled aggregate concrete. Construction and building materials 2013; 49: 852-862.
[14] Ammar, Younes., Emad, Elbeltagi., Aboelkasim, Diab., Giulia, Tarsi., Fam, Saeed., Cesare, Sangiorgi. (2023). Incorporating coarse and fine recycled aggregates into concrete mixes: mechanical characterization and environmental impact. Journal of Material Cycles and Waste Management, doi: 10.1007/s10163-023-01834-1
[15] Sara, F., Mohammed., Abdulamir, Atalla, Karim., Ansam, Z., Thamer. (2023). Experimental Study on The Effect of Size, Type, and Replacement Ratio of Recycled Aggregate on The Mechanical Properties of Reactive Powder Concrete. doi: 10.61263/mjes.v2i1.43
[16] Malak, H., Dawood., Ali, K., Al-Asadi. (2023). Mechanical properties and flexural behaviour of reinforced concrete beams containing recycled concrete aggregate. Scientific Review Engineering and Environmental Sciences, doi: 10.22630/srees.4250
[17] Mohammed, Bentahar., Moulai, Arbi, Youcef., N., Mahmoudi., Habib, Benzaama. (2024). Numerical modeling of the stress-strain behaviour of a viscoelastic material in a 2D model by FEM. Studies in Engineering and Exact Sciences, doi: 10.54021/seesv5n1-070

IOP Conf. Series: Materials Science and Engineering **1323** (2024) 012015 doi:10.1088/1757-899X/1323/1/012015

FEM Parameters for Earthquake Analysis of Retaining Walls

Maria Topalska[1] and Lena Mihova[1*]

[1] Geotechnics, University of Architecture, Civil Engineering and Geodesy, Sofia, Bulgaria

*E-mail: mihova_fte@uacg.bg

Abstract. The paper presents the main characteristics and features of numerical finite element (FE) models for the study of the seismic response of soil-retaining wall. Since walls are linear structures, 2D plane strain models are suitable for conducting of the analysis. The models need to include a finitely sized soil domain, whose boundary conditions are defined according to the way earthquake loading is defined, respectively as pseudo-static impact or dynamic impact as a function of time. The FE mesh must meet the criteria of sufficient density in order to ensure the necessary accuracy of the solution. With the use of multi-nodal finite elements the displacements function in the model is improved. Retaining walls as well as other structural elements and devices are modelled with 2D plate and 1D linear elements defined in terms of a full or limited set of stresses. To simulate the behaviour in the contact of the structural elements and soil, interface elements are introduced modelling the gapping and slipping in the contact. The material parameters are determined depending on the selected constitutive models. In dynamic earthquake analyses, parameters shall be introduced to take into account the reduction of the soil stiffness. Dynamic loading is usually introduced with an accelerogram whose selection is made on the basis of an analysis of the parameters of strong motion records.

1. Introduction

Retaining walls are structures whose mechanical behaviour occurs in 2D plane strain. A simplified engineering approach for the seismic analysis of retaining walls is to apply pseudo-static solutions in limit equilibrium theory. The most widely used formulae in engineering practice for determining pseudo-static active and passive earth pressure on retaining walls during an earthquake are the Mononobe-Okabe formulae [1], [2], [3]. Dynamic earthquake analyses of geotechnical structures are performed using software most often developed on the basis of discretisation mathematical methods. These are numerical time history analyses where the seismic impact is introduced as a function of time, usually through an accelerogram. The accuracy of the analyses is determined by the completeness and precision of the mathematical formulations for describing the mechanical behaviour of the materials by the dependencies for the discretisation and the boundary conditions of the soil domain, as well as by the accuracy of the algorithms used for obtaining the solution. Some of the geotechnical software products used for performing 2D dynamic seismic analysis are PLAXIS, FLAC, EERA, STRATA, QUAKE/W (GEOSTUDIO) and SWANDYNE. Commentary on the capabilities of various software products, as well as comparative analyses can be found in the sources [4], [5].

IOP Conf. Series: Materials Science and Engineering **1323** (2024) 012015 doi:10.1088/1757-899X/1323/1/012015

This paper presents the main parameters and features of the FE numerical models for dynamic earthquake analysis of the retaining wall-soil interaction, with the software PLAXIS 2D being used when building these models [6], [7], [8].

2. FE formulation for dynamic site response analysis

The matrix equation of the seismic response of the soil has the following form

$$\boldsymbol{M}\,\ddot{\boldsymbol{u}} + \boldsymbol{C}\dot{\boldsymbol{u}} + \boldsymbol{K}\boldsymbol{u} = \boldsymbol{F}\ , \tag{1}$$

where: $\boldsymbol{M}$ is the mass matrix; $\boldsymbol{C}$ – the damping matrix; $\boldsymbol{K}$ – the stiffness matrix; $\boldsymbol{F}$ – the dynamic impact force vector; $\boldsymbol{u}$ – the displacement vector; $\dot{\boldsymbol{u}}$ – the velocity vector; $\ddot{\boldsymbol{u}}$ – the acceleration vector.

The damping is a result of the dissipation of seismic energy due to the friction between particles and the occurrence of plastic deformations in the soil. The damping matrix $\boldsymbol{C}$ is calculated by Rayleigh as a linear combination of the matrices $\boldsymbol{M}$ and $\boldsymbol{K}$ through the coefficients α_R and β_R according to the equation

$$\boldsymbol{C} = \alpha_R \boldsymbol{M} + \beta_R \boldsymbol{K}\ . \tag{2}$$

Equation (1) is solved in PLAXIS 2D using the Newmark implicit integration scheme, i.e.

$$\left[c_0\boldsymbol{M} + c_1\boldsymbol{C} + \boldsymbol{K}\right]\Delta\boldsymbol{u} = \boldsymbol{F}_{ext}^{t+\Delta t} + \boldsymbol{M}\left(c_2\dot{\boldsymbol{u}}^t + c_3\ddot{\boldsymbol{u}}^t\right) - \boldsymbol{F}_{int}^t\ , \tag{3}$$

where c_0 .... c_3 are the Newmark coefficients, which are functions of the integration parameters α and β and the time step Δt.

The parameters α and β need to fulfil the following conditions

$$\beta > 0.5, \qquad \alpha \geq \frac{1}{4}\left(\frac{1}{2} + \beta\right)^2, \tag{4}$$

where the following values of $\alpha = 0.25$ and $\beta = 0.5$ being recommended (the so-called average acceleration method).

To carry out a mathematically correct solution, it is necessary to fulfil the numerical discretisation criteria. The minimum distance between two nodes of a finite element l_{min} and the maximum interval in the time domain Δt_{max} must fulfil the conditions respectively [9]

$$l_{min} \leq \frac{\lambda_{min}}{8} = \frac{V_{s,min}}{f_{p,max}}\ , \tag{5}$$

$$\Delta t_{max} < \Delta t_{critical} = \frac{l_{min}}{V_{s,max}}\ , \tag{6}$$

where: λ_{min} is the minimum length of the seismic wave; $V_{s,min}$ and $V_{s,min}$ – the minimum and the maximum wave velocities in the soil deposit, respectively; $f_{p,max}$ – the maximum frequency of the motion.

3. FE discretization and boundary conditions of soil domain

3.1 Soil domain geometry and discretization

In finite element models, a finitely sized soil domain of the ground is assumed. For the discretisation of the soil area in plane strain models, plate finite elements with a thickness of 1 and node parameters of displacements u_x and u_y in the direction of the coordinate axes x and y are used (Figure 1-a). The nodal displacements u_x and u_y are directly determined by the solution of the global system of equations of the method. The software PLAXIS 2D works with 6-node and 15-node triangular finite elements (Figure 1-b, c), with polynomial interpolation of the 2nd and 4th order, respectively, for the displacements within the finite element.

3.2 Finite elements for modelling of retaining walls

Three methods, and accordingly, three types of finite elements can be used in PLAXIS 2D for the discretisation of a continuous retaining wall. The first method involves using plate elements. This method is suitable for retaining walls with a large thickness such as heavy concrete walls as it allows for the accurate input of the cross-sectional geometry of the wall into the numerical model (Figure 2-a, b). The solution results used for the design of the wall are the displacements and stresses that occur in the plane of the cross-section of the wall. The disadvantage of this modelling method is the inability to directly obtain internal forces such as bending moments, axial and shear forces from the solution, which are usually set as values in the design formulae.

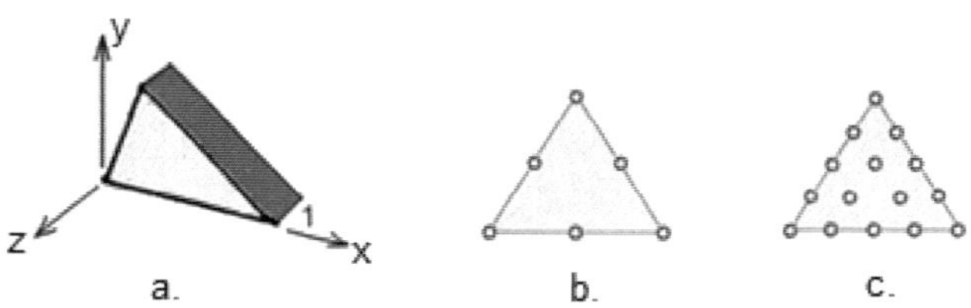

Figure 1. Plate finite elements in the software PLAXIS 2D.

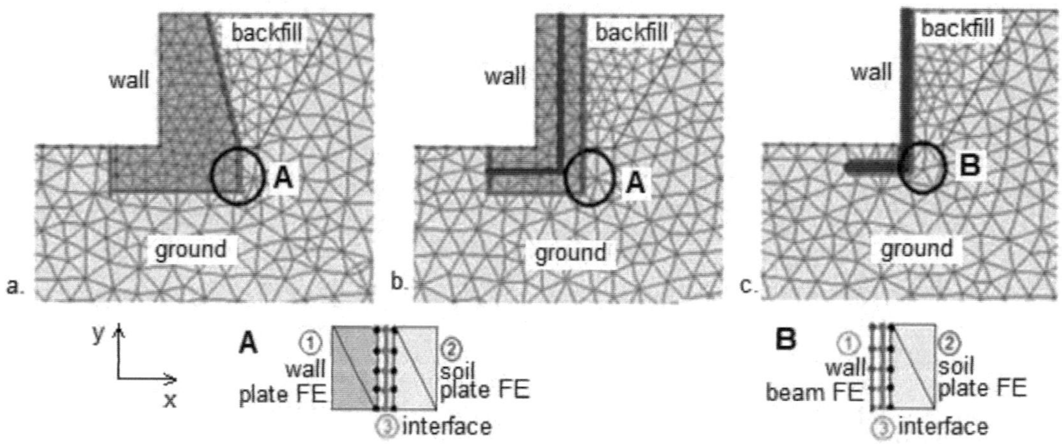

Figure 2. Modelling of the retaining wall-soil system.

Thin continuous walls are suitably modelled in PLAXIS 2D with 3-node or 5-node beam elements that has a dimension of 1 in the *z*-direction (the direction perpendicular to the wall cross section). In a plane strain model, the beam element is represented as a linear element (Figure 2-c). It must have the length and number of nodes equal to those of the adjacent soil plate element. In addition to the displacements u_x and u_y, plate elements also have rotation ϕ_z in the *x-y* plane as a nodal parameter. The material properties of this element are linear stiffness *EA* and bending stiffness *EI*. Then direct results are obtained for bending moment, axial force (tension, compression) and shear force in the *x-y* plane.

The contact between the continuous retaining wall and the soil is modelled with interface elements. The interface elements have zero or very small thickness and are defined with normal stiffness k_n and tangential stiffness k_s. This concept of modelling the contact, known as Goodman's stiffness method [10], is associated with the operation of two springs - a normal and a shear spring (Figure 3). The stresses that occur in these elements are

$$\begin{aligned} \sigma &= k_n \delta_n \\ \tau &= k_s \delta_s \end{aligned}, \qquad (7)$$

where δ_n is the relative displacement across the contact surface - gapping, and δ_s is the relative tangential displacement in the contact surface - slipping.

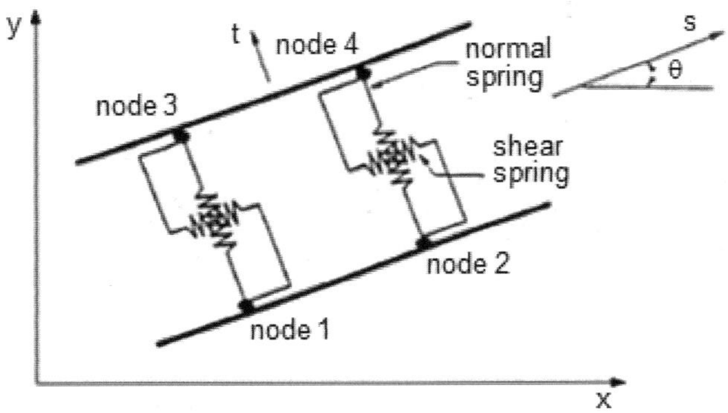

Figure 3. Interface element components [9].

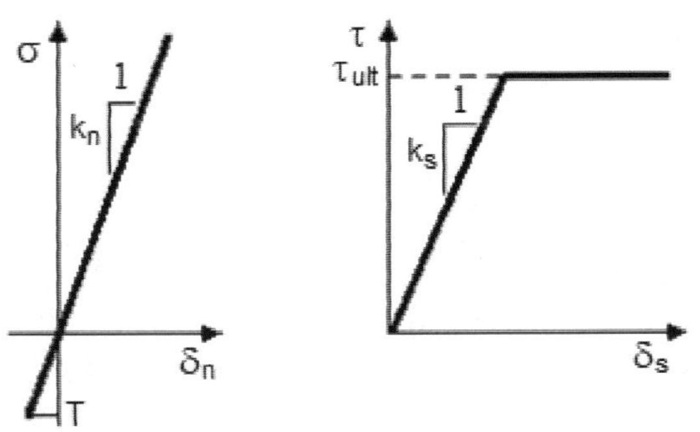

Figure 4. Dependencies for an interface element in the retaining wall-soil interaction.

The theoretical dependencies for stresses and deformations in the contact surface are presented in Figure 4. The contact area is characterised by elastoplastic behaviour, which is accounted for by defining the strength parameters of the interface element. The maximum tensile stress of the interface element is determined by the tensile strength T of the interface. In the soil-structure interaction, the tensile strength is excluded, i.e. $T = 0$. The maximum shear stress τ_{ult} of the interface element is determined by the shear strength of the contact defined by the Coulomb equation

$$\tau_{ult} = \sigma\ tg\varphi_{int} + c_{int} \ , \tag{8}$$

where φ_{int} and c_{int} are the contact friction angle and the contact adhesion, respectively. The φ_{int} and c_{int} parameters in PLAXIS 2D are derived from the internal friction angle φ and cohesion c of the soil in the contact area using a reduction factor R_{int}, i.e.

$$\begin{aligned} tg\varphi_{int} &= R_{int}\ tg\varphi \\ c_{int} &= R_{int}\ c \end{aligned} \ . \tag{9}$$

The plate elements are not suitable for modelling retaining walls constructed from discrete structural elements such as piles (embedded beams) since the latter interact with the surrounding soil environment spatially. To model such walls in PLAXIS 2D, a 2D embedded beam element is formulated. The concept of this element is schematically presented in Figure 5.

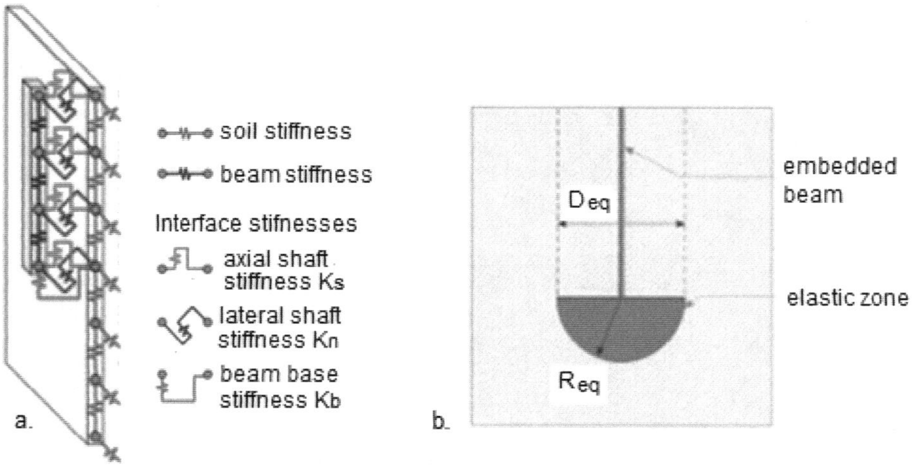

Figure 5. 2D embedded beam element: a. Interface elements; b. Elastic zone under the beam base [7], [11].

The following three types of interface elements are introduced for the contact of the embedded beam with the soil: (1) an element with stiffness in axial direction of the embedded beam K_s; (2) an element with stiffness in direction perpendicular to the axial direction K_n; (3) an element with stiffness at the base of the embedded beam K_b. The stiffnesses K_s, K_n, and K_b depend on the stiffness of the surrounding soil, defined by the shear modulus G_{soil}, the spacing between embedded beams $L_{spasing}$, and the diameter D_{eq} (or the radius R_{eq}, respectively) of the elastic zone under the base of the embedded beam. They are determined by the following formulae [7]

$$K_s = ISF_{RS} \frac{G_{soil}}{L_{spacing}}, \qquad ISF_{RS} = 2.5\left(\frac{L_{spacing}}{D_{eq}}\right)^{-0.75}$$
$$K_n = ISF_{RN} \frac{G_{soil}}{L_{spacing}}, \qquad ISF_{RN} = 2.5\left(\frac{L_{spacing}}{D_{eq}}\right)^{-0.75} . \tag{10}$$
$$K_b = ISF_{RB} \frac{G_{soil} R_{eq}}{L_{spacing}}, \qquad ISF_{RB} = 25\left(\frac{L_{spacing}}{D_{eq}}\right)^{-0.75}$$

The diameter D_{eq} is related to the bearing capacity of the ground and it defines the zone under the base of the embedded beam, without the presence of plastic deformations (Figure 3-b). It is determined as a function of the linear stiffness *EA* and the bending stiffness *EI* of the embedded beam according to the formula [9]

$$D_{eq} = \sqrt{12EI / EA} . \tag{11}$$

To model linear structural elements that act as supports for retaining walls such as bars, struts and anchors (Figure 6), PLAXIS 2D introduces 5-node and 3-node linear finite elements, which work for axial force and friction with the soil if they are in contact with the soil. For modelling anchor ropes, 2-node finite elements are defined, which transmit the tensile force between two points without activating friction with the soil. The supporting structural elements can alternatively be modelled as springs (Figure 6) with linear stiffness, which is determined by the formula $k = EA / L$, where L denotes the length of the structural element.

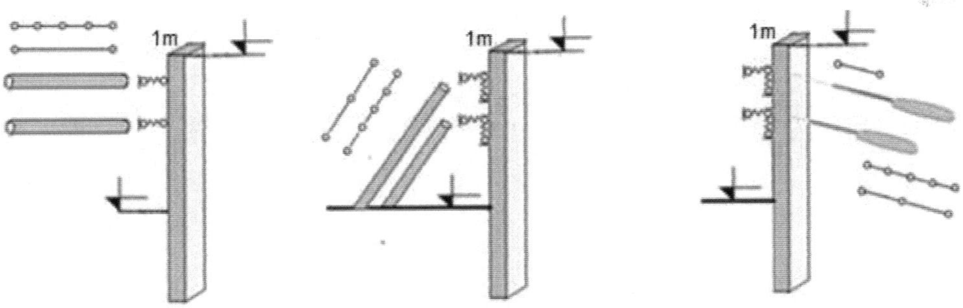

Figure 6. Modelling of linear supporting elements of retaining walls.

3.3 Boundary conditions of the soil domain
It is necessary to define appropriate boundary conditions for the bounded soil area in the FE numerical model in order to simulate the infinity of the ground. The fixed supports at the boundary nodes of the model nullify displacements (Dirichlet boundaries) (Figure 7-a). Fixed supports are used only in the analysis of the initial in situ state because in dynamic analysis they will cause the presence of reflected waves within the soil domain. Therefore, in dynamic analysis boundary conditions are defined to simulate the actual unbounded propagation of waves, eliminating their reflection at the model boundaries

To absorb the reflected waves, authors Lysmer and Kuhlemeyer [12] defined viscous boundaries, which consist of dampers in the horizontal and vertical directions. The dampers fully absorb the stresses caused by dynamic impact, causing the boundary to move. Such boundaries are suitable for use when the source of impact is within the considered soil domain, defined as the main domain. However, when the dynamic impact is applied at the boundary of the main domain, e.g., an accelerogram at the bottom side of the main domain, introducing viscous boundaries along the sides of the soil area is not suitable because the dampers would "absorb" the impact itself. In such cases, free field boundaries are introduced on the left and right sides of the domain, and a compliant base boundary is introduced at the bottom side (Figure 7-b).

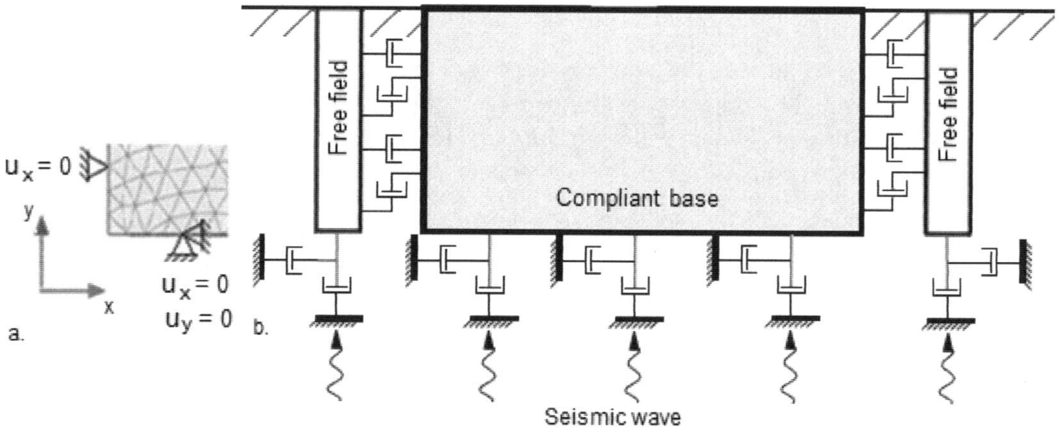

Figure 7. Boundary conditions: a. For static analyses; b. For dynamic earthquake analyses [13].

The free field boundary elements consist of a 1D element ("column") to which dampers are attached. The 1D element has material properties identical to those in the main domain. The free field boundaries combine the free field motion with the absorption of incoming waves reflected from structures within the main domain. In the absence of structures, the same free field motion occurs in the main domain and in the boundary elements, and the dampers are not activated. In the free field boundaries, normal and shear stresses arise as reactions, which are calculated using the following expressions

$$\begin{aligned}\sigma_n &= -\rho\, V_p \left(\dot{u}_x^m - \dot{u}_x^{ff}\right)\\ \tau &= -\rho\, V_s \left(\dot{u}_y^m - \dot{u}_y^{ff}\right)\end{aligned}, \qquad (12)$$

where: ρ is the soil density, V_p and V_s – the compressional and shear wave velocity, respectively; $\dot{u}^m$ and $\dot{u}^{ff}$ – the particle velocities in the main domain and in the free field boundary, respectively.

The compliant base boundary [14] is a combination of a viscous boundary, which absorbs the incoming waves, and a boundary where the dynamic input is applied. Earthquake input in PLAXIS 2D is introduced as a prescribed displacement history, which is transferred into a load history.

The removal of the boundary fixities in the dynamic analysis induces the movement of these boundaries under action of in situ stresses, and the displacements of the boundaries are not equal to zero at the end of the time history loading (Figure 8-a). For zeroing the final displacements (Figure 8-b), a “drift correction” procedure is applied by defining a low-frequency function (polynomial or periodic), which is added to the time history function.

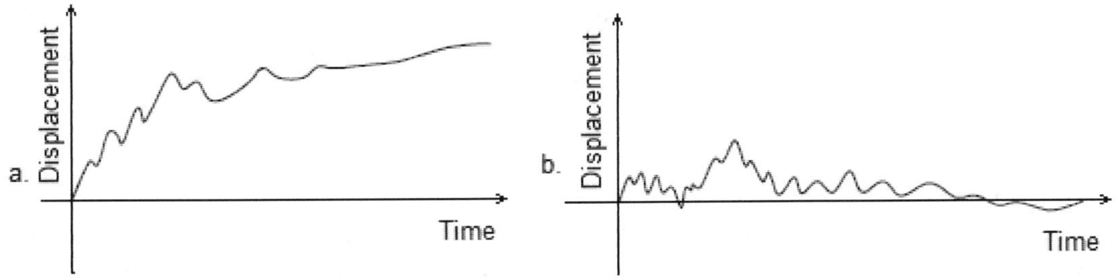

Figure 8. Displacement history: a. Before the "drift correction" procedure; b. After the "drift correction" procedure [13].

3.4 Derivation of design accelerograms

The use of information in the strong motion databases allows for conducting dynamic analyses with records of real earthquakes. Accelerograms are most commonly used. The selection of suitable accelerograms for analysis of retaining walls is carried out based on a set of criteria [15].

The frequency content of the strong motion is a key feature that influences the dynamic response of the ground and engineering structures. The frequency content is fully described by the Fourier Amplitude Spectrum and the Acceleration Response Spectrum. Several scalar parameters are defined based on these spectra [16], [17]. The most indicative and widely used frequency parameters are the mean period T_m, obtained from the Fourier Amplitude Spectrum, and the predominant period T_p, defined as the period at the maximum acceleration in the Acceleration Response Spectrum. The mean period T_m is defined for the range of discrete frequencies 0.25 Hz $\leq f_i \leq$ 20 Hz by the expression

$$T_m = \frac{\sum_i C_i^2 (1/f_i)}{\sum_i C_i^2} \quad \text{with} \quad \Delta f \leq 0.005\text{Hz}, \tag{13}$$

where C_i are the Fourier Amplitude coefficients and Δf is the frequency interval.

The comparison between the mean period of the strong ground motion T_m and the fundamental period of the soil deposit T_s enables the assessment of the risk of resonance or enhanced response of the soil deposits and adjacent structures. The fundamental period of the soil deposit T_s is determined by the formula $T_s = 4H / V_s$, where H is the thickness of the soil deposits and V_s is the shear wave velocity. In the case of equality between periods T_m and T_s, a state of resonance occurs that leads to substantial increasing in amplitudes of the ground motion. The closer the period T_m to the period T_s, the more the strong ground motion record is appropriate for the dynamic analysis.

Eurocode 8 (BDS EN 1998-5:2005) [3] introduces an additional criterion based on the assessment of the coincidence of the strong ground motion Response Spectrum and a target Response Spectrum. According to the prescriptions of Eurocode 8, a ground motion record is appropriate if its spectral accelerations $S_{a,SGM}$ exceed 90% of the target spectral accelerations $S_{a,target}$. This requirement needs to be fulfilled in the period interval ($0.2T_s \div 2T_s$). When the requirement is not satisfied, a procedure of spectral matching has to be performed. It can be performed alternatively in the time domain or in the frequency domain. The matching procedure

IOP Conf. Series: Materials Science and Engineering **1323** (2024) 012015 doi:10.1088/1757-899X/1323/1/012015

in the time domain does not change the frequency content of the motion. It consists in uniformly scaling of spectral accelerations of the motion $S_{a,SGM}$ up or down so a complete match has to be accomplished for the natural period of foundation soil T_s. The scale factor F_s is computed by the expression $F_s = S_{a,target} / S_{a,SGM}$ at $T = T_s$. The goal is to achieve $S_{a,SGM}$ values equal to the $S_{a,target}$ or maximum 10% smaller than the $S_{a,target}$. Some authors suggest the scale factor interval $1/2 < F_s < 2$ and accelerograms, which need a scale factor F_s out of this interval, are considered as inappropriate for the analysis [18].

The earthquake ground motions are usually recorded on the ground surface at free-field conditions. The direct use of such records in FE analyses, where the seismic impact is applied at the bottom boundary of the model, does not correct. In order to define an input accelerogram appropriate for the FE analysis, a numerical deconvolution procedure is required. The deconvolution consists in obtaining the accelerogram at a specified depth of the ground that reproduces the recorded free-field accelerogram at the ground surface when the seismic waves pass through the soil layers. The deconvolution can be carried out by two methods.

The first method is related to a frequency domain analysis (Figure 9-a). It can be carried out only when the constitutive model of the soil is linear elastic or equivalent linear elastic, i.e. the behaviour of the soil is not stress-path dependent [19]. The software SHAKE [20] for 1D seismic site response analysis is the most commonly used for performing this type of deconvolution. The target free field time history function (accelerogram) is presented by Fourier series of harmonic functions. Each term of the Fourier series is multiplied by an appropriate value of a transformation function. This function defines how the signal is modified passing through the ground layers from the terrain to a specific depth (base of the soil domain in the FE model). At last, an inverse Fourier transformation is performed to present the input motion at the base of the soil domain in the FE model as accelerogram.

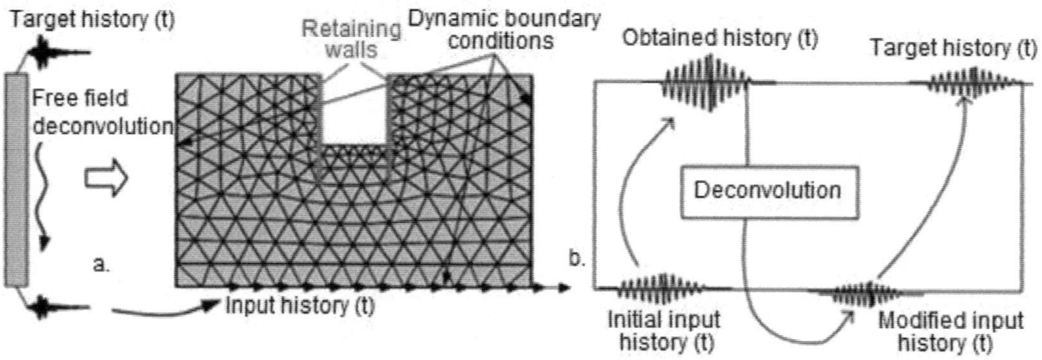

Figure 9. Deconvolution procedures: a. Frequency domain analysis; b. Time domain analysis. (using Figures in [21], [22])

The second method of deconvolution is related to a time domain analysis that takes into account nonlinear elastoplastic stress path dependent soil behaviour (Figure 9-b). The method consists of iterative procedure. Each iteration includes a convolution nonlinear analysis applying an initial motion at the base of the soil domain. From the analysis, a resultant motion at the ground surface is obtained. The Fourier Amplitude Spectra of derived and target motion at the ground surface are compared and correction factors for each frequency are calculated as ratios of corresponding Fourier amplitudes. After using the correction factors to the initial Fourier Amplitude Spectrum, a new Fourier Amplitude Spectrum at the base is obtained. Then an Inverse

Fast Fourier transform is performed in order to obtain the new time history function at the base. This function is applied at the base and a new convolution site response analysis is performed. The procedure is repeated until the required criterion for matching of derived and target motion at the ground surface is satisfied. The method of deconvolution based on matching the Fourier amplitudes of an input and a target motion is the most appropriate for determining an unfavourable input ground motion for analysis of earthquake response of retaining walls. The spectral matching technique allows to set a frequency range of interest and optimize the dynamic analysis applying a procedure of frequency filtration of the input motion. The latter is directly related to the finite element dimensions (Equation 5).

Specifying the target motion in terms of Response Spectrum is the widely used version of time domain deconvolution method [21], [22], [23]. In this case, correction factors are computed for the Response Spectrum ordinates in a range of frequencies. This approach usually is used when the engineering systems are designed according to a Response Spectrum given in codes or derived from microzonation of the construction site.

3.5 Constitutive modelling of the soil

The HSSmall constitutive model for soils [24], [25] developed in the software PLAXIS 2D is suitable for dynamic analyses. This model involves over 10 material parameters that are determined experimentally. The stiffness of the soil is described through the nonlinear Duncan-Chang stress-strain relationship [26] by defining the modules E_i, E_{50} and E_{ur} (Figure 10-a). The degradation of the soil stiffness due to hysteresis behaviour of the soil under cyclic loading is described by a dependence of the shear modulus on the shear strains (Figure 10-b). To generate this dependence in PLAXIS 2D, the initial shear modulus G_0 at shear strain $\gamma = 0$ and the shear strain $\gamma_{0.72}$ at reduced shear modulus $G = 0.72\ G_0$ are defined.

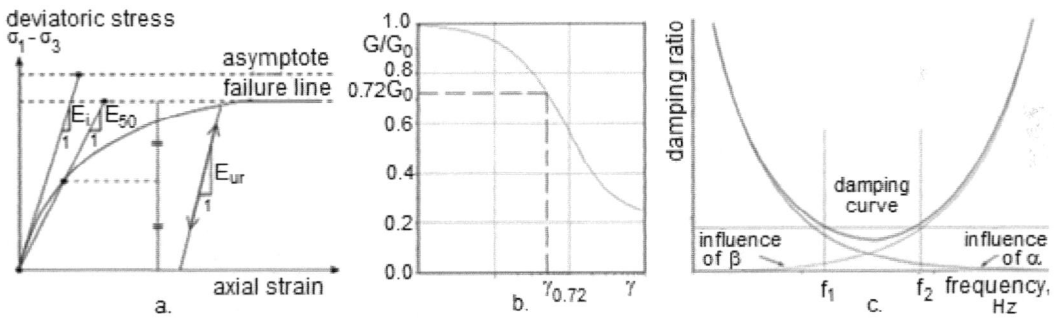

Figure 10. a. Stress-strain curve from soil drained triaxial test; b. Stiffness reduction curve; c. Influence of the Rayleigh damping coefficients [8].

The HSSmall model partially takes into account the soil attenuation at small shear strains γ (in the range $\gamma = 0 \div 10^{-5}$). Therefore, a viscous attenuation formulated by equation (2) according to Rayleigh theory is introduced. According to this theory, the damping ratio ξ is related to the frequency f by the following equation

$$\xi = \frac{\alpha_R}{4\pi f} + \beta_R \pi f \ , \tag{14}$$

IOP Conf. Series: Materials Science and Engineering **1323** (2024) 012015 doi:10.1088/1757-899X/1323/1/012015

where α_R and β_R are Rayleigh coefficients. They are computed after substitution in equation (14) two control frequencies f_1 and f_2, and a target damping ratio ξ. There are different approaches for derivation of the frequencies f_1 and f_2 using the fundamental frequency of the soil deposits f_s and the predominant frequency of the earthquake impact f_p in the Fourier Amplitude Spectrum [5], [8]. The Figure 10-c presents the damping curve. It shows that within the range of control frequencies ($f_1 \div f_2$) the damping ratio is less than the target damping ratio and, out of this range, the damping ratio is higher than the target damping ratio.

4. Case study

The presented methodology is applied to create a model for seismic analysis of a retaining wall. The wall was built in Sofia city, Geo Milev district. It has a total height of 6.5 m and an embedding depth of 1.5 m. The ground at construction site is built of Quaternary deposits with basic parameters shown in Table 1. The seismic impact is an accelerogram of the earthquake Loma Prieta'89 (LP EQ) choosed after applying selection criteria. A plane-strain FE model for analysis is created using the software PLAXIS 2D.

Table 1. Soil parameters.

Layer	thickness m	γ kN/m³	φ deg	c kPa	$E_{oed} = E_{50}$ MPa	E_{ur} MPa	G_0 MPa	$\gamma_{0.72}$ %
3 – Clayey sand	6.0	20.0	34	16	26	80	150	0.015
4 – Silty clay	10.0	18.5	21	27	20	60	120	0.050
4a – Clay	>15.0	21.0	20	35	30	90	130	0.040

4.1 FE modelling

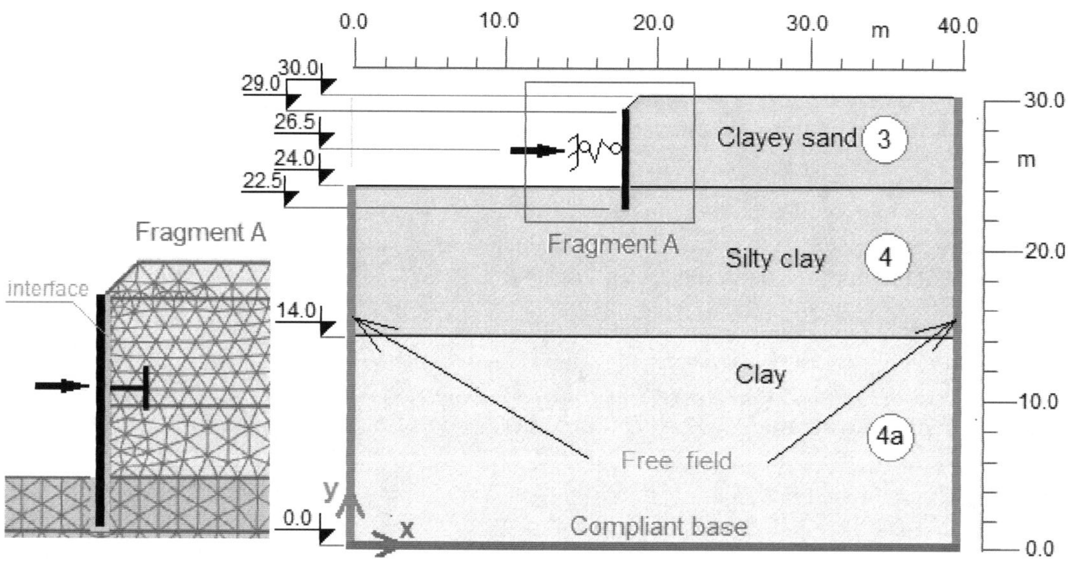

Figure 11. Plane-strain FE model.

The geometry and the boundary conditions of the FE model are shown in Figure 11. A soil domain with dimensions of 40 m /30 m is considered. The wall is braced by a strut, which is simulated by a spring support.

The soil domain is discretized using 15-node triangular elements (Figure 1-c) and the wall is discretized using 9-node linear beam elements (Figure 2-c). An elasto-plastic interface element defined by Equations (9) is applied for the modelling of retaining wall-soil interaction. The value 0.5 is accepted for the reduction factor R_{int}.

The mechanical behaviour of the soil is described by the constitutive model HSSmall, The shear strains $\gamma_{0.72}$ of the soil layers corresponding to $0.72G_0$ (72% reduction of the initial shear modulus) are shown in Table 1 and they are determined by the empirical correlations of Vucetić and Dobri [27]. The reduction curves ($G/G_0 - \gamma$) of the soil layers are presented in Figure 12, where by G_t and G_s are denoted tangent shear modulus and secant shear modulus, respectively.

The Rayleigh coefficients α_R and β_R are specified according to the procedure proposed in the Reference Manual [8]. The target frequencies f_1 and f_2 are calculated using both the fundamental frequency of the soil deposits f_s = 1.35 Hz and the predominant frequency in the Fourier Amplitude Spectrum of the seismic impact f_p = 1.56 Hz (see items 4.2 and 4.3).

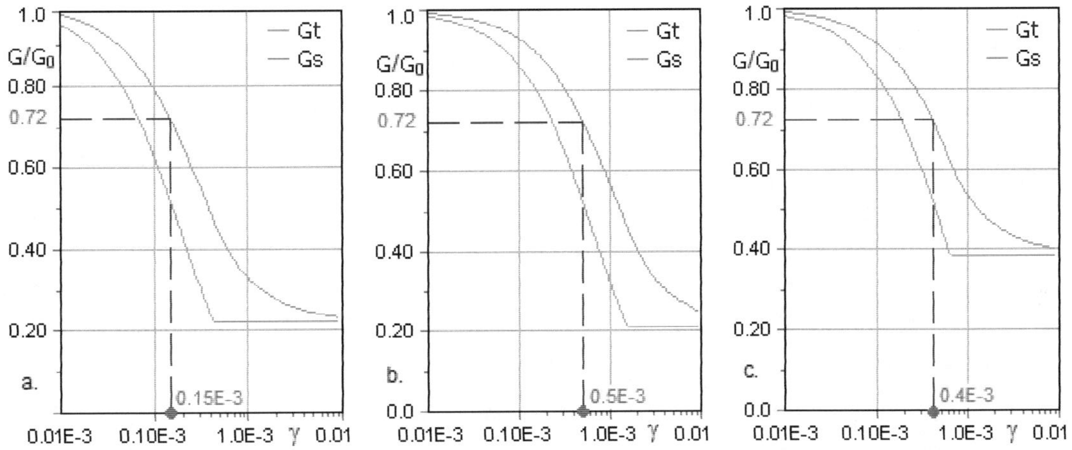

Figure 12. *G*-modulus reduction curves of the soil layers: a. Layer 3; b. Layer 4; c. Layer 4a.

4.2 Frequency characteristics of the soil deposits

The location of the construction site is shown in the map of Sofia city in Figure 13, where the isolines of the ground fundamental period up to 30 m depth is presented. The map was created on the basis of seismic microzoning carried out in 2001 [28]. According to this map, the ground fundamental period at the construction site is T_s = 0.75 s, therefore, the ground fundamental frequency is f_s =1.35 Hz. These values were confirmed by in situ microtremor measurements and an application of the HVSR (Horizontal-to-Vertical Spectral Ratio) technique [29].

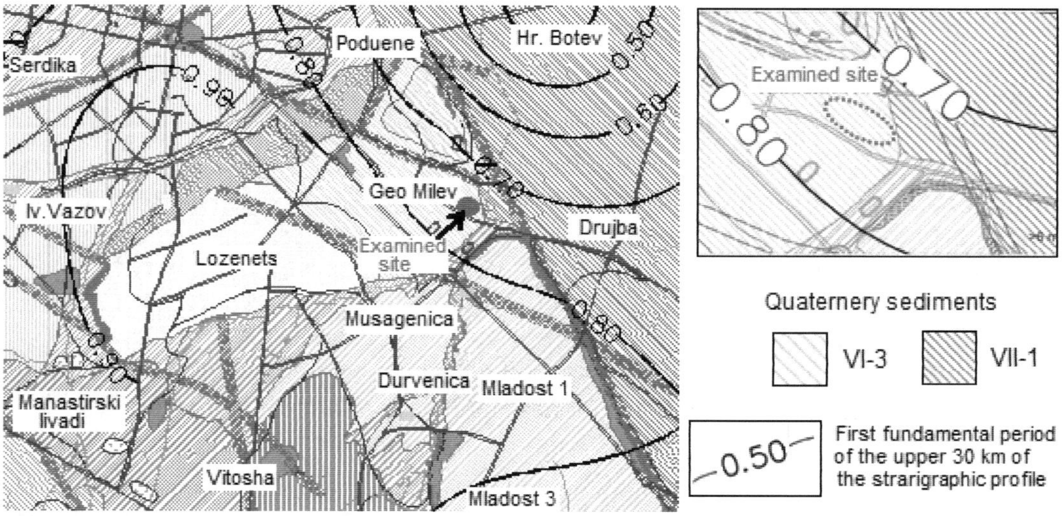

Figure 13. Map of Sofia city: The fundamental period of the soil deposits up to 30 m depth [28].

4.3 Application of the accelerogram selection criteria

The appropriate accelerogram selection is made using the set of accelerograms in a public database of strong ground motion records.

At first, the earthquake scenario is defined by the magnitude M and the focal depth d. In the present study M > 6 and d < 35 km are assumed. As a second step, the frequency characteristics of the records are considered. The Loma Prieta'89 earthquake (LP EQ) has been selected because it has an average period T_m = 0.638 s, which is very closed to the fundamental period T_s of the soil deposits up to 30 m depth at the examined site. The difference between T_m and T_s is 17%.

Some basic parameters of the Loma Prieta'89 earthquake as magnitude M, focal depth d, horizontal peak ground acceleration PHA, predominant period T_p, mean period T_m, and significant duration t_s are shown in Table 2.

Table 2. Parameters of the Loma Prieta'89 earthquake.

M -	d km	PHA g	T_p s	T_m s	t_s s
6.9	19	0.13	0.38	0.638	24.02

The next step in the strong ground motion evaluating is the comparison between Response Spectrum (RS) of the strong ground motion and a target Response Spectrum. In this study, as a target spectrum is assumed the spectrum defined in the National Annex of Eurocode 8 (BDS EN 1998-5:2005/NA:2012/Ed.1:2015) [3] for ground type C and importance class II. The spectra are shown in Figure 14-a. It can be noted that the difference between two spectra exceeds the tolerance of 10%. Therefore, a simple scaling is applied to the Loma Prieta'89 earthquake accelerations using a scale factor 1.63, so that the equality of spectral accelerations at T_s is achieved (Figure 14-b).

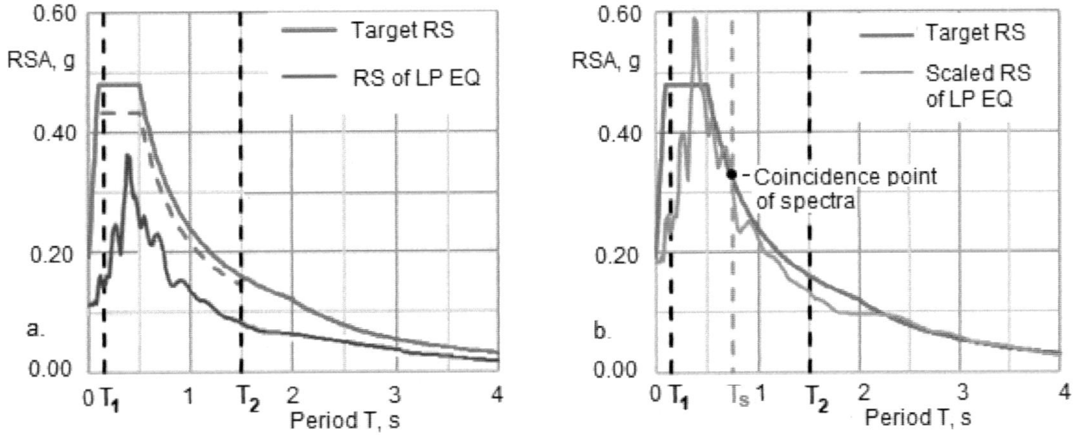

Figure 14. Comparison between target RS and: a. RS of the Loma Prieta'89 earthquake; b. Scaled RS of the Loma Prieta'89 earthquake.

4.4 Accelerogram deconvolution

The deconvolution procedure with an accelerogram of the Loma Prieta'89 earthquake is performed with the help of the software SHAKE [20]. The original accelerogram and its Fourier Spectrum are compared with those obtained after deconvolved procedure (Figure 15). The results show that when the seismic waves reach the ground surface, passing through the soil deposits 30 m thick, the seismic signal significantly amplifies. The maximum amplitude of acceleration amplifies 2 times (Figure 16-a). The highest amplification factor of the Fourier Amplitude Spectrum is 7.7 and it is recorded at a frequency of 1.6 Hz ~$1\ /\ T_m$ = 1.57 Hz (Figure 16-b), i.e. the amplification of the Fourier Amplitude Spectrum is greatest at the frequency closest in value to the natural frequency of the soil deposits.

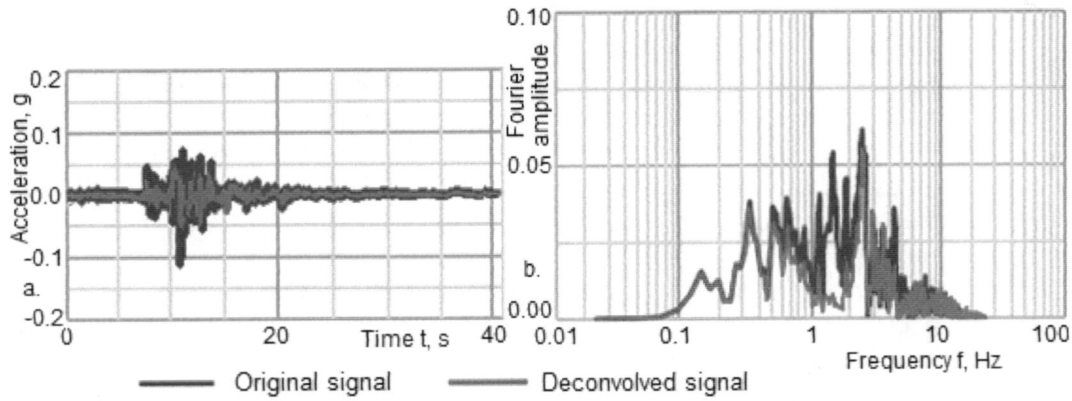

Figure 15. The Loma Prieta'89 earthquake: a. Accelerograms of the free-field signal and the deconvolved signal to 30 m depth; b. Fourier Spectra of the original free-field signal and the deconvolved signal to 30 m depth.

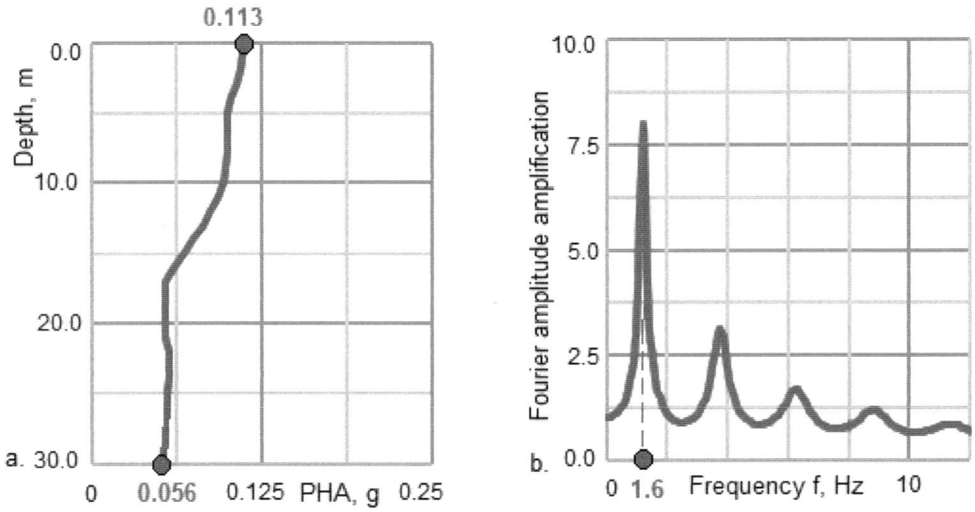

Figure 16. The Loma Prieta'89 earthquake: a. PHA varying through the soil deposits; b. Amplification factor of the Fourier Spectrum.

5. Conclusion

The performance of FE time history seismic analysis of a wall-ground system requires detailed knowledge about the soil dynamic behaviour and the interaction mechanism of the soil and structural elements, as well as the softwares capabilities for modelling. The software PLAXIS 2D uses the advanced multiparametric constitutive model HSSmall, which describes the complex elasto-plastic behaviour of the soil and its stiffness degradation under dynamic impact. The different types of finite elements developed in the software in terms of dimensionality, nodal parameters and degree of approximation enable alternative approaches for creation of numerical models. The introduction of dynamic boundaries (compliant base boundary and free field side boundaries) of the soil domain in the FE model allows simulating the infinite soil half-space. The time history function (accelerogram) for dynamic analysis is applied at the base of the FE model. For derivation of input function from a strong ground motion record made on the ground surface, a deconvolution procedure is carried out. The discretization of the soil domain and the time domain in FE models must obey mathematical criteria ensuring the convergence and accuracy of the numerical solution.

The presented methodology is illustrated by the selection of parameters for FE seismic analysis of a retaining wall. The ground of the construction site consists of Quaternary deposits with fundamental period of T_s = 0.75 s. The earthquake impact is set by the Loma Prieta'89 earthquake accelerogram characterized with a peak acceleration of PHA = 0.113 g and a mean period of the Fourier Amplitude Spectrum T_m = 0.638 s. The choice of accelerogram is based on the fulfillment of criteria for unfavourable seismic impact. The deconvolution procedure performed in order to obtain the design accelerogram in the FE model confirms that when the seismic waves pass through soil deposits, they are amplified.

IOP Conf. Series: Materials Science and Engineering **1323** (2024) 012015 doi:10.1088/1757-899X/1323/1/012015

References

[1] Okabe S 1926 General Theory of Earth Pressures *Journal of Japan Society of Civil Engineers* **12** 1 pp 123-134
[2] Mononobe N and Matsuo M 1929 On the Determination of Earth Pressures during Earthquakes *Proc. World Engineering Congress* (Tokyo) **9** 388 pp 179-187
[3] EN 1998 2003 Eurocode 8: Design of Structures for Earthquake Resistance, Part 5: Foundations, Retaining Structures and Geotechnical Aspects (CEN European Committee for Standardization Bruxelles Belgium)
[4] Amorosi A, Boldini D and Elia G 2010 Parametric Study on Seismic Ground Response by Finite Element Modeling *Computers and Geotechnics* **37** 4 pp 515-528
[5] Chiaradonna A 2022 Defining the Boundary Conditions for Seismic Response Analysis – A Practical Review of Some Widely-Used Codes *Geosciences* **12** 83
[6] PLAXIS 2D Scientific Manual 2018
[7] PLAXIS 2D Material Model Manual 2018
[8] PLAXIS 2D Reference Manual 2018
[9] Kuhlemeyer L and Lysmer J 1973 Finite Element Method Accuracy for Wave Propagation Problems *Journal of the Soil Mechanics and Foundations Division* **99** 5 pp 421-427
[10] Goodman R E, Taylor R L and Brekke T L 1968 A Model for the Mechanics of Jointed Rock *Journal of Soil Mechanics and Foundations Division* **94** 3 pp 637-659
[11] Sluis J 2012 *Validation of embedded pile row in Plaxis 2D* Master's thesis
[12] Lysmer J and Kuhlemeyer R 1969 Finite Dynamic Model for Infinite Media *Journal of the Engineering Mechanics Division* **95** 4 pp 859-877
[13] FLAC2D/FLAC3D Itasca Software 2024 *Dynamic Modeling Considerations*
[14] Joyner W B, Chen A T F 1975 Calculation of Nonlinear ground Response in Earthquake *Bulletin of Seismological Society of America* **65** pp 1315-1336
[15] Topalska M and Kouteva-Guentcheva M 2024 Design Accelerograms Derivation for Dynamic Analysis of Retaining Walls *Proc of the 3th International Conference on Environmental Protection and Disaster Risks and 12th Annual CMDR COE Conf on Crisis Management and Disaster Response* (Sofia)
[16] Rathje E M et al 1998 Simplified Frequency Content Estimates of Earthquake Ground Motions *Journal of the Geotechnical Engineering Division* **124** 2 pp 150-159
[17] Rathje E M et al 2004 Empirical Relationships for Frequency Content Parameters of Earthquake Ground Motions *Earthquake Spectra* **20** 1 pp 110-144
[18] Genovese F et al 2019 A Procedure for the Selection of Input Ground Motion for 1D Seismic Response Analysis *Associazione Geotecnica Italiana* pp 2591-2598
[19] Reimer R B 1973 Deconvolution of seismic response for linear systems *Tech. Rep. UCB/EERC*-**73/10** (Earthquake Engineering Research Center University of California Berkeley)
[20] Ordonez G A 2012 *Shake 2000 A Computer Program for the 1-D Analysis of Geotechnical Earthquake Engineering Problems User`s Manual* (GeoMotions LLC Lacey Washington USA)
[21] Manica A M et al 2023 Time Domain Deconvolution in Nonlinear Elastoplastic Soil Deposits *Soil Dynamic and Earthquake Engineering* **172** 107993
[22] Sooch G S and Bagchi 2012 An Effect of Seismic Wave Scattering on the Response of Dam-Reservoir-Foundation Systems *15th World Conference on Earthquake Engineering* (Lisbon Portugal)
[23] Pinzon L A et al 2023 Time-Domain Deconvolution Procedure for Elastoplastic Materials: Application to the Treasure Island Site during the 1989 Loma Prieta Earthquake *MethodsX* **11** 102386
[24] Schanz T, Vermeer P A and Bonnier P G 1999 The Hardening-Soil Model: Formulation and Verification (In R B J Brinkgreve *Beyond 2000 in Computational Geotechnics*) (Balkema Rotterdam) pp 281-290
[25] Benz T 2006 *Small-Strain Stiffness of Soil and its Numerical Consequences* PhD thesis (Institut für Geotechnik der Universität Stuttgart)
[26] Duncan J M and Chang C-Y 1970 Nonlinear Analysis of Stress and Strain in Soil *Journal of the Soil Mechanics and Foundations Division* **SM5** pp 1629-1653
[27] Vucetic M and Dorby R 1991 Effect of Soil Plasticity on Cyclic Responses *Journal of geotechnical engineering* **117** GT1 pp 89-107
[28] Paskaleva et al 2001 Seismic microzoning of Sofia city
[29] Popova M et al 2021 Experimental Evaluation of the Dynamic Parameters f_0 and A0 using the Method HVSR *11th Congress of the Balkan Geophysical Society* (Bucharest)

IOP Conf. Series: Materials Science and Engineering **1323** (2024) 012016 doi:10.1088/1757-899X/1323/1/012016

Seismic design of an existing six story administrative building

Emad Abdulahad [1] and Aleks Grancharov [2]

[1] Department of Reinforced Concrete Structures, Faculty of Structural Engineering, University of Architecture, Civil Engineering and Geodesy (UACEG), Sofia, Bulgaria.
[2] University of Architecture, Civil Engineering and Geodesy (UACEG), Sofia, Bulgaria.

*E-mail: alex_gr@abv.bg

Abstract. The occurrence of inadmissible limit states is essentially a phenomenon that may appear during the construction or opreration of a given structure and manifests as a defect, failure or accident. The restoration of the bearing capacity is related to the future operation and the category of the building according to the current regulations and is an integral part of the overall process of seismic design of the structures. This report is related to a public six-story building with two underground levels, calculated according to the Bulgarian national regulations. The bearing structure is reinforced concrete, monolithic, the floors are flat slab and the seismic resistance design consists of three reinforced concrete cores in the center of the structure and reinforced concrete walls along the stairwell. A visual and instrumental inspection documents that the defects and damages shown are of an unacceptable nature. Based on the conclusion of the structural assessment and the nature of the mentioned failures, measures of intervention should be designed.

1. Introduction

The construction of the building has began in 1993, but after the completion of the foundation, it is stopped for 3 years. Subsequently, the construction is resumed, and the structure is completed in 1997. The bearing system for horizontal loads is classified as a "shear wall system" and consists of three reinforced concrete cores in the center of the structure and reinforced concrete shear walls around the staircase, as shown in Figure 1. The slabs are 25 cm thick. The floor height is 3.6 meters. In the basement of the building, there is a bomb shelter with a slab 70 cm thick at an elevation of -2.35 meters and walls 50 cm thick. It is located from axis 1 to 7. From axis 7 to 10, there is an underground floor used as a technical room, which is located on a reinforced concrete slab at an elevation of -3.50 meters on backfill. The mat foundation is 60 cm thick at an elevation of -7.00 meters.

In 2013, the building suffered damage in the form of a punching failure of a column at axes 5 and G. A strenghtening project was implemented, which did not take into account the seismic resistance of the building and involved the addition of reinforced concrete walls along axes 4, 5, 6, and 7 between axes V and G (Figure 2). The columns along axis G at axes 4, 5, 6, and 7 are strenghtened with a reinforced concrete jacket. Steel beams were placed in the spans between the new reinforced concrete walls to unload the columns and limit the deflection of the slab.

Reversed reinforced concrete beams were constructed along the facade at axis G between axes 4, 5, 6, and 7. A reinforced concrete jacket is applied to the basement walls.

The most recent inspection of the building, conducted in 2023,has documented extremely low concrete strength characteristics, as well as numerous defects and damages. The concrete strength is found to correspond to class C12/15, and the reinforcement was class AIII(fy=410MPa). The slabs have unacceptable deflections, which have reached up to 7 cm in almost all spans between longitudinal axes A and B. Deficiencies are discovered in the execution of the 2013 project, including poor connections between the slabs and the newly constructed basement walls. The ground under the foundation has undergone differential settlement, reaching 15 cm along axis 10 and axis G.See the separate guidelines document for further information.

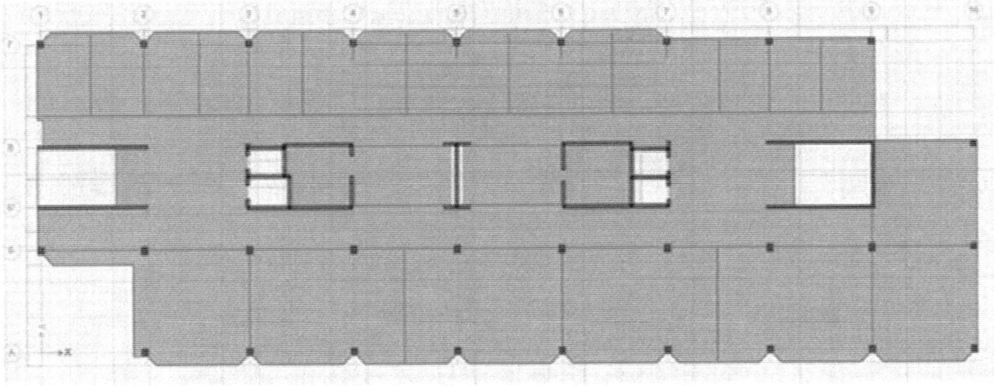

Figure 1. Layout of the vertical load-bearing structure in the plan before 2013

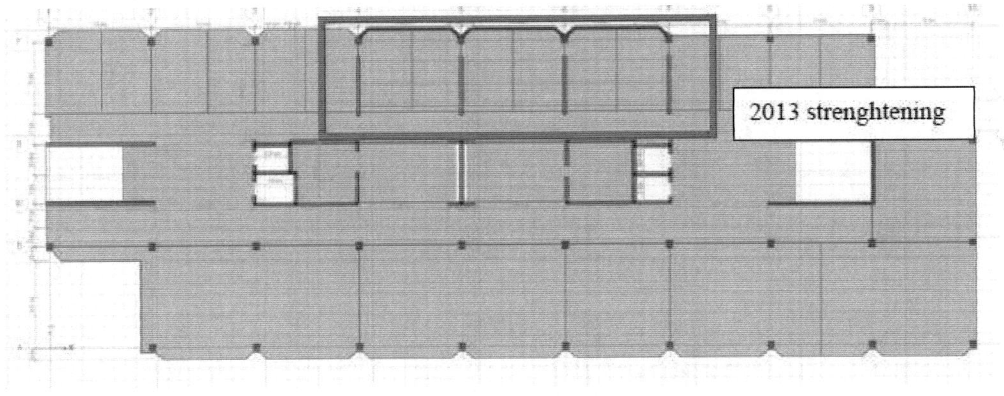

Figure 2. Layout of the vertical load-bearing structure in the plan after 2013

2. Initial analysis of the existing structure

Initially, an analysis of the structure in its current state was conducted using the spectral analysis method [1][2][9][10][13], with the strength characteristics of the materials from the 2023 inspection and additional detailed structural survey [15]. The results indicate that the building is torsionally flexible (Fig. 5). The building is irregular and has significant eccentricities between the centers of mass and the centers of rigidity at various floor levels, particularly at the

IOP Conf. Series: Materials Science and Engineering **1323** (2024) 012016 doi:10.1088/1757-899X/1323/1/012016

roof and the slab at elevation -2.35 meters (Table 1). The inter-story displacements (Fig. 4) are unacceptable, and the forces in the seismic walls exceed their load-bearing capacity. Many of the elements do not meet the requirements for local ductility.

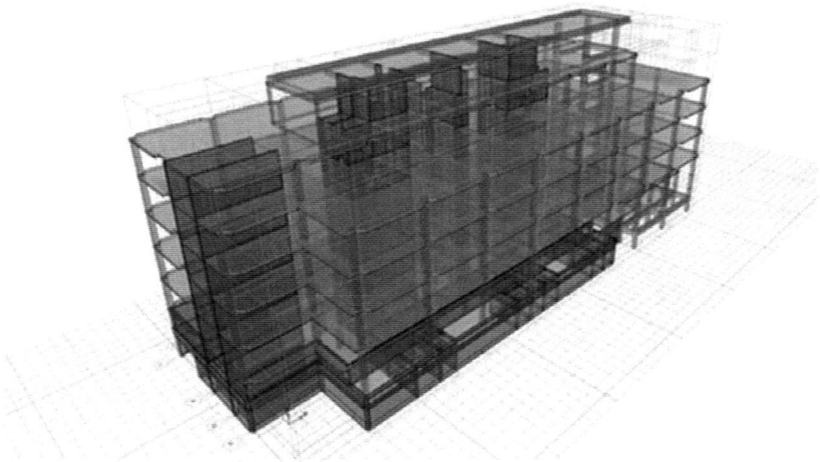

Figure 3. Three-dimensional model of the existing building.

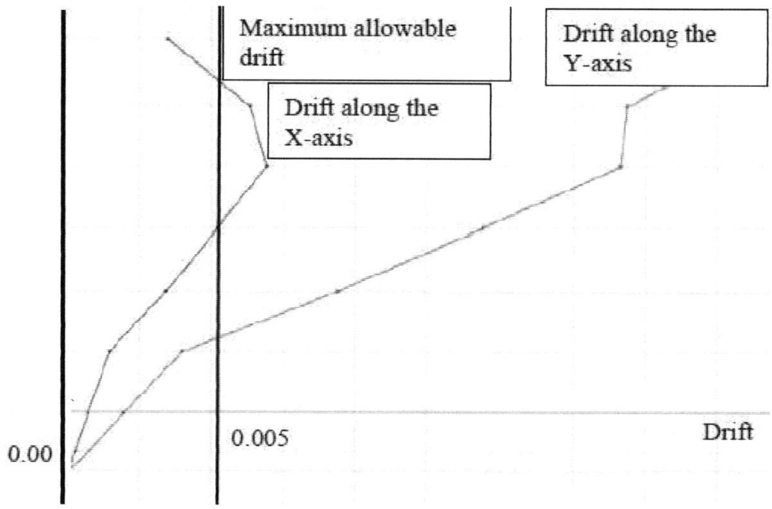

Figure 4. Inter-story displacements (drifts)

IOP Conf. Series: Materials Science and Engineering **1323** (2024) 012016 doi:10.1088/1757-899X/1323/1/012016

Table 1. Eccentricities between the center of mass and the center of rigidity

	X (m)	Y (m)	X (%)	Y (%)
Roof	4.867	0.836	7%	4%
+17,85	0.140	1.630	0%	7%
+14,3	1.956	-1.074	3%	-5%
+10,7	2.007	-1.467	3%	-7%
+3,5	1.792	-1.277	3%	-6%
-0,10	2.049	-1.445	3%	-7%
-2,35	7.080	0.290	11%	1%

The large eccentricity at the roof and the slab at elevation +17.85 meters is due to the reduction of the dimensions of the last two floors. The eccentricity at elevation -2.35 meters is caused by the presence of a bomb shelter with a 70 cm thick slab between axes 1 and 7, while a reinforced concrete slab on backfilling is present between axis 7 and axis 10 at elevation -3.50 meters.

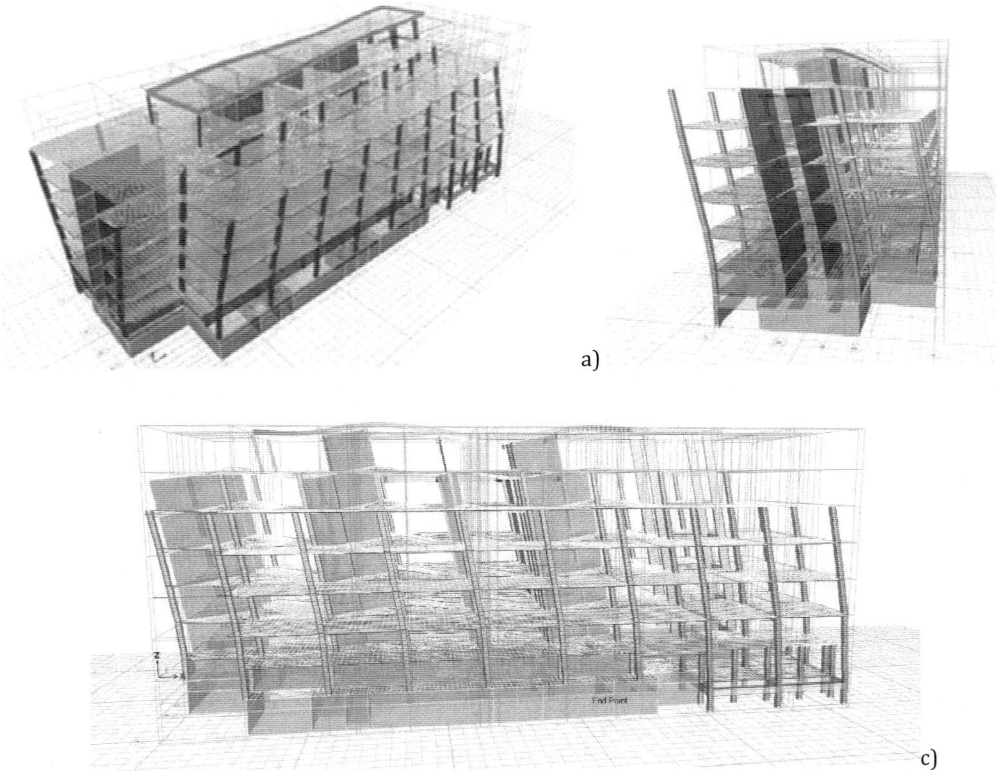

Figure 5. . Natural frequencies of initial structure a)First form T=1.136s; b)Second form T=0.742s; c)Third form T=0.563s

3. Seismic strengthening of the building

A structural survey and additional investigation of the load-bearing structure are conducted to achieve the necessary completeness of information regarding geometry, details, and materials. According to the requirements of Eurocode 8.3 [9], section 3.3 - Level of Information and Table 3.3 (Fig. 6), a confidence factor CF_{KL3} = 1.0 is determined.

Knowledge Level	Geometry	Details	Materials	Analysis	CF
KL1	From original outline construction drawings with sample **visual** survey ***or*** from **full** survey	Simulated design in accordance with relevant practice ***and*** from **limited** *in-situ* inspection	Default values in accordance with standards of the time of construction ***and*** from **limited** *in-situ* testing	LF- MRS	CF_{KL1}
KL2		From incomplete original detailed construction drawings with **limited** *in-situ* inspection ***or*** from **extended** *in-situ* inspection	From original design specifications with **limited** *in-situ* testing ***or*** from **extended** *in-situ* testing	All	CF_{KL2}
KL3		From original detailed construction drawings with **limited** *in-situ* inspection ***or*** from **comprehensive** *in-situ* inspection	From original test reports with **limited** *in-situ* testing ***or*** from **comprehensive** *in-situ* testing	All	CF_{KL3}

Figure 6. Table 3.1 from section 3.3 of Eurocode 8.3 [9]

The non-compliance of the structure with the Eurocode regulations, specifically the lack of conditions for local ductility, necessitates its evaluation during the elastic stage when strengthening. This, combined with the high importance class of the building (Category III), implies an additional increase in forces, and thus a need for larger dimensions of the elements bearing seismic loads. The seismic assessment and strengthening program is categorized as active according to Eurocode 8.3 [9].

The second solution involves both strengthening the existing elements and adding new reinforced concrete elements [7][9][10][14] to improve the structure's rotation performance and reduce the eccentricity between the centers of mass and rigidity. A connection has been realized between the slabs at elevations -2.35 and -3.50 meters, which is possible because of the large dimensions of the bomb shelter elements, which improves the eccentricity at elevation -2.35 meters. However, the forces in the individual reinforced concrete walls remain high, or those in the newly added walls are excessively large to meet the load-bearing capacity requirements. The building's architecture, including the need for natural light on much of the facades, makes it impossible to place sufficient reinforced concrete shear walls. Addressing the issue by constructing a frame system is ineffective due to their lower rigidity compared to reinforced concrete shear walls. Furthermore, frames are effective mostly at high levels of ductility [5][6], whereas the specifics of the building do not permit their use.

4. Final solution for seismic strengthening of the existing building

To preserve the architectural requirements and ensure the proper functioning of the building's spaces, the decision is made to install additional composite steel and reinforced concrete braces [8][9][11][12][14][16] along axes 1 and 10 externally to the facade, with connections to the existing structure achieved through chemical anchors. The columns and beams are composite steel and reinforced concrete, while the diagonals are steel.

The combination of the existing rigid reinforced concrete walls with the braces creates a high degree of static indeterminacy. In the case of plastic deformation of the steel diagonals, the redistribution of forces relies on the reinforced concrete cores and shear walls, rather than solely on the frame action of the braces themselves. Combining steel and reinforced concrete in the columns and beams limits the loss of stability in the steel and allows for the absorption of large tensile forces.

On the long facades (axes A and G), steel braces have been installed in three spans (between axes 2 and 3, 5 and 6, and 8 and 9) to cover two floors [4][8][9][11][14][16]. This aims to limit deformations and forces in the longitudinal direction and reduce the forces in the existing elements. Several reinforced concrete walls have also been designed inside the building (Fig. 7) – around the staircase between axes 8 and 9, along axis 6 between axes A and B, and along axis B between axes 3 and 4, 5 and 6, and 9 and 10, to bring the centers of mass and rigidity as closer as possible.

Existing elements have been strengthened with a reinforced concrete jacket [2][3][7][9][10] to achieve the necessary load-bearing capacity and to repair the damages and defects. This significantly improves the building's behavior, with the first two modes being translational and the third being rotational (Fig. 8, Fig. 9). The checks for plan regularity and inter-story slab displacement have been satisfied. The new seismic-resistant elements and the reinforced concrete jackets for the old elements have been designed. A decision was made to strengthen the slabs by implementing a steel beam grid [11][12]. The foundation has been reinforced with micropiles [17] and an additional layer of concrete on the slab to achieve the required load-bearing capacity [7][10].

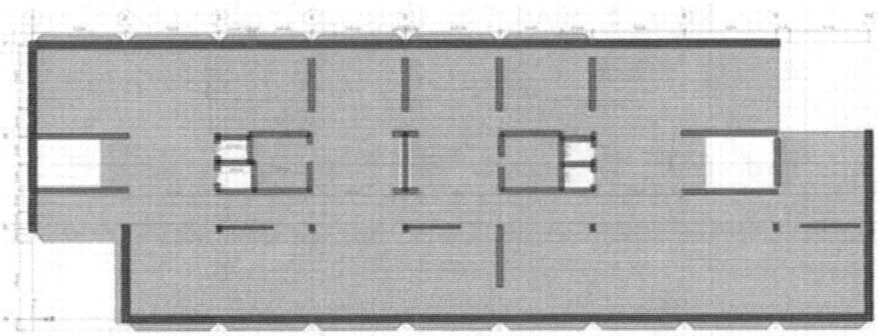

Figure 7. Plan of the vertical elements of the reinforced structure

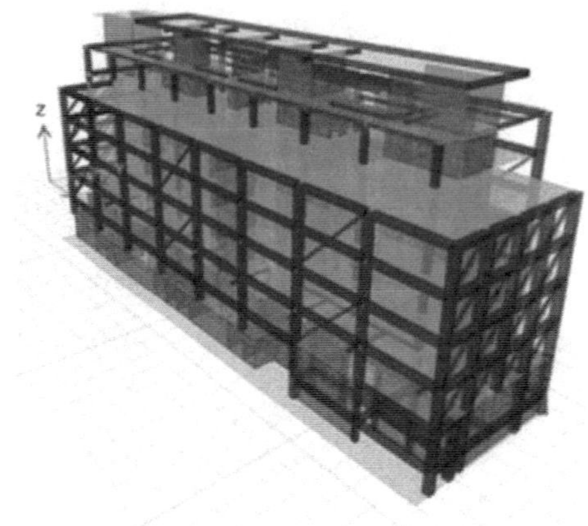

Figure 8. Three-dimensional structural model of the reinforced structure

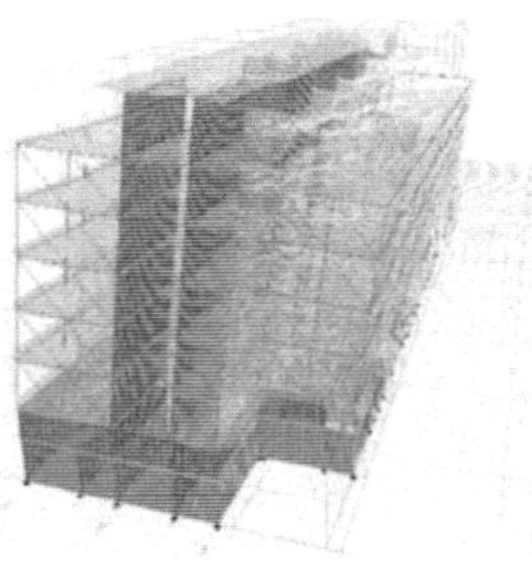

a)

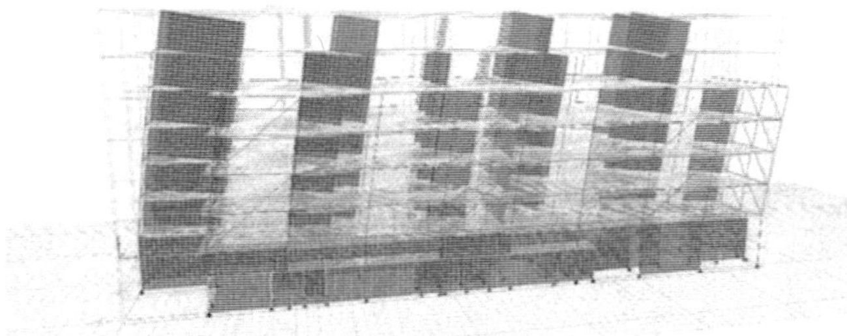

b)

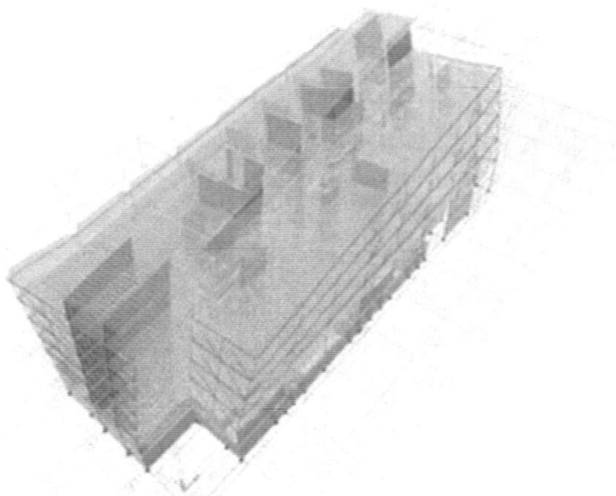

c)

Figure 9.. Natural frequences of strengthened structure a)First form T=0,602s; b)Second form T=0,494s; c)Third form T=0.459s

IOP Conf. Series: Materials Science and Engineering **1323** (2024) 012016 doi:10.1088/1757-899X/1323/1/012016

5. Conclusion

The choice of a behavior factor of 1.5 aims to meet the requirements of Eurocode 8.3 due to the lack of local ductility.

From an economic perspective, to ensure the existing building's seismic resilience, additional detailed investigations and comprehensive structural survey are required to obtain complete information and achieve a confidence factor $CF_{KL} = 1.0$.

In some cases, depending on the building's configuration and architecture, strengthening the existing reinforced concrete walls and adding new ones does not improve the building's performance. For the specific case presented in the report, the forces in the existing reinforced concrete walls and some of the newly added walls are substantial, and the chosen solutions are economically and architecturally unfeasible.

The installation of braces on the facade fully satisfies the checks according to Eurocode, reduces the forces in the existing elements, and improves the building's performance, with the first two modes of behavior being translational.

References

[1] Gancheva, R. Design of Reinforced Concrete Structures for Seismic Actions, 2013
[2] Neshev, H. Design of Reinforced Concrete Wall Structures Subjected to Seismic Action for Intermediate Ductility Level (DCM) According to Eurocode, 2022
[3] Kardzhiev, V., Neshev, H. Methodology for Creating an Iterative Diagram for Sizing Rectangular Symmetric Reinforced Concrete Sections Using Microsoft Office Excel, 2016
[4] Georgiev, Ts., Raicheva, L. Two-Story Vertical Connections with Crossed Diagonals Intersecting at a Floor Beam. Design Considerations for Seismic Actions. Annual of the university of architecture, civil engineering and geodesy ISSN 2534-9759
[5] Kardzhiev, V., Georgiev, E. Specific requirements in design of beams from reinforced concrete frame structures for seismic action at high level of total ductility (DCH). Annual of the University of Architecture, Civil Engineering and Geodesy, Vol. 56, Issue 3 - International Jubilee Scientific Conference "80th Anniversary of UACEG", UACEG, Sofia, 2023, ISSN 2534-9759, p. 1219÷1230, 1828/ p.
[6] Kardzhiev, V., Georgiev, E. Specific requirements in design of columns from reinforced concrete frame structures for seismic action at high level of total ductility (DCH). Jubilee Scientific Conference "80th Anniversary of UACEG", UACEG, Sofia, 2023, ISSN 2534-9759, p. 1231÷1242, 1828/ p.
[7] Traykova M., Rehabilitation and Strengthening of Buildings- problems, structural solutions, practical recommendations, 2014
[8] Elghazouli, A. Seismic design procedures for concentrically braced frames. Proceedings of the Institution of Civil Engineers Structures&Buildings 156 Issue SB4, Paper 13066, 2002.
[9] BDS EN 1990 – Basis of structural design
[10] BDS EN 1992-1-1 – Design of concrete structures: General rules and rules for buildings
[11] BDS EN 1993-1-1 – Design of steel structures - General rules and rules for buildings
[12] BDS EN 1994-1-1 – Design of composite steel and concrete structures - General rules and rules for buildings
[13] BDS EN 1998-1 – Design of structures for earthquake resistance - General rules, seismic actions and rules for buildings
[14] BDS EN 1998-3 - Design of structures for earthquake resistance – Assessment and retrofitting of buildings
[15] Panichkov, D. Inspection and Testing of Structural Components and Facilities
[16] Chen, C., Mahin, A. Performance-Based Seismic Demand Assessment of Concentrically Braced Steel Frame Buildings. PEER Report 2012/103, Pacific Earthquake Engineering Research Center, Headquarters at the University of California, Berkeley, 2012
[17] BDS EN 1997-1 – Geotechnical design – General rules

IOP Conf. Series: Materials Science and Engineering **1323** (2024) 012017 doi:10.1088/1757-899X/1323/1/012017

Dynamic behavior of soil-tunnel interaction system in transversely isotropic half-space

Sonia Parvanova[1*], Amar Pashov[2] and Georgi Vasilev[1]

[1] Department of Structural Engineering, University of Architecture, Civil Engineering and Geodesy, Sofia, Bulgaria
[2] KU Leuven, Department of Civil Engineering, Kasteelpark Arenberg 40, B-3001 Leuven, Belgium

*E-mail: slp_fce@uacg.bg

Abstract. Dynamic behavior of lined tunnels buried in a transversely isotropic half-space with surface relief is the focus of this paper. Different computational techniques are employed to model the soil-structure system subjected to time-harmonic point loads. Firstly, the direct Boundary Element Method is employed, using frequency-dependent fundamental solution of the governing equation in two-dimensional anisotropic elastodynamics to model the underground structure in a plane-strain state. In addition, the built-in options of the commercial software package ANSYS are utilized to compute the three-dimensional wave-field response. As an alternative to the 3D models, a computationally efficient two-and-a-half-dimensional approach has been developed for a transversely isotropic medium. It is based on the Finite Element Method combined with Perfectly Matched Layers. Comparisons between the results obtained by the three approaches are presented, and some recommendations regarding their application are provided.

1. Introduction

The current study focuses on the dynamic soil-structure interaction of lined tunnels buried in a nonhomogeneous transversely isotropic (TI) half-space with surface relief. Transverse isotropy accounts for the stiffness differences in horizontal and vertical directions, due to the long-term geological sedimentation processes. Numerous numerical techniques have been developed for modeling soil-tunnel interaction problems. We only relate to the applications of the Boundary Element Method (BEM) and the Finite Element Method (FEM) combined with Perfectly Matched Layers (PMLs) to tunnel structures mainly in anisotropic media.

The BEM is an attractive technique for modeling wave propagation in semi-infinite media because its formulation via boundary integral equations is equivalent to the original Boundary-Value Problem (BVP), described by partial differential equations. The analytically derived fundamental solution for an anisotropic continuum ensures a high level of accuracy by satisfying Sommerfeld's radiation condition in infinity. Three-dimensional (3D) and two-dimensional (2D) time-domain elastodynamic fundamental solutions (Green's functions) for linearly elastic anisotropic materials are initially obtained by the Radon transform in Wang and Achenbach [1]. An excellent work on the application of BEM in 2D anisotropic electrodynamics in the frequency domain could be found in Garcia-Sanchez [2]. Ba *et* al. [3] investigated the scattering and diffraction of plane qP- and qSV-waves by a canyon in a multi-layered transversely isotropic (TI) half-space in both frequency and time domains. More recently, Ba et al. [4,5] investigated the dynamic impedance of rigid foundations embedded in anisotropic layered soil using indirect BEM. Parvanova and Dineva [6] addressed a 2D dynamic problem in frequency and time domains for a flexible foundation resting in an elastic anisotropic half-plane with arbitrary geometry using direct BEM.

The classical 2D in-plane and out-of-plane strain states are applicable to numerous problems related to invariant structures, but they also have several limitations. For example, they cannot cover 3D wave propagation in case of moving loads. For structures such as tunnels, which exhibit spatial invariance and infinite length along one direction, a two-and-a-half dimensional (2.5D) approach can be employed. In this approach, the 3D wave field in the soil-tunnel system is represented through 2D discretization in the wavenumber-frequency domain. The solution involves computing a sequence of 2D problems across a range of spatial wavenumbers. Subsequently, through inverse Fourier transform the 3D field is synthesized.

Pedersen et al. [7] employed the indirect boundary element method (IBEM) to analyze the 3D seismic response of 2D topographies using full-space Green's functions for a harmonic point force moving along a direction parallel to the generatrix of the topography. Stamos and Beskos [8] used the BEM to evaluate the 3D dynamic response of lined tunnels in a half-space. The problem was reduced to a 2D one by a coordinate transformation and appropriate integration of the full space dynamic fundamental solution along the direction of the tunnel axis. In Tadeu and Kausel [9] the derivation of 2.5D Green's functions for 3D elastodynamic problems using integral transform methods and/or boundary elements is presented. A 2.5D Green function for a given wave number along the infinite direction is given in [10]. Francois et al. [11] applied a 2.5D coupled finite element–boundary element methodology to investigate the dynamic interaction between a layered soil and longitudinally invariant structures. The prediction of 3D wave field in longitudinally invariant soil-structure interaction problems by a 2.5D coupled FEM-BEM-Method of Fundamental Solutions (MFS) methodology was used by Liravi *et* al. [12].

An alternative to the BEM for modeling unbounded domains are the Perfectly Matched Layers (PMLs), originally introduced for absorbing electromagnetic waves by Berenger [13] extended for structural elastodynamics by Basu and Chopra [14]. In this work, we employ the 2.5D FEM-PMLs formulation proposed by Francois et al. [15], derived in a weak form for an isotropic medium. A recent application of the complex frequency shifted PMLs with the thin layer method for modelling wave propagation in a layered TI medium could be found in Li at al. [16]. There the authors observed that traditional PMLs may lead to exponentially increasing backpropagation waves, attributed to material anisotropy when simulating TI media. However, in the current paper, no such computational instabilities are observed, despite the similarities in the characteristics of TI mediums.

Based on the brief state-of-the-art review, it could be concluded that there is a lack of results considering the dynamic soil-tunnels interaction accounting for material anisotropy. This motivates the authors to concentrate on the dynamic analysis of soil-tunnel interaction, considering both lined and unlined tunnels, along with soil topography, material anisotropy, and the characteristics of the dynamic source. For verification purposes, FEM solutions are performed using the built-in options of ANSYS. These solutions employ the concept of PMLs for 3D problems in both isotropic and anisotropic mediums, and infinite finite elements for 3D isotropic continua. The computational efficiency and accuracy of the infinite elements are discussed in [17].

The paper is organized as follows: The problem statement together with its reformulation via boundary integral equations and via FEM combined with PMLs are given in Section 2. What follows is the verification of the developed computational scheme in Section 3. Subsequently, we examine the dynamic wave field of the BVP formulated in Section 2.1 and summarize the respective numerical results in Section 4, illustrating the effects of soil anisotropy. Finally, the concluding remarks are summarized in Section 5.

2. Problem outline

2.1 Boundary value problem

Consider an elastic anisotropic heterogeneous half-space Ω_{hs} in a Cartesian coordinate system $Ox_1x_2x_3$. The heterogeneity arises from the presence of surface or subsurface relief, as well as embedded underground tunnels, Ω_t. Two lined or unlined tunnels, infinite along Ox_2 axis, with equal embedment depth with respect to the flat free surface, and located at a prescribed distance one another, are considered (Fig. 1). The wave propagation is generated by a point source embedded in the half-space. The dynamic source is represented by a time-harmonic body force $f_i = f_{0i}\,\delta(\mathbf{x},\mathbf{x}_0)\exp(\mathrm{i}\omega t)$ of magnitude f_{0i} and source position vector $\mathbf{x}_0 = (x_{01}, x_{02}, x_{03})$, ω is the excitation frequency.

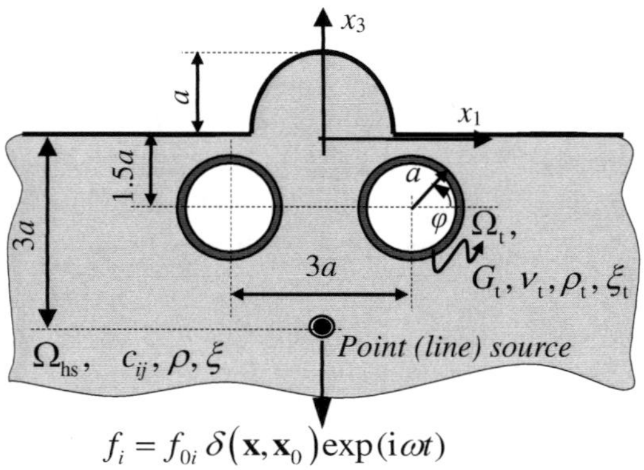

Figure 1. Problem geometry.

The governing equation of motion in frequency domain is expressed as:

$$\sigma_{ij,j}(\mathbf{x},\omega) + \rho\omega^2 u_i(\mathbf{x},\omega) = f_{0i}\,\delta(\mathbf{x},\mathbf{x}_0,\omega), \quad \mathbf{x}(x_1,x_2,x_3) \in \Omega_{hs}, \Omega_t. \tag{1}$$

Here σ_{ij} are the stress components; $\mathbf{x}(x_1,x_2,x_3)$ is the position vector of the observer point; ρ is the mass density; $u_i(\mathbf{x},\omega), i = 1,2,3$ are displacement components, related with the strain tensor ε_{ij} by the geometry equation $\varepsilon_{ij} = (u_{i,j} + u_{j,i})$, $i, j = 1,2,3$. Since the response is time-harmonic, the common multiplayer $\exp(\mathrm{i}\omega t)$ is suppressed. Comma subscripts denote partial differentiation with respect to the spatial coordinates, while the summation convention over repeated indices is implied.

The constitutive stress-strain equation in a TI medium, where the stiffness tensor is expressed in the contracted notation of Voigt [18], reads as:

$$\begin{Bmatrix} \sigma_{11} \\ \sigma_{22} \\ \sigma_{33} \\ \sigma_{12} \\ \sigma_{23} \\ \sigma_{31} \end{Bmatrix} = \begin{bmatrix} c_{11} & c_{11}-2c_{66} & c_{13} & & & \\ c_{11}-2c_{66} & c_{11} & c_{13} & & & \\ c_{13} & c_{13} & c_{33} & & & \\ & & & c_{66} & & \\ & & & & c_{44} & \\ & & & & & c_{44} \end{bmatrix} \begin{Bmatrix} \varepsilon_{11} \\ \varepsilon_{22} \\ \varepsilon_{33} \\ 2\varepsilon_{12} \\ 2\varepsilon_{23} \\ 2\varepsilon_{31} \end{Bmatrix} \tag{2}$$

where c_{ij} are the components of the stiffness tensor $\mathbf{C}$ of the soil medium. In case of TI medium $c_{11}=c_{22}$, $c_{44}=c_{55}$, $c_{13}=c_{23}$ and $c_{12}=c_{11}-2c_{66}$. Five independent parameters define such materials: horizontal Young's modulus $E_h=E_1=E_2$; vertical Young's modulus $E_v=E_3$; vertical shear modulus $G_v=c_{44}=c_{55}$; horizontal Poisson ratio $\nu_h=\nu_{12}$; vertical-horizontal Poisson's ratio $\nu_{vh}=\nu_{31}=\nu_{32}$. The horizontal shear modulus is $c_{66}=E_h/\left(2\left(1+\nu_h\right)\right)$. These parameters are related to c_{11}, c_{13} and c_{33} stiffness tensor components by the following expressions, [5]:

$$c_{11}=\frac{E_h\left(1-\frac{E_h}{E_v}\nu_{vh}^2\right)}{\left(1+\nu_h\right)\left(1-\nu_h-2\frac{E_h}{E_v}\nu_{vh}^2\right)}; \quad c_{13}=\frac{E_h\nu_{vh}}{1-\nu_h-2\frac{E_h}{E_v}\nu_{vh}^2}; \quad c_{33}=\frac{E_h\left(1-\nu_h\right)}{1-\nu_h-2\frac{E_h}{E_v}\nu_{vh}^2}. \tag{3}$$

The boundary conditions along the free surface are tractions free. For the soil-tunnel interfaces the traction equilibrium and displacement compatibility are satisfied. Finally, the traction free boundary conditions are fulfilled along the internal tunnel wall interfaces. In the case of unlined tunnels, the respective cavity contours are traction free.

2.2 Reformulation of the posed problem via boundary integral equations in plane-strain state

For this BVP a verified authors' BEM software is used, based on the frequency-dependent fundamental solution (FS) derived by Radon transform for 2D general anisotropic continua [6]. Consider an elastic anisotropic heterogeneous half-plane in a Cartesian coordinate system $Ox_1x_2x_3$ in plane-strain state with $x_2=0$. The mechanical model defined in frequency domain via governing Eq. (1) can be reformulated by a system of displacement boundary integral equations (4) along the domain boundaries, see [19]:

$$C_{ij}u_j(\mathbf{x},\omega)=\int_{\Gamma_\Omega} U_{ij}^*(\mathbf{x},\xi,\omega)\,t_j(\xi,\omega)\,d\Gamma(\xi)-\int_{\Gamma_\Omega} T_{ij}^*(\mathbf{x},\xi,\omega)\,u_j(\xi,\omega)\,d\Gamma(\xi)+f_{0j}U_{ij}^*(\mathbf{x},\mathbf{x}_0,\omega), \tag{4}$$

where C_{ij} is the jump term depending on the local geometry at the source point $\mathbf{x}=(x_1,x_3)$, the couple $\mathbf{x},\xi$ presents the position vectors of the source and receiver point $\xi=(\xi_1,\xi_3)$, respectively; $U_{ij}^*(\mathbf{x},\xi,\omega)$ is the displacement fundamental solution of the governing elliptic partial differential equation (1) for in-plane wave motion in general anisotropic continuum, see Garcia-Sanchez (2005); $T_{ij}^*(\mathbf{x},\xi,\omega)=C_{ijql}U_{qk,l}^*(\mathbf{x},\xi,\omega)n_k$ is the corresponding traction FS, C_{ijql} are the components of the stiffness tensor from Eq. (2), n_k is the outward pointing unit normal vector to the respective surface. The boundary Γ_Ω is the contour of the domain under consideration Ω_{hs} or Ω_{t}. In 2D plane-strain state the time-harmonic body force is $f_i = f_{0i}\,\delta(\mathbf{x},\mathbf{x}_0)\exp(\mathrm{i}\omega t)$, $i=1,3$ and line source position vector $\mathbf{x}_0 = (x_{01}, x_{03})$.

2.3 Reformulation of the posed problem via FEM combined with PMLs

A computationally efficient two-and-a-half dimesional (2.5D) approach is employed. The 2.5D methodology exploits the invariance of the underground structure in longitudinal Ox_2 direction [11]. This allows for Fourier transformation of the coordinate along the track, resulting in a solution where the original 3D problem is substituted by 2D mesh for each wavenumber, k_{x_2}. The soil-tunnels system is modeled by 2D plane finite elements with additional degree of freedom in longitudinal direction. To prevent spurious wave reflection at the boundaries of Ω_{hs} domain, PMLs are installed, see Fig. 2b. We follow the formulation developed by Francois *et* al. [15] extended for TI media. A MATLAB software code has been developed where the BVP described in section 2.1 is reformulated in 2.5D FEM context as [15]:

$$\left(\mathbf{K}_0 - \omega^2\mathbf{M} - \mathrm{i}k_{x_2}\mathbf{K}_1 - k_{x_2}^2\mathbf{K}_2\right)\mathbf{u}\left(k_{x_2},\omega\right) = \mathbf{f}\left(k_{x_2},\omega\right), \tag{5}$$

where $\mathbf{K}_0 = \iint_A \lambda_1\lambda_3 \mathbf{B}_1^{\mathrm{T}}\mathbf{C}\mathbf{B}_1 \mathrm{d}A$ represents the stiffness matrix for a 2D in-plane problem, A is the area of the particular finite element; $\mathbf{C}$ is the stiffness material tensor from Eq. (2); $\mathbf{M}$ is the mass matrix. Matrices $\mathbf{K}_1 = \iint_A \lambda_1\lambda_3\left(\mathbf{B}_1^{\mathrm{T}}\mathbf{C}\mathbf{B}_2 - \mathbf{B}_2^{\mathrm{T}}\mathbf{C}\mathbf{B}_1\right)\mathrm{d}A$ and $\mathbf{K}_2 = \iint_A \lambda_1\lambda_3\mathbf{B}_2^{\mathrm{T}}\mathbf{C}\mathbf{B}_2\mathrm{d}A$ are stiffness matrices responsible for 3D wave propagation combining both in-plane and out-of-plane wave motion. Matrices $\mathbf{B}_1$ and $\mathbf{B}_2$ relate the stain and displacement vectors in space-frequency domain $\boldsymbol{\varepsilon} = \mathbf{B}_1\mathbf{u}(x_2,\omega) + \mathbf{B}_2\partial\mathbf{u}(x_2,\omega)/\partial x_2$, $\mathbf{u}(\mathbf{x},\omega)$ is the displacement vector of size 3 that contains the degrees of freedom of the 2D mesh, k_{x_2} is the wavenumber in longitudinal x_2 direction. Finally, λ_1 and λ_3 are stretch functions attenuating propagating (p) and evanescent (e) waves, see Basu & Chopra [14]:

$$\lambda_i(s_i) = 1 + f_{s_i0}^{e}\left(\frac{s_i}{L}\right)^m - \mathrm{i}\frac{f_{s_i0}^{p}}{a_0}\left(\frac{s_i}{L}\right)^m. \tag{6}$$

In Eq. (6) L is the thickness of the PMLs, $s_i, i=1,3$ stands for the direction of the attenuated wave, $a_0 = \omega L / C_{\text{S}}$ is a dimensionless frequency, C_{S} is the shear wave velocity. The attenuation coefficients $f_{s_i0}^{e}$ and $f_{s_i0}^{p}$ are chosen $f_{s_10}^{p} = f_{s_30}^{p} = 20$ and $f_{s_10}^{e} = f_{s_30}^{e} = C_s\pi/(l_{FE}\omega) - 1$, where l_{FE} is the size of the FE in the PMLs. The exponent *m* is set to 2, as recommended in Basu & Chopra [14]. On the interface between the PMLs and the elastic domain the stretch functions have unit values, i.e. $s_i = 0, i=1,3$, see Fig. 2b. We note that for the elastic domain $\lambda_1 = \lambda_3 = 1$.

IOP Conf. Series: Materials Science and Engineering **1323** (2024) 012017 doi:10.1088/1757-899X/1323/1/012017

3. Numerical scheme verification

3.1 Two circular tunnels entirely embedded in transversely isotropic half-plane with surface relief in the shape of semi-circular hill

The first validation example is a TI soil medium containing two circular embeded unlined tunnels of radius a. A semi-circular hill of the same radius a forms the surface relief. The distance between both the tunnels is $3a$, the burial depth is $1.5a$ (Fig. 1).

Three different soil mediums are considered, summarized in *Table 1*. The first two material types are TI with Young's moduli ratio $E_h / E_v = 0.5$ and $E_h / E_v = 2$, respectively. The third material type is isotropic with $E_h / E_v = 1$, shear modulus $G = 40500\,\text{kN/m}^2$ and Poisson's ratio 0.25. The mass density of all material types is $\rho = 1.8\,\text{t/m}^2$. The hysteretic frequency independent rate damping $\xi = 5\%$ is applied. It is modeled through complex stiffness tensor $c_{ij}^* = c_{ij}\left(1 + 2\mathrm{i}\xi\right)$. The dynamic excitation is presented by an embedded line source of magnitude $f_i = (0, -f_{03})$ located at point $\mathbf{x}_0 = \left(0, -3a\right)$, see Fig. 1.

Table 1. Material properties.

Mat. type	stiffness tensor components, [kN/m²]
(TI-1) $E_h / E_v = 0.5$	$c_{11} = 76090.9$; $c_{13} = 24545.45$; $c_{33} = 147272.7$; $c_{44} = 40500$; $c_{66} = 27000$ $\left(E_h = 67500;\ E_v = 135000;\ G_v = 40500;\ \nu_h = \nu_{vh} = 0.25\right)$
(TI-2) $E_h / E_v = 2$	$c_{11} = 189000$; $c_{13} = 67500$; $c_{33} = 101250$; $c_{44} = 40500$; $c_{66} = 54000$ $\left(E_h = 135000;\ E_v = 67500;\ G_v = 40500;\ \nu_h = \nu_{vh} = 0.25\right)$
(ISO-3) $E_h / E_v = 1$	$c_{11} = 121500$; $c_{13} = 40500$; $c_{33} = 121500$; $c_{44} = c_{66} = 40500$

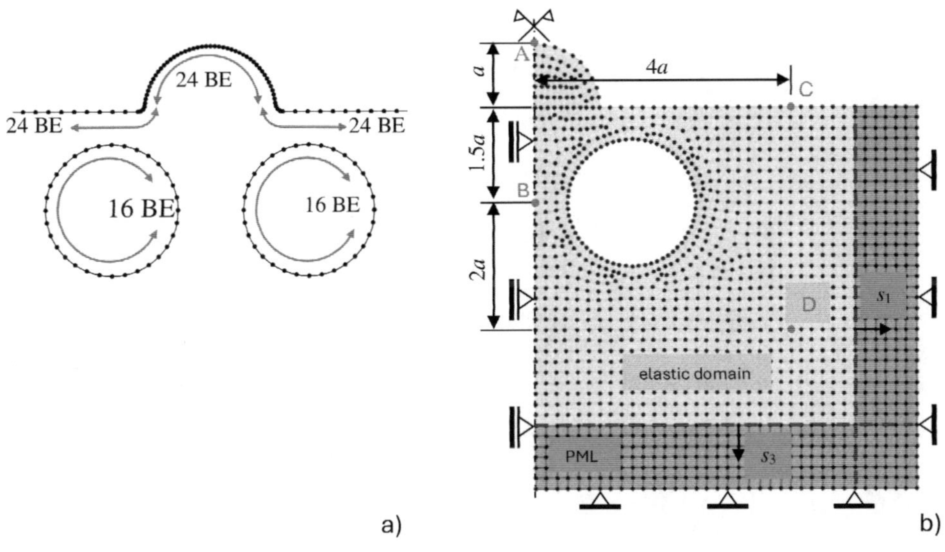

Figure 2. Numerical models: a) BEM mesh; b) FEM-PMLs mesh.

IOP Conf. Series: Materials Science and Engineering **1323** (2024) 012017 doi:10.1088/1757-899X/1323/1/012017

The BEM mesh consists of 104 quadratic boundary elements (BE), 72 of which are used for the free surface of the half-plane discretizing length equal to +/-10a with respect to the coordinate system origin. Each of the cavity contours is modelled by 16 BE as shown in Fig. 2a. The FE mesh is shown in Fig. 2b where the symmetry is exploited to reduce the computational effort. The FEM mesh comprises of 1297 quadratic 8-noded finite elements (FE) with 4109 nodes. The thickness of the PMLs strip equals the radius of the canyon a, discretized by 6 FE. By evaluating the 2.5D solution at wavenumber $k_{x_2} = 0$ we obtain the 2D plane-strain case.

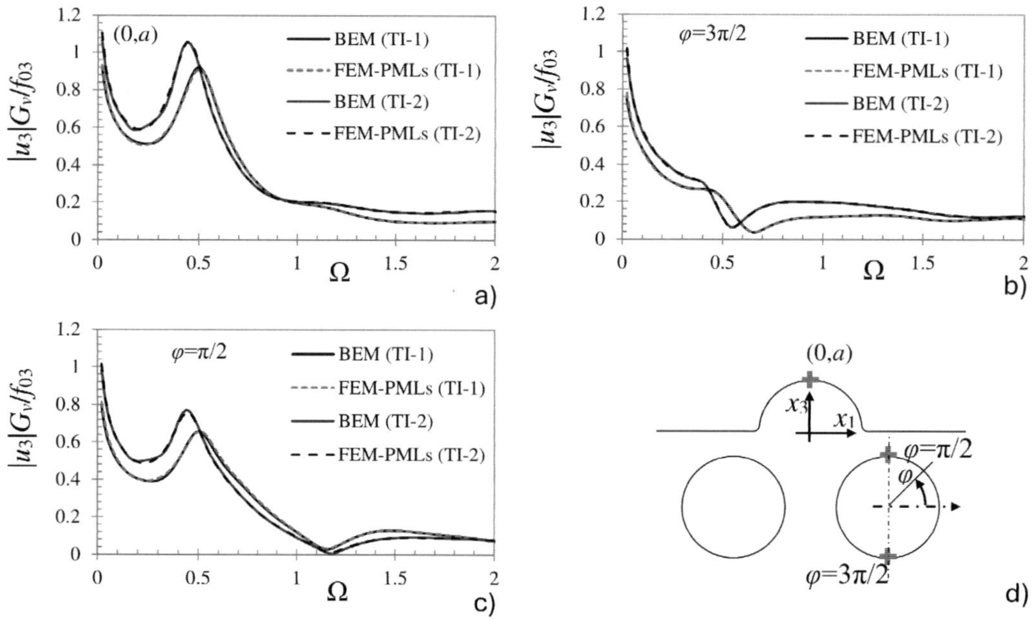

Figure 3. Amplitudes of vertical displacements versus normalized frequency for TI soil medium: a) top of the hill $(0,a)$; b) bottom of the right hole with polar angle φ=3π/2; c) top of the right hole with polar angle φ=π/2; d) scheme of the representative points.

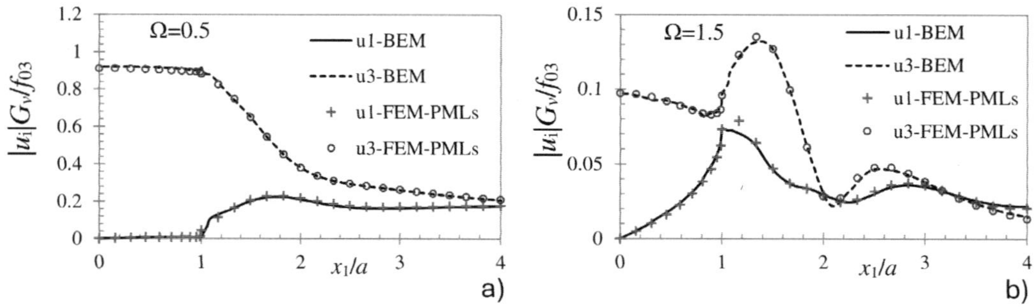

Figure 4. Amplitudes of normalized displacement components along the free surface of TI soil medium with $E_h / E_v = 0.5$ for fixed normalized frequency: a) Ω=0.5; b) Ω=1.5.

The vertical displacement components of three representative points versus normalized frequency $\Omega = \omega a / C_S$ and both TI soil mediums are shown in Fig. 3. The representative points are top of the hill with coordinates $(0,a)$ and two nodes of the right cavity contour with polar angles $\varphi=3\pi/2$ and $\varphi=\pi/2$. The results are obtained by direct BEM and 2.5D FEM combined with PMLs.

The normalized displacement components along the free surface for two fixed normalized frequencies are plotted in Figs. 4-5 for both TI material types. The first normalized frequencies, in Figs. 4 and 5 respectively, are $\Omega=0.50$ and $\Omega=0.46$, and correspond to the first peaks observed in Fig. 3a. The second normalized frequency is chosen $\Omega=1.50$ in both the figures.

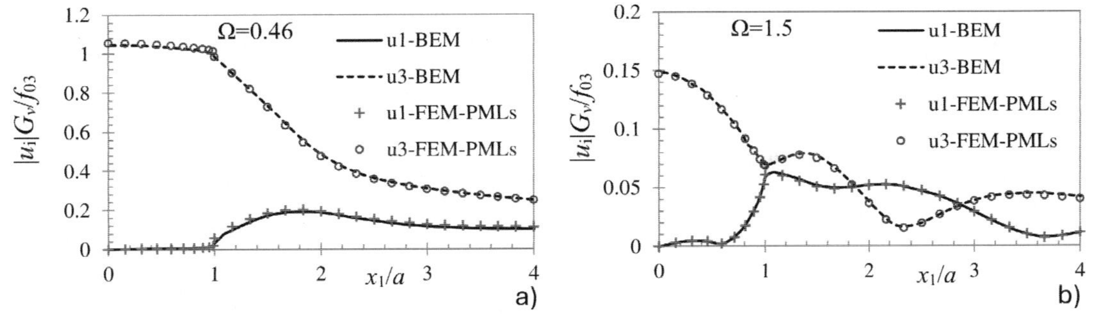

Figure 5. Amplitudes of normalized displacement components along the free surface of TI soil medium with $E_h / E_v = 2$ for fixed normalized frequency: a) Ω=0.46; b) Ω=1.5.

3.2 Two cylindrical unlined tunnels entirely embedded in TI half-space with surface relief in the shape of semi-cylindrical hill under elastic harmonic wave radiated by an embedded point source of magnitude f_{0i}, $i=1,2,3$. Comparisons between 2.5D FEM-PMLs and 3D FEM (ANSYS) solutions.

The geometry of the current example which consists of two unlined cylindrical tunnels embedded in a TI half-space, with surface relief represented by semi-cylindrical hill of radius $a=1$, is shown in Fig. 1. The current 2.5D FEM-PMLs solution is conducted using the numerical model and corresponding mesh outlined in test example 1 and depicted in Fig. 2b. The 3D FEM solution modeled in ANSYS software considers 1/4 of the domain by applying appropriate boundary conditions. In the longitudinal x_2 direction the tunnel's length is $5a$. Three material soil types are considered, see Table 1. In the case of isotropic soil medium, two distinct methodologies for absorbing boundary conditions are utilized. In the first one the infinite nature of the half-space is reproduced by infinite FE (INFIN257), while in the second one with PMLs. For TI soil type, the simulation is confined solely by PMLs, as infinite FE are limited to isotropic materials. The elastic domain is modelled with solid FE (SOLID187).

Comparisons are made based on displacements in spatial coordinates for fixed frequencies. The 2.5D results are transformed from wavenumber k_{x_2} to space x_2 domain by inverse Fourier transform. Due to the symmetry of the displacements caused by a unit point load only the positive wavenumbers are considered. The displacement components in the space domain are retrieved as follows:

$$u_i(x_2,\omega)=\frac{1}{\pi}\int_0^{\infty} u_i(k_{x_2},\omega)e^{-ik_{x_2}x_2}dk_{x_2}, \tag{7}$$

where $i=1,3$ refers to the displacement component. The wavenumber sampling in the current simulations is defined in terms of the dimensionless wavenumber $\bar{k}_{x_2}=k_{x_2}C_S/\omega$. The dimensionless wavenumber ranges 0 to 3, sampled with 301 equidistant points.

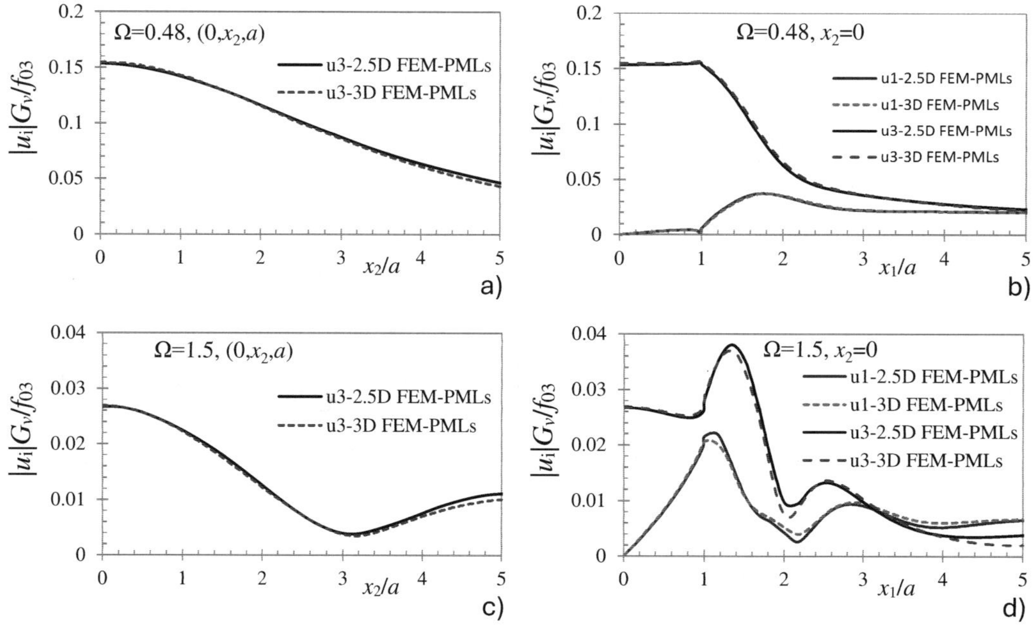

Figure 6. Amplitudes of normalized displacement components of elastic TI half-space with $E_h / E_v = 0.5$: for fixed normalized frequency Ω=0.48 and a) top of the hill along x_2 axis, b) free surface for x_2=0; and Ω=1.5 c) top of the hill along x_2 axis, d) free surface for x_2=0.

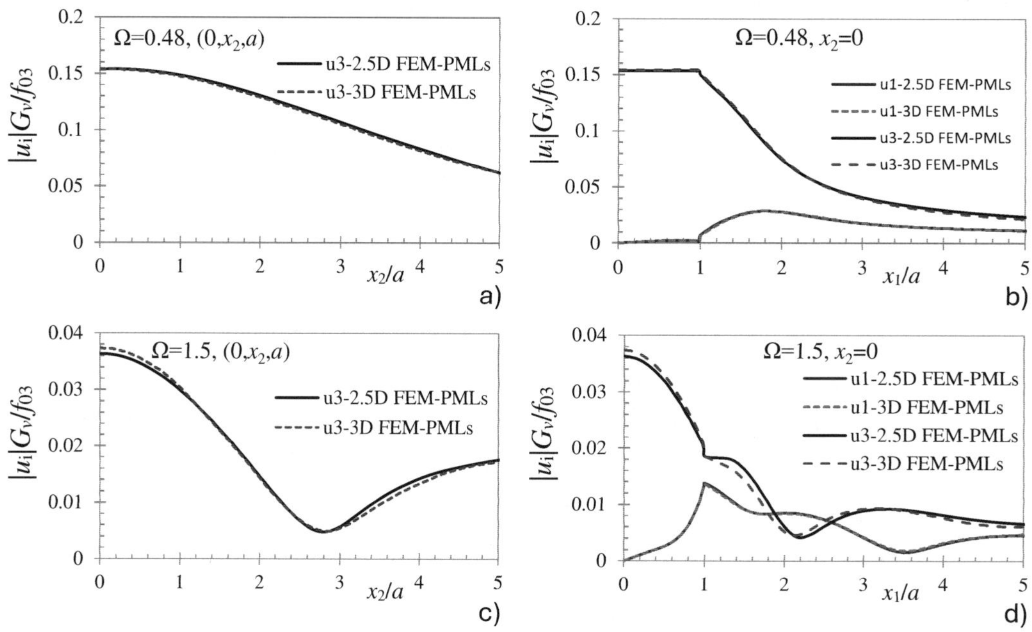

Figure 7. Amplitudes of normalized displacement components of elastic TI half-space with $E_h / E_v = 2$: for fixed normalized frequency Ω=0.48 and a) top of the hill along x_2 axis, b) free surface for x_2=0; and Ω=1.5 c) top of the hill along x_2 axis, d) free surface for x_2=0.

IOP Conf. Series: Materials Science and Engineering **1323** (2024) 012017 doi:10.1088/1757-899X/1323/1/012017

Results for all soil types listed in Table 1 are compared. The displacement components for TI medium with $E_h / E_v = 0.5$ and $E_h / E_v = 2$ are shown in Fig. 6 and Fig. 7 respectively. Those for isotropic soil medium are presented in Fig. 8. The displacements components for the top of the hill or point with coordinates $(0, x_2, a)$ for fixed dimensionless frequencies $\Omega = 0.48$ and $\Omega = 1.5$ are shown in Figs. 6a), 7a) and 8a) and Figs. 6c), 7c) and 8c), respectively. The normalized displacement components along the free surface for the plane $x_2 = 0$ are presented in Figs. 6b), 7b) and 8b) and Figs. 6d), 7d) and 8d) for both normalized frequencies. The results are obtained by authors' 2.5D FEM-PMLs approach and 3D FEM solution performed by ANSYS software with PMLs capabilities. In the case of isotropic material, the results are confirmed by 3D FEM ANSYS solution with infinite FE instead of absorbing layers.

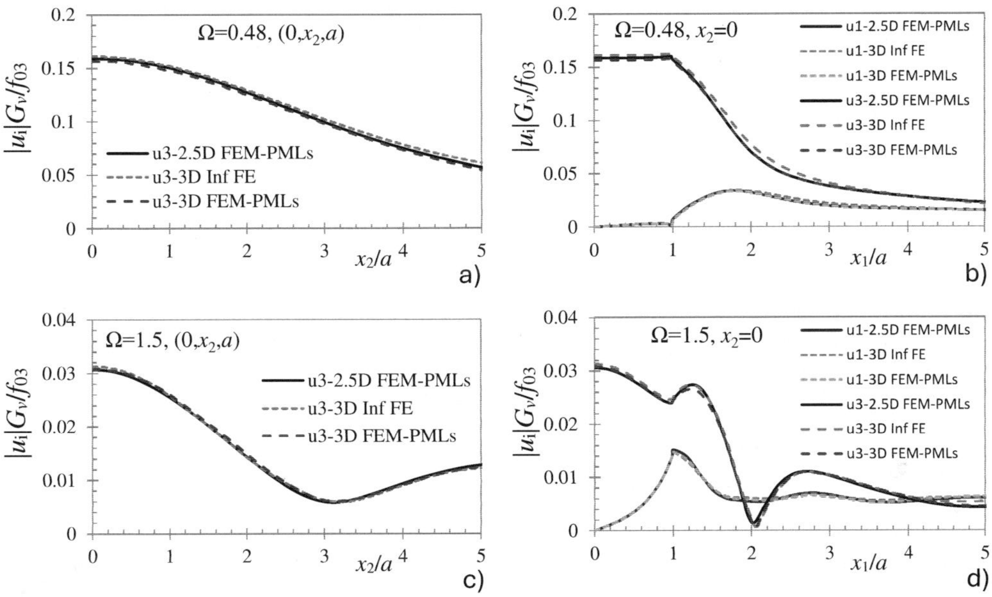

Figure 8. Amplitudes of normalized displacement components of elastic isotropic half-space with $E_h / E_v = 1$: for fixed normalized frequency Ω=0.48 and a) top of the hill along x_2 axis, b) free surface for x_2=0; and Ω=1.5 c) top of the hill along x_2 axis, d) free surface for x_2=0.

4. Numerical results

The methodology, described in section 2.3, and verified in Section 3, is now applied for modelling of two embedded lined tunnels in TI half-space as shown in Fig. 1. As before the burial depth is $1.5a$, the distance separating both tunnels is $3a$, where a is the external radius of the tunnel wall and radius of the semi-circular hill. In the current simulations $a = 1$ m the wall thickness is assumed 0.1 m. The mechanical properties of the walls correspond to a concrete material with shear modulus G_t=3.1e7 kN/m^2, Poisson's ratio ν_t=0.2 and mass density ρ_t=2.5 t/m^3. The hysteretic damping ratio in the tunnel wall is 0.1%. The dynamic excitation is presented by a unit

IOP Conf. Series: Materials Science and Engineering **1323** (2024) 012017 doi:10.1088/1757-899X/1323/1/012017

vertical point force of magnitude $f_i = (0,0,-1\,kN)$ concentrated at the bottom surface of both the tunnels.

The FEM mesh consists of 1361 quadratic finite elements (FE) with 4301 nodes. Each of the tunnel walls is discretized by 64 quadratic FE. The thickness of the PMLs strip is equal to the radius of the canyon a, discretized by 6 FE. As before the three material types are considered listed in Table 1.

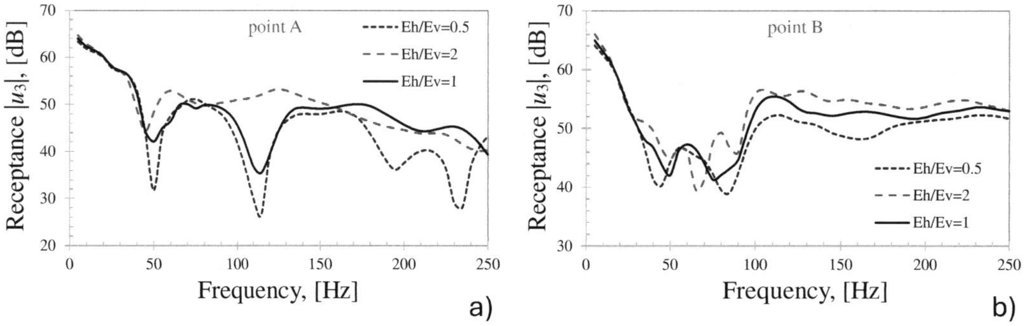

Figure 9. Receptance versus frequency: a) point A; b) point B.

The obtained results are shown in terms of receptance at representative points in horizontal and vertical direction. The evaluation point in longitudinal x_2 direction is x_{02} where the longitudinal distribution of load is $\delta(x_2 - x_{02})$. The results are presented in dB based on reference $10^{-9}\,m/kN$. The representative points are: the top of the hill A(0, x_{02},1a); the point between both the tunnels B(0, x_{02},-1.5a); point C from the free surface with coordinates (4a, x_{02},0); point from the near field D(4a, x_{02},-3.5a). The representative points are depicted in Fig. 2b. Vertical receptance obtained by the present methodology for points A and B (see Fig. 2b) are portrayed in Fig. 9. Horizontal and vertical receptance at points C and D are depicted in Fig. 10 and Fig. 11, respectively. In all the figures the effect of the soil anisotropy is clearly pronounced.

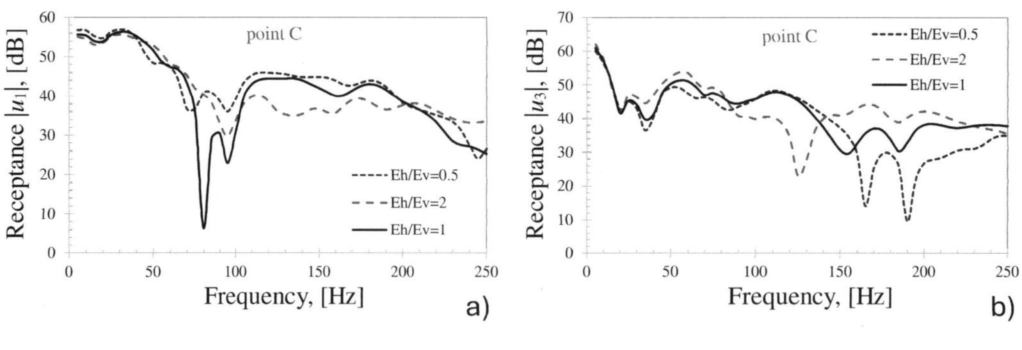

Figure 10. Receptance in point C versus frequency: a) x_1 component; b) x_3 component.

All the results in this section are obtained by MATLAB software code based on 2.5D FEM-PMLs in wavenumber-frequency domain. The dimensionless wavenumber ranges from 0 to 3, sampled with 301 equidistant points.

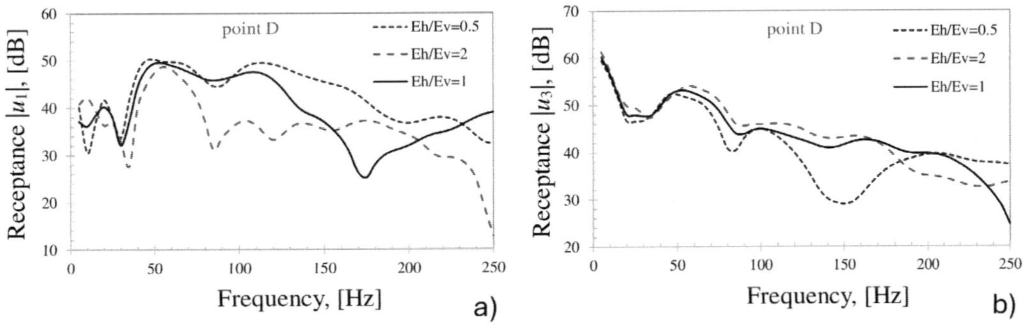

Figure 11. Receptance in point D versus frequency: a) x_1 component; b) x_3 component.

5. Conclusions

A mechanical model for 2.5D elastic wave propagation in a TI half-space with topographic relief of arbitrary geometry containing two unlined/lined tunnels and a point source of excitation is developed in this paper. The core numerical technique is computationally efficient 2.5D FEM combined with PMLs accounting for the 3D wave propagation. The solutions are verified and compared by the direct boundary integral equation method in frequency domain in case of 2D TI problems and by the commercial FEM software ANSYS in case of 3D soil-tunnel interaction systems.

Table 2. Analysis type application.

Space	Medium	Analysis type	ANSYS Inf. FE	ANSYS PMLs	BEM present	2.5FEM-PMLs present
2D	Iso	Time-harmonic	✓		✓	✓
		Transient	✓		✓*	✓*
	TI	Time-harmonic			✓	✓
		Transient			✓*	✓*
3D	Iso	Time-harmonic	✓	✓		✓
		Transient	✓			✓*
	TI	Time-harmonic		✓		✓
		Transient				✓*

*Only linear transient analysis after IFFT

The areas of applicability of the methods considered in this paper are summarized in Table 2. For 3D elastodynamics in TI mediums the present 2.5 FEM-PMLs approach and built-in options of ANSYS with PMLs are applicable. The present 2.5D technique is more efficient in treatment of longitudinally invariant structures due to the reducing storage requirements with respect to the classical 3D solutions. Since 3D ANSYS solution is time-harmonic, the treatment of moving loads would be more complicated and requires longer discretization length in longitudinal direction which considerably increases the computational time and memory requirements. Contrary, the moving loads may be easily implemented in the soil-tunnel system in which the 3D wave field is obtained by 2D discretization in wavenumber-frequency domain in the framework of the 2.5D FEM.

Acknowledgments: The authors wish to acknowledge the support provided by the Research, Consultancy and Design Centre (RCDC) of the University of Architecture Civil Engineering and Geodesy - Sofia through Grant No. BN-298/24.

References

[1] Wang CY, Achenbach JD 1994 Elastodynamic fundamental solutions for anisotropic solids *Geophys. J. Int.* **118** 384–392

[2] Garcia-Sanchez FG 2005 *Numerical study of fracture problems in elastic anisotropic and piezoelectric solids* Dissertation, University of Sevilla

[3] Ba Z, Lee VW, Liang J, Yan Y 2017 Scattering of plane qP- and qSV-waves by a canyon in a multi-layered transversely isotropic half-space *Soil Dyn. Earthq. Eng.* **98** 120–140

[4] Ba Z, Liang J, Lee VW, Hu L 2018 IBEM for Impedance Functions of an Embedded Strip Foundation in a Multi-Layered Transversely Isotropic Half-Space *J. Earthq. Eng.* **22** 1415–46

[5] Ba Z, Fu J, Wang F, Wang Y 2022 Three-Dimensional Dynamic Response Analysis of Rigid Foundation Embedded in Layered Transversely Isotropic Half-Space *Journal of Earthquake Engineering* **26(16)** 8611–28

[6] Parvanova S, Dineva P 2022 Dynamic BEM analysis of elastic foundation on anisotropic half-plane *Archive of Applied Mechanics* **92** 2617–36

[7] Pedersen HA, FSánchez-Sesma J, Campillo M 1994 Three-dimensional scattering by two-dimensional topographies *Bulletin of the Seismological Society of America* **84(4)** 1169–83

[8] Stamos AA, Beskos DE 1996 3-D seismic response analysis of long lined tunnels in half-space *Soil Dynamics and Earthquake Engineering* **15(2)** 111–118

[9] Tadeu AJB, Kausel E 2000 Green's Functions for Two-and-a-Half-Dimensional Elastodynamic Problems *Journal of Engineering Mechanics* **126(10)** 1093–97

[10] Jean P, Guigou C, Villot M 2004 A 2.5D BEM Model for Ground-Structure Interaction *Building Acoustics* **11(3)** 157–173

[11] François S, Schevenels M, Galvín P, Lombaert G, Degrande G, 2010 A 2.5D coupled FE–BE methodology for the dynamic interaction between longitudinally invariant structures and a layered halfspace *Computer Methods in Applied Mechanics and Engineering* **199(23–24)** 1536–48

[12] Liravi H, Arcos R, Ghangale D, Noori B, Romeu J 2021 A 2.5D coupled FEM-BEM-MFS methodology for longitudinally invariant soil-structure interaction problems *Computers and Geotechnics* **132** 104009

[13] Berenger J 1994 A perfectly matched layer for the absorption of electromagnetic waves *Journal of Computational Physics* **41** 115–135

[14] Basu U, Chopra A 2003 Perfectly matched layers for time-harmonic elastodynamics of unbounded domains: theory and finite-element implementation *Computer Methods in Applied Mechanics and Engineering* **192(11-12)** 1337–1375

[15] Francois S, Schevenels M, Lombaert G, Degrande G 2012 A two-and-a-half-dimensional displacement-based PML for elastodynamic wave propagation *Numerical Methods in Engineering* **90(7)** 819–837

[16] Li H, He C, Gong Q, Zhou S, Li X, Zou C 2024 TLM-CFSPML for 3D dynamic responses of a layered transversely isotropic half-space *Computers and Geotechnics* **168** 106131

[17] Kazakov K 2013 On the Total Dynamic Response of Soil-Structure Interaction System in Time Domain Using Elastodynamic Infinite Elements with Scaling Modified Bessel Shape Functions *American Journal of Computational Mathematics* **3** 104–109

[18] Voigt W 1906 *Lehrbuch der kristallphysik* (Leipzig: Springer)

[19] Dominguez J 1993 *Boundary Elements in Dynamics* (WIT Press/Computational Mechanics Publications Southampton)

IOP Conf. Series: Materials Science and Engineering **1323** (2024) 012018 doi:10.1088/1757-899X/1323/1/012018

Dynamic response of a beam on an elastic Winkler foundation subjected to time-harmonic load

Petar Shiskov[1], German Stoyanov[1], Amar Pashov[2] and Sonia Parvanova[1*]

[1] Department of Structural Engineering, University of Architecture, Civil Engineering and Geodesy, Sofia, Bulgaria
[2] KU Leuven, Department of Civil Engineering, Kasteelpark Arenberg 40, B-3001 Leuven, Belgium

*E-mail: slp_fce@uacg.bg

Abstract. This study investigates the dynamic behavior of a beam on an elastic Winkler foundation under time-harmonic excitation. The frequency dependent ordinary differential equation of a beam with prescribed bending stiffness is analytically solved by using initial parameters' method and Krylov's functions for beams of medium length, and by Zimmermann functions in case of infinitely long beams. The obtained results are validated by an authors' numerical solution based on the Finite Element Method. The dynamic response of numerical models, falling between medium and infinitely long beams, is validated by both analytical approaches. Conclusions and recommendations regarding their application are provided. Finally, a method accounting for moving point loads is presented.

1. Introduction

Various models exist for the simplified modeling of beams resting on elastic half-space, starting with one parameter Winkler model [1] followed by two-parameters model of Pasternak [2], and to higher-order models such as those proposed by Kerr [3]. Winkler's assumption that the soil reaction at a point is linearly proportional to the deflection of the beam at that point is the basis of many soil-structure interaction problems and has a number of applications in geotechnical engineering. The very first studies on a beam on an elastic Winkler foundation can be traced back to the pioneering works of Zimmermann [4] and Krylov [5], who presented analytical solutions to numerous static problems. Through the years, these problems have been augmented by many researchers. For instance, a closed-form static solution by a variational formulation based on the minimum of the total potential energy functional is developed in [6]. Based on the theory of beams on an elastic foundation, laterally loaded piles with linearly distributed subgrade reaction coefficients are investigated in [7]. A beam with variable cross section resting on an elastic foundation, characterized by two bed coefficients, is solved in [8].

The dynamic behavior of a beam on an elastic foundation subjected to moving point loads is related to models describing railways, continuous highways, or airport pavements. With the rapid development of high-speed cars and trains the requirements for more realistic numerical models producing more accurate and reliable results increase. Among the first closed-form solutions of an infinite Euler–Bernoulli beam on Winkler spring foundation subjected to a moving point load with a constant velocity is that of Kenney [9]. Thompson [10] and Fryba [11] further developed similar analytical solutions, which have been used as a basis for more precise and complicated models by many researchers. More recently Sun [12] obtained a closed-form solution of the steady-state response of a beam on a viscoelastic subgrade subjected to a moving constant point load for different combinations of damping and speed. In addition to the analytical solutions, long established and well-known numerical models based on the Finite Element Method (FEM) are used to simulate the response of a beam on an elastic support [13-15].

In this paper, the steady-state response of finite and infinite beams resting on a viscoelastic Winkler foundation subjected to a point load is examined. Closed-form time-harmonic solutions for beams of medium and infinite lengths are proposed, and the results are confirmed by the FEM. Next, a procedure based on Fourier transform is proposed which accounts for the moving point load with constant velocity.

The paper is organized as follows: The problem statement together with analytical solutions of the governing differential equation are presented in Section 2. The FEM formulation of the posed problem is outlined in Section 3. The results regarding beam deflection, bending moment and shear force for different beam length, obtained analytically and numerically, are discussed in Section 4. Section 5 concerns moving point load with a constant speed along the beam achieved through discrete forward and inverse Fourier transforms. Concluding remarks are summarized in Section 6.

2. Problem statement

In the following, we consider a model of a beam resting on an elastic foundation. The beam is modeled as an Euler-Bernoulli beam with bending stiffness EI , where E is the Young's modulus of the beam material, I is the moment of inertia. The mass per unit length is $m=\rho A$, where ρ is the mass density, and A the cross-sectional area, see Fig. 1. The elastic foundation has a stiffness k, and viscous damping c. The longitudinal beam's horizontal axis is denoted as x_1, while the positive vertical direction is assumed downwards along the x_2 direction. The sign convention for internal forces is depicted in Fig. 1b.

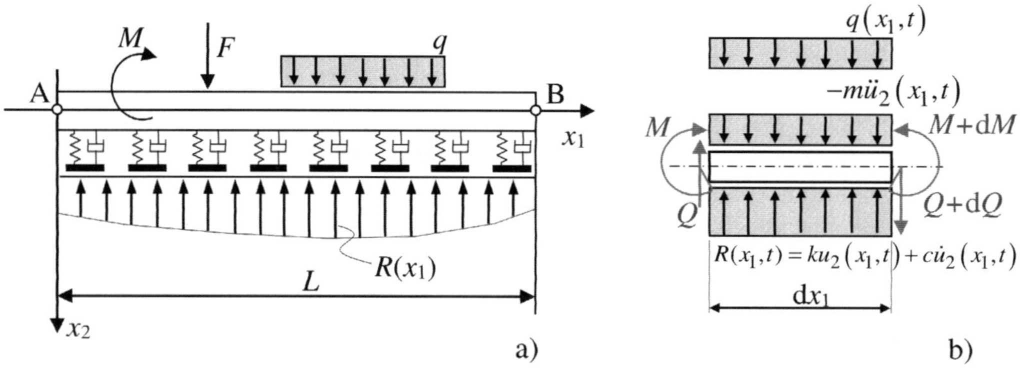

Figure 1. Beam on an elastic foundation: a) Problem geometry; b) differential element of the beam.

The partial differential equation governing the dynamic equilibrium of the beam can be derived by examining the equilibrium of the differential element shown in Fig. 1b. This part of the beam is subjected to: (1) distributed load of intensity $q(x_1,t)$; (2) inertial or Dalambert's forces of intensity $-m\ddot{u}_2(x_1,t)$, where $\ddot{u}_2$ is the acceleration at a point with coordinate x_1, being the second derivative with respect to the time of vertical displacement $u_2(x_1,t)$; (3) vertical reaction forces of the foundation proportional to the vertical displacement via the stiffness coefficient and to the velocity $\dot{u}_2$ via the damping coefficient $R(x_1,t) = ku_2(x_1,t) + c\dot{u}_2(x_1,t)$; and to the (4) internal forces as shown in Fig. 1b. The vertical equilibrium is obtained as:

$$\mathrm{d}Q / \mathrm{d}x_1 = ku_2(x_1,t) + c\dot{u}_2(x_1,t) + m\ddot{u}_2(x_1,t) - q(x_1,t), \tag{1}$$

where Q is the shear force, t is the time variable.

Considering the equilibrium of moments along the left section of the differential element, neglecting the components of order $\mathrm{d}x_1^2$ and higher, we attain $\mathrm{d}M / \mathrm{d}x_1 = Q$. Using the well know elastic curve equation $-EI\left(\mathrm{d}^2u_2 / \mathrm{d}x_1^2\right) = M$ and replacing it in Eq. 1, we reach the governing equation of the dynamic equilibrium in the time domain with respect to the vertical displacement:

$$\frac{\mathrm{d}^4u_2}{\mathrm{d}x_1^4} + \frac{1}{EI}\left(ku_2(x_1,t) + c\dot{u}_2(x_1,t) + m\ddot{u}_2(x_1,t)\right) = \frac{q(x_1,t)}{EI}. \tag{2}$$

Assume that the beam is subjected to time-harmonic loads $\bar{q}e^{\mathrm{i}\omega t}$ of amplitude $\bar{q}$ and excitation frequency ω, we get:

$$u_2 = \bar{u}_2e^{\mathrm{i}\omega t}; \quad \dot{u}_2 = \mathrm{i}\omega\bar{u}_2e^{\mathrm{i}\omega t}; \quad \ddot{u}_2 = -\omega^2\bar{u}_2e^{\mathrm{i}\omega t}. \tag{3}$$

Here $\mathrm{i}=\sqrt{-1}$ is the imaginary unit, $\bar{u}_2$ is the vertical displacement amplitude. Replacing Eq. (3) in Eq. (2) and suppressing the common multiplier $e^{\mathrm{i}\omega t}$ we obtain the differential equation of dynamic equilibrium in frequency domain:

$$\begin{gathered}\frac{\mathrm{d}^4\bar{u}_2}{\mathrm{d}x_1^4} + \frac{1}{EI}\left(k + \mathrm{i}\omega c - \omega^2 m\right)\bar{u}_2 = \frac{\bar{q}}{EI};\\ \bar{u}_2^{IV} + 4\alpha^4\bar{u}_2 = \bar{q} / EI; \quad \alpha = \sqrt[4]{\left(k + \mathrm{i}\omega c - \omega^2 m\right) / (4EI)}.\end{gathered} \tag{4}$$

In Eq. (4) the parameter α includes the flexural rigidity of the beam as well as the elasticity of the foundation, the damping coefficient, and the excitation frequency. In the static analysis $\omega = 0$, α has dimension 1/ length and in that respect $1/\alpha$ is called characteristic length.

The solution of Eq. (4) could be expressed as $\bar{u}_2(x_1,\omega) = \bar{u}_{2,0}(x_1,\omega) + \bar{u}_{2,F}(x_1,\omega)$ where $\bar{u}_{2,0}(x_1,\omega)$ is the solution of the homogeneous differential equation and $\bar{u}_{2,F}(x_1,\omega)$ is a particular integral corresponding to the applied loads. The roots of the characteristic equation of the differential Eq. (4) are:

$$r_1 = (\alpha + \mathrm{i}\alpha), \quad r_2 = (\alpha - \mathrm{i}\alpha), \quad r_3 = (-\alpha + \mathrm{i}\alpha), \quad r_4 = (-\alpha - \mathrm{i}\alpha). \tag{5}$$

The respective solution of the homogeneous equation holds:

$$\bar{u}_{2,0}(x_1,\omega) = B_1e^{(\alpha+\mathrm{i}\alpha)x_1} + B_2e^{(\alpha-\mathrm{i}\alpha)x_1} + B_3e^{(-\alpha+\mathrm{i}\alpha)x_1} + B_4e^{(-\alpha-\mathrm{i}\alpha)x_1}, \tag{6}$$

where B_1, B_2, B_3, B_4 are integrational constants which are determined from the boundary conditions. Using Euler's expressions $e^{\mathrm{i}\alpha x_1} = \cos\alpha x_1 + \mathrm{i}\sin\alpha x_1$ and $e^{-\mathrm{i}\alpha x_1} = \cos\alpha x_1 - \mathrm{i}\sin\alpha x_1$ Eq. (6) is rewritten as:

$$\bar{u}_{2,0}(x_1,\omega) = D_1e^{\alpha x_1}\sin\alpha x_1 + D_2e^{\alpha x_1}\cos\alpha x_1 + D_3e^{-\alpha x_1}\sin\alpha x_1 + D_4e^{-\alpha x_1}\cos\alpha x_1. \tag{7}$$

IOP Conf. Series: Materials Science and Engineering **1323** (2024) 012018 doi:10.1088/1757-899X/1323/1/012018

Further replacing $e^{\alpha x_1} = \operatorname{ch}\alpha x_1 + \operatorname{sh}\alpha x_1$ and $e^{-\alpha x_1} = \operatorname{ch}\alpha x_1 - \operatorname{sh}\alpha x_1$ an alternative form of the same solution becomes:

$$\bar{u}_{2,0}(x_1,\omega) = \operatorname{ch}\alpha x_1\left[C_1\cos\alpha x_1 + C_2\sin\alpha x_1\right] + \operatorname{sh}\alpha x_1\left[C_3\cos\alpha x_1 + C_4\sin\alpha x_1\right]. \quad (8)$$

In the above D_1, D_2, D_3, D_4 and C_1, C_2, C_3, C_4 are integrational constants which will be obtained by satisfying the boundary conditions.

2.1 Solution of beams of medium length by initial parameters method

Beams of medium length are beams with $0.5 \le \bar{\alpha}L \le 5$, where $\bar{\alpha} = \sqrt[4]{k/4EI}$, and L is the length of the beam. For such beams the loads applied at the one end of the beam have a finite and not negligible influence on the other end. For these beams the method of initial parameters is suitable.

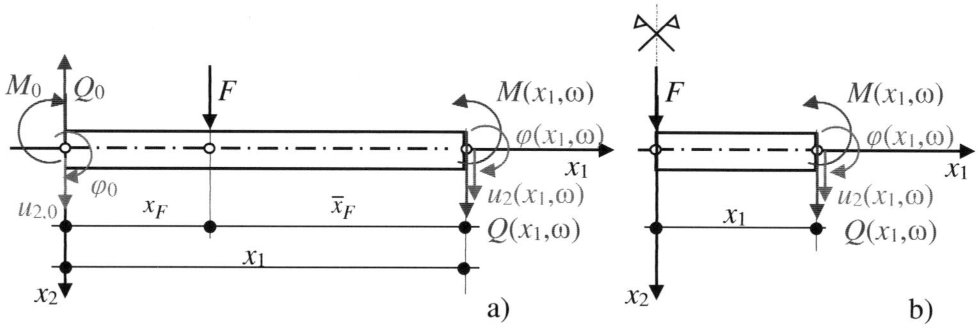

Figure 2. Internal forces: a) beam of medium length; b) infinite beam on elastic foundation.

Initial parameters of the beam on an elastic foundation are: initial vertical displacement $u_{2,0}$, initial rotation φ_0, initial bending moment M_0, and initial shear force Q_0, positioned at $x_1 = 0$ (Fig. 2a). The integrational constants C_1, C_2, C_3, C_4 from Eq. (8) are expressed by the initial parameters satisfying the following boundary conditions:

$$U_{2,0} = EIu_{2,0}(0,\omega);\quad \Phi_0 = EI\varphi_0 = EIu'_{2,0}(0,\omega);\quad M_0 = -EIu''_{2,0}(0,\omega);\quad Q_0 = -EIu'''_{2,0}(0,\omega), \quad (9)$$

where $U_{2,0}$ and Φ_0 are EI multiple values of initial vertical displacement and initial rotation, respectively. Eq. (9) constitutes a system of four equations with four unknowns. Solving this system yields the integration constants C_1, C_2, C_3, C_4 from Eq. (8) in terms of initial parameters. After substituting the expressions for C_1, C_2, C_3, C_4 in Eq. (8) and subsequently differentiating, the following expressions are obtained:

$$\begin{Bmatrix} U_2(x_1,\omega) \\ \Phi(x_1,\omega) \\ M(x_1,\omega) \\ Q(x_1,\omega) \end{Bmatrix} = \begin{bmatrix} A & B/\alpha & -C/\alpha^2 & -D/\alpha^3 \\ -4\alpha D & A & -B/\alpha & -C/\alpha^2 \\ 4\alpha^2 C & 4\alpha\ D & A & B/\alpha \\ 4\alpha^3 B & 4\alpha^2 C & -4\alpha D & A \end{bmatrix} \begin{Bmatrix} U_{2,0} \\ \Phi_0 \\ M_0 \\ Q_0 \end{Bmatrix} + \begin{Bmatrix} U_{2,F} \\ \Phi_F \\ M_F \\ Q_F \end{Bmatrix}. \quad (10)$$

Eq. (10) provides the vertical displacement, rotation, bending moment and shear force at any section of the beam with respect to the initial parameters. Two of these parameters are always known from the boundary conditions at $x_1 = 0$, while the other two could be obtained from boundary conditions at $x_1 = L$. With all initial parameters available, we can determine all the required parameters at any section of the beam. The functions A, B, C and D are known as Krylov's functions [5], for which the following substitutions hold true:

$$\begin{aligned} &A(\alpha x_1) = \operatorname{ch}\alpha x_1 \cos\alpha x_1; \quad B(\alpha x_1) = (\operatorname{ch}\alpha x_1 \sin\alpha x_1 + \operatorname{sh}\alpha x_1 \cos\alpha x_1)/2; \\ &C(\alpha x_1) = \operatorname{sh}\alpha x_1 \sin\alpha x_1 / 2; \quad D(\alpha x_1) = (\operatorname{ch}\alpha x_1 \sin\alpha x_1 - \operatorname{sh}\alpha x_1 \cos\alpha x_1)/4. \end{aligned} \tag{11}$$

The last vector in Eq. (10) is the particular integral due to a concentrated point force at x_F. The distance from the point load to the section under consideration is $\bar{x}_F = x_1 - x_F$, see Fig. 2a. This vector could be determined using the following expressions:

$$U_{2,F}(x_1,\omega) = \frac{D(\alpha\bar{x}_F)}{\alpha^3}F; \; \Phi_F(x_1,\omega) = \frac{C(\alpha\bar{x}_F)}{\alpha^2}F; \; M_F(x_1,\omega) = -\frac{B(\alpha\bar{x}_F)}{\alpha}F; \; Q_F(x_1,\omega) = -A(\alpha\bar{x}_F)F \tag{12}$$

2.2 Solution of beams of infinite length

The beam is considered as infinitely long when $\bar{\alpha}L > 5$. For such beams the counter effect of the end conditioning terms (forces and displacements) on each other is negligible. When investigating one end of the beam we can assume that the other end is infinitely far away. To this end, the Krylov's functions $A(\alpha L)$, $B(\alpha L)$, $C(\alpha L)$ and $D(\alpha L)$ are practically zero which greatly simplifies the computations.

An infinite beam resting on an elastic foundation loaded by point force is shown in Fig. 2b. The coordinate origin is positioned at the point of force application. Due to the symmetric impact, both symmetric and anti-symmetric responses occur, hence only the positive part of x_1 is considered. The boundary conditions for derivation of D_1, D_2, D_3, D_4 from Eq. (7) are as follows:

$$\begin{aligned} &\varphi(0,\omega) = 0; \quad Q(0,\omega) = -F/2; \\ &u_2(\infty,\omega) = 0; \quad u_2'(\infty,\omega) = 0. \end{aligned} \tag{13}$$

From the second row of Eq. (13) it follows that $D_1 = D_2 = 0$. To fulfill the condition of zero rotation, at $x_1 = 0$, one obtains $D_3 = D_4 = \bar{D}$. Finally, $\bar{D}$ is evaluated as:

$$-EIu_2'''(0,\omega) = -EI4\bar{D}\alpha^3 = -F/2 \quad \Rightarrow \quad \bar{D} = F/(8EI\alpha^3). \tag{14}$$

After substitution of Eq. (14) in Eq. (7) and its consecutive differentiation the following yields:

$$\begin{aligned} &U_2(x_1,\omega) = EIu_2(x_1,\omega) = \frac{F}{8\alpha^3}Z_2(\alpha x_1); \quad \Phi(x_1,\omega) = EIu_2'(x_1,\omega) = -\frac{F}{4\alpha^2}Z_3(\alpha x_1); \\ &M(x_1,\omega) = -EIu_2''(x_1,\omega) = \frac{F}{4\alpha}Z_4(\alpha x_1); \quad Q(x_1,\omega) = -EIu_2'''(x_1,\omega) = -\frac{F}{2}Z_1(\alpha x_1). \end{aligned} \tag{15}$$

In the above $Z_1 - Z_4$ are periodic decaying functions known as Zimmerman functions [4]:

$$\begin{aligned} &Z_1(\alpha x_1) = e^{-\alpha x_1}\cos\alpha x_1; \quad Z_2(\alpha x_1) = e^{-\alpha x_1}(\sin\alpha x_1 + \cos\alpha x_1); \\ &Z_3(\alpha x_1) = e^{-\alpha x_1}\sin\alpha x_1; \quad Z_4(\alpha x_1) = e^{-\alpha x_1}(-\sin\alpha x_1 + \cos\alpha x_1). \end{aligned} \tag{16}$$

3. FEM formulation for beams on an elastic foundation

The boundary value problem described in section 2 is reformulated in FEM context as:

$$(\mathbf{K} + \mathbf{K}_W + i\omega\mathbf{C} - \omega^2\mathbf{M})\mathbf{d} = \mathbf{f}, \tag{17}$$

where $\mathbf{K}$ and $\mathbf{K}_W$ are the global stiffness matrices of the beam and the foundation, respectively; $\mathbf{C}$ is the damping matrix; $\mathbf{M}$ is the mass matrix. The nodal displacement vector is $\mathbf{d}$, and $\mathbf{f}$ is the vector of nodal forces. The global matrices are assembled by element ones, composed as:

$$\mathbf{k}_W^e = \int_0^l \mathbf{N}^{\mathrm{T}}\mathbf{N}dx_1 k; \quad \mathbf{m}^e = \int_0^l \mathbf{N}^{\mathrm{T}}\mathbf{N}dx_1 m; \quad \mathbf{c}^e = \int_0^l \mathbf{N}^{\mathrm{T}}\mathbf{N}dx_1 c. \tag{18}$$

In the above $\mathbf{N}$ is the matrix containing shape functions related with the approximation of the displacements with the nodal values. The integration is over the element length l. A two-node frame element with 6 degrees of freedom, two translational and one rotational at each joint, is employed in the MATLAB FEM code with standard polynomial approximation.

4. Numerical examples

Two numerical examples are considered to demonstrate the capabilities of the methodologies outlined in the previous section. An infinite beam loaded by a unit point force, resting on a viscoelastic foundation is the first numerical example. The second one is a beam of finite length $L = 20\ \mathrm{m}$ subjected to a unit point force at $L/2$. For this beam $\bar{\alpha} = \sqrt[4]{k/4EI} = 0.5\ \mathrm{m}^{-1}$, respectively $\bar{\alpha}L = 10$ or the initial parameters have negligible influence on the section of the point force as $\bar{\alpha}L/2 = 5$. This means that the beam could be considered as a finite beam of medium length or infinitely long. Parameters of the soil-beam system are as follows: $EI = 80000\ \mathrm{kN\,m^2}$, $m = 8\ \mathrm{t/m}$, $k = 20000\ \mathrm{kN/m/m}$, $c = 100\ (200)\ \mathrm{kN\,s/m^2}$. The excitation frequency ω varies from zero to 50 Hz.

The vertical deflection and internal forces along the infinite beam are obtained by Eq. (15-16), while for the beam with finite length by Eq. (10-12), for the considered frequency interval. For both the beams numerical FE solution is performed. In the first case, the numerical model represents a beam with a length $L = 200\ \mathrm{m}$ discretized into 100 two-node finite elements. The FE length increases progressively from the point of force application to the ends of the beam. The largest to smallest element length ratio is set to 4. The second beam, with finite length, is discretized by 20 equal sized FE of length 1 m. The results for the vertical deflection of the beam at the point of application of the unit force versus frequency are depicted in Fig. 3. More specifically in Fig. 3a the amplitude of the vertical deflection for a damping coefficient $c = 100\ \mathrm{kN\,s/m^2}$ is plotted, obtained for infinite and finite beam by both analytical and numerical solutions. In Fig. 3b the same set of results for $c = 200\ \mathrm{kN\,s/m^2}$ is depicted. The FEM solution is applicable to both beams, regardless of the length. Fig. 3 shows that FEM solutions have no plotting difference with respect to the analytical ones for both the beams and both damping coefficients. The results for the finite beam are very close to these of the infinite one up to the cut-on frequency $f = \sqrt{k/m}/2\pi = 50/2\pi\ \mathrm{Hz}$. The larger damping coefficient, the smaller oscillations of the finite beam deflection with respect to the infinite one.

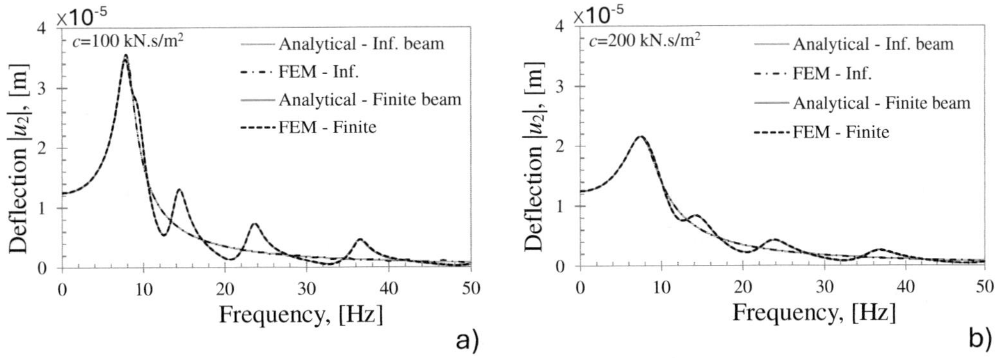

Figure 3. Amplitude of vertical deflection at point of application of the force versus frequency for damping coefficient: a) c=100 kN.s/m^2; b) c=200 kN.s/m^2.

In Fig. 4 the internal forces of both the beams, obtained by both analytical solutions, are compared for a fixed frequency $f = 7.45$ Hz corresponding to the peak values observed in Fig. 3b. Fig. 4a illustrates the real and imaginary variation of the bending moment along the beam axis, depicted on the tension side for infinite and finite length beams. The real and imaginary shear force along the beam axis is portrayed in Fig. 4b. As observed in the vertical deflections up to the cut-on frequency finite and infinite beam analytical solutions for the internal forces are almost identical.

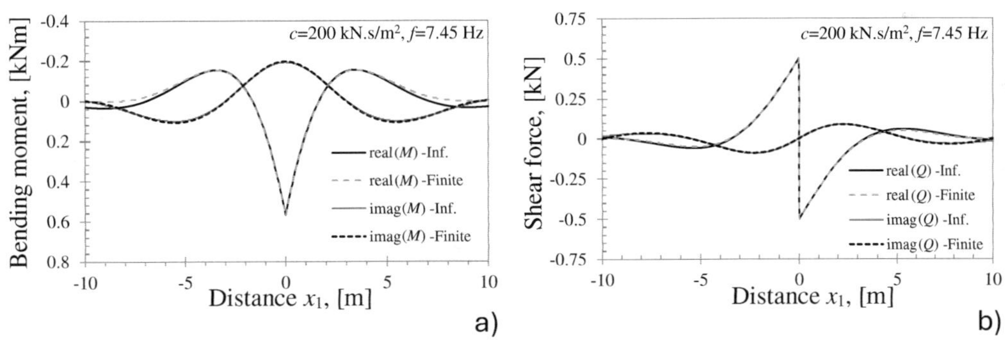

Figure 4. Internal forces along the beam axis for fixed frequency f=7.45 Hz: a) bending moment, M; b) shear force, Q.

5. Beam subjected to a moving point load with constant velocity

The procedures described in Sections 2 and 3 are solutions of beams on viscoelastic foundation in space-frequency domain. In what follows we propose a methodology which allows to obtain the response of the beam due to a constant moving point load with a given velocity v_F. This requires the representation of u_2 to be in the wavenumber-frequency domain, which is obtained

IOP Conf. Series: Materials Science and Engineering **1323** (2024) 012018 doi:10.1088/1757-899X/1323/1/012018

by a forward Fourier transformation from the space variable x_1 to the wavenumber k_{x_1}, defined as:

$$u_2\left(k_{x_1},\omega\right)=\int_{-\infty}^{+\infty}u_2\left(x_1,\omega\right)e^{\mathrm{i}k_{x_1}x_1}\mathrm{d}x_1\,. \tag{19}$$

The displacement components $u_2\left(\omega/v_F\right)$ corresponding to $k_{x_1}=\omega/v_F$ are extracted from the above transformation at each angular frequency ω. For a constant harmonic load, the solution of vertical displacement in the spatial-frequency domain is obtained from the wavenumber domain representation as $u_2\left(x_1,\omega\right)=\dfrac{1}{v_F}u_2\left(\omega/v_F\right)e^{-\mathrm{i}(\omega/v_F)x_1}$. An inverse Fourier transformation is applied to transfer the frequency ω to time variable t, defined as:

$$u_2(x_1,t)=\frac{1}{2\pi}\int_{-\infty}^{+\infty}u_2(x_1,\omega)e^{\mathrm{i}\omega t}\mathrm{d}\omega\,. \tag{20}$$

Eq. (19) and (20) are applied in the current computations by discrete Fast Fourier Transform (FFT) and inverse FFT, respectively. The sampling used in the simulations follows the steps: the considered length of the beam should be $L=n_f v_F/f_N$ where n_f is the number of frequency steps, f_N is the Nyquist frequency. The space increment is $\Delta x_1=L/(2n_f)$ resulting in $2n_f+1$ discrete values of the displacements. The Nyquist wavenumber becomes $k_{x_1N}=2\pi/(2\Delta x_1)$, the wavenumber step is $\Delta k_{x_1}=2\pi/\left(2n_f\Delta x_1\right)$. This increment is adjusted to provide discrete values of the wavenumber interval equal to ω/v_F, without a need of interpolation. Based on the performed solutions for numerous elastic supports and beam characteristics, we recommend Nyquist frequency to be set up to 20 times the cut-on frequency, and less than 500 equal frequency steps.

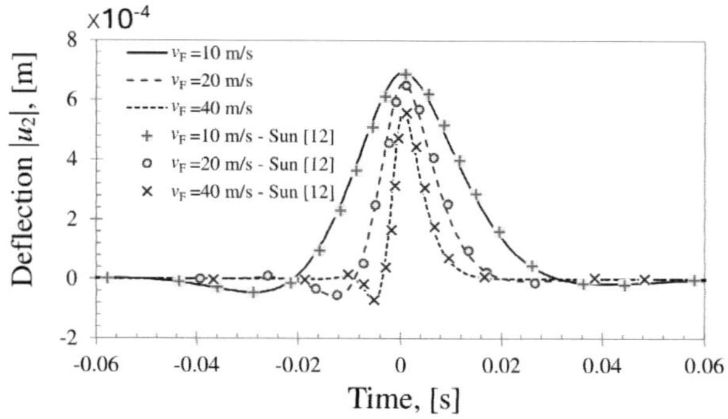

Figure 5. Vertical deflection at x_1=0 versus time. Comparison with Sun [12].

The first numerical example in this section is a validation of the proposed methodology. Consider an infinite beam laying on viscoelastic foundation solved by Sun [12] with the following mechanical properties: $EI=2.3\,\mathrm{kN\,m^2}$, $m=48.2\,kg/\mathrm{m}$, $k=68.9\,\mathrm{MPa}$, $c=230\,\mathrm{kN\,s/m^2}$, and the amplitude of the point force is $F=10.5\,\mathrm{kN}$. The vertical deflection versus time for a receiver at

$x_1 = 0$ m and different force velocities, v_F = 10, 20 and 40 m/s, is plotted in Fig. 5. The reference plots of Sun [12] are reproduced directly from their own figures digitally.

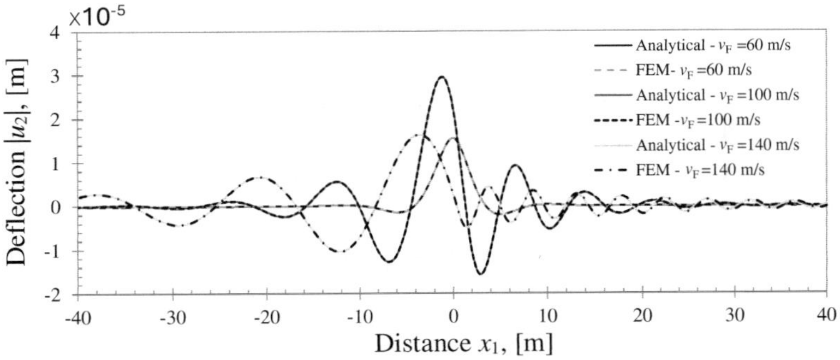

Figure 6. Vertical deflection at t=0 along the beam axis for different point load velocity.

The last set of results concerns the deflection of the infinite beam considered in Section 4 due to a moving point load with constant velocity. The critical velocity of the soil-beam system is $v_{cr} = \sqrt[4]{4EIk/m^2} = 100$ m/s [16]. The velocity of the point force is set to below, at, and above the critical velocity, with values v_F = 60, 100 and 140 m/s. Vertical displacement along the beam on elastic foundation at $t = 0$ s due to a constant moving point load is depicted in Fig. 6. We observe at subsonic speed, the displacement pattern is similar to the one obtained by a non-moving load. At supersonic speed the deflections are widely spread. The largest displacements are indeed obtained for the critical velocity of 100 m/s. The plotted results are derived analytically and numerically by FEM. The response obtained in space-frequency domain is further transformed in space-time domain by applying Fourier transformations given in Eq. 19-20. The numerical solution considers a beam with length L=440 m, discretized by 200 FE with progressively increasing size from application of the point load to the beam ends. Largest to smallest FE length ratio is 4. The cut-on frequency is 50 rad/s, Nyquist frequency is 1000 rad/s and the frequency step is 2 rad/s. The discrete values in the FEM solution, corresponding to the requested sampling for the different load velocities, are interpolated from the solution with L=440 m, which is the necessary length for the highest velocity. Fig. 6 shows that the numerical solutions correspond well to the analytical ones.

6. Conclusions

In this paper the steady-state response of beams resting on an elastic Winkler foundation is examined. Analytical solutions for both the beams of medium and infinite lengths subjected to a concentrated point load are evaluated and verified numerically by FEM approach. The obtained results show that beams with $\bar{\alpha}L/2 \geq 5$ have identical dynamic behavior due to a point load in the center of the beam up to the cut-on frequency. It could be concluded that when the ratio of length to characteristic length of the beam is larger than 10 the finite beam could be examined as infinite one.

IOP Conf. Series: Materials Science and Engineering **1323** (2024) 012018 doi:10.1088/1757-899X/1323/1/012018

A procedure which allows to consider the moving point load with constant velocity based on Fourier transform is proposed. A sampling is recommended, and the obtained results are verified and confirmed against a closed-form solution of beam on viscoelastic subgrade subjected to moving loads proposed by Sun [12]. The suggested procedure for treatment of moving point loads is applicable for both analytical and numerical solutions. Due to the space limitation only one numerical simulation is presented which expectedly shows that deflection response of a beam with damping is asymmetric and the maximal deflection occurs behind the moving load. The largest vertical displacement is achieved for moving loads with critical velocity and the deflection decreases for speeds above or below the critical one.

Acknowledgments: The authors wish to acknowledge the support provided by the Research, Consultancy and Design Centre (RCDC) of the University of Architecture Civil Engineering and Geodesy - Sofia through Grant No. BN-298/24.

References

[1] Winkler E 1867 *Die Lehre von der Elastizität und Festigkeit* (Prague)
[2] Pasternak PL 1954 *Fundamentals of a new method of analyzing structures on an elastic foundation by means of two foundation moduli* (Gosudarstvennoe Izdatelstro Liberaturi po Stroitelstvu i Arkhitekture, Moscow) (in Russian)
[3] Ker AD 1964 Elastic and viscoelastic foundation models *Journal of Applied Mechanics, Transactions ASME* **31(3)** 491–498
[4] Zimmerman H 1888 *Die Berechnung des Eisenbahn-oberbaues* (Berlin)
[5] Krylov A N 1931 *On the calculation of beams lying on an elastic foundation* (Leningrad: Publishing House of the USSR Academy of Sciences) (in Russian)
[6] Dinev D 2012 Analytical solution of beam on elastic foundation by singularity functions *Engineering MECHANICS* **19(6)** 381–392
[7] Chen J, Feng Y, Shu W 2016 An improved solution for beam on elastic foundation using quintic displacement functions *KSCE J Civ Eng* **20** 792–802
[8] Poliakov AM, Shtanko PK, Pakhaliuk VI 2019 Calculation of a variable cross-section beam on an elastic foundation with two coefficients of compliance *HIRM-2019, IOP Publishing, Journal of Physics: Conference Series* **1353** 012110
[9] Kenney JT 1954 Steady-state vibrations of beam on elastic foundation for moving load *Journal of Applied Mechanics* **21** 359–364
[10] Thompson WE 1962 *Analysis of Dynamic Behavior of Roads Subject to Longitudinally Moving Loads* Report No. VJ-1620-V-1 (Cornell Aeronautical Laboratory, Inc., New York
[11] Fryba L 1972 *Vibration of Solids and Structures under Moving Loads* (Noordhoff International Publishing)
[12] Sun L 2002 A closed-form solution of beam on viscoelastic subgrade subjected to moving loads *Computers & Structures* **80** 1–8
[13] Thambiratnam D, Zhuge Y 1996 Dynamic analysis of beams on an elastic foundation subjected to moving loads *Journal of Sound and Vibration* **198(2)** 149–169
[14] Wu JS, Shih PY 2000 The dynamic behaviour of a finite railway under the high-speed multiple moving forces by using finite element method *Communications in Numerical Methods in Engineering* **16** 851–866
[15] Andersen L, Nielsen SRK, Kirkegaard PH 2001 Finite element modeling of infinite Euler beams on Kelvin foundations exposed to moving loads in convected co-ordinates *Journal of Sound and Vibration* **241(4)** 587–604
[16] Kim SM 2004 Vibration and stability of axial loaded beams on elastic foundation under moving harmonic loads *Engineering Structures* **26** 95–105

IOP Conf. Series: Materials Science and Engineering **1323** (2024) 012019 doi:10.1088/1757-899X/1323/1/012019

Predicting the parameters of reinforced concrete elements by neural network

Vladimir Yakov[1*] and **Petar Hristov**[2]

[1] Department of Computer-Aided Engineering, Faculty of Structural Engineering, University of Architecture, Civil Engineering and Geodesy /UACEG/, Sofia, Bulgaria
[2] Department of Reinforced Concrete Structures, Faculty of Structural Engineering, University of Architecture, Civil Engineering and Geodesy /UACEG/, Sofia, Bulgaria

*E-mail: yakov_fce@uacg.bg

Abstract. Obtaining predictive results at planning and conceptual design phase often represents a serious problem on which the choice of design options depends to a significant extent This paper seeks solutions to the problem using neural networks and artificial intelligence and illustrates the idea by simple examples of pre-trained neural network models.

1. Introduction

In the context of rapidly emerging new technologies utilizing nonlinear programs and systems, modern construction and construction investment planning and design are adopting numerous new ideas aimed at increasing efficiency. There is increasingly frequent and strong discussion about the implementation of Building Information Modeling (BIM) [1][2], and efforts are being made to find generalized indicators for the condition and capacity of building structures [3][4].

This is a modern approach that could, to some extent, replace specialized studies, such as those concerning the consumption of reinforcement and concrete in reinforced concrete structures in [5], [6]

The design of reinforced concrete structures in the Republic of Bulgaria has been carried out according to Bulgarian national normative documents since 1927, and since the year 2000, according to the Eurocodes system (EC0, EC1, EC2, and EC8), with the corresponding national applications.

Both in modern Bulgarian standards and in the Eurocodes (EC), the limit state method is used. The trend is to fully transition to designing according to EC in the near future, which makes previous systematic studies on the costs of steel and concrete outdated and difficult to apply in practice. The idea is to use neural networks at the planning and preliminary design stages (conceptual phase) of investment projects to achieve a relatively accurate estimate of expected material costs (concrete, reinforcement, etc.).

This problem could be largely solved with the creation of software products using neural networks and artificial intelligence approaches. The idea is to create an appropriate neural network, trained with real or generated data from individual elements, which would provide, based on the input parameters, quantities of reinforcement and suitable cross-sections for a certain set of structural elements. This, in turn, would allow for an objective assessment of future material costs.

2. Application of artificial intelligence, particularly artificial neural networks, in civil engineering

Nowadays, methods of artificial intelligence, particularly artificial neural networks, are increasingly being applied in civil engineering.

In [7], the application of artificial neural networks for sizing a rectangular reinforced concrete section subjected to bending with double reinforcement is evaluated.

3. Demonstrative example

For the purposes of this report, a simple demonstrative example has been developed based on a hypothetical small four-story building with a monolithic reinforced concrete load-bearing structure consisting of rectangular four-sided supported slabs, beams, and columns.

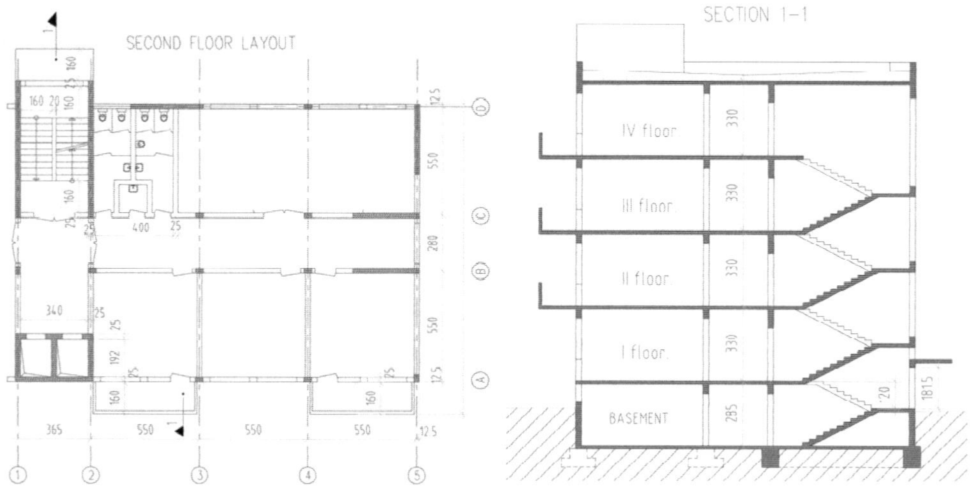

Figure 1. Second Floor Layout and section 1-1

An ANN has been built, which, given the input parameters of the inter-floor structure, results in the quantity of reinforcement in the beam elements.

The network has been trained with a certain amount of real data obtained from designs according to the EC system, and additionally, results generated from a simple sizing program created for the purposes of this study.

In the current study, The training of the neural network used the usual range of parameters for residential and administrative buildings. Loading is given implicitly by the width and length of the cross-reinforced slabs, respectively transversely and longitudinally to the beam. The width of the beams remains unchanged—25 centimeters, but their height varies from 30 to 60 centimeters.

The input parameters are:

- l_L - the length of the span parallel to the beam
- 1_C - the length of the span perpendicular to the beam
- h - the height of the beam
- h_f - the thickness of the slab.

Correspondingly, the output parameters are two:

- AS - the quantity of longitudinal reinforcement
- AS' - the quantity of transverse reinforcemen

The ANN consists of an outer layer with 4 neurons—one for each input parameter, one output layer respectively with two neurons for the longitudinal and transverse reinforcement of the beam, and an internal (hidden) layer with 10 neurons.

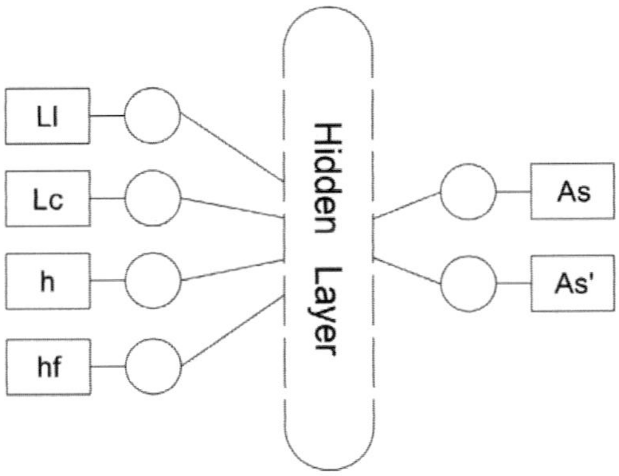

Figure 2. Diagram of Artificial Neural Network used in this study

In [7], a similar task is discussed, and a detailed study is conducted on the influence of the number of internal layers—either one or two—and the number of neurons in each on the accuracy of the training results. Some of the results are shown in Figures 3, which is taken from [7].

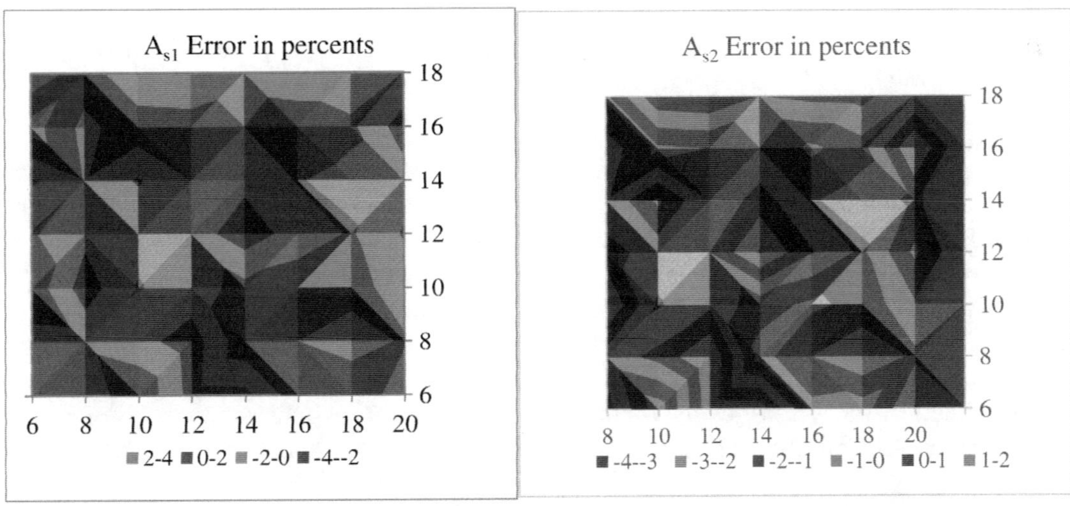

Figure 3. Error in predicted values in relation to the number of neurons in the function.

The results of the study indicate that there is no strong dependency, and on the other hand, the data is relatively limited compared to the previous study where the dataset comprised 3400 sets, whereas here it includes only 300. For this reason, 10 neurons in a single hidden layer were chosen. The outcomes of the training, validation, and testing confirm this choice.

Figure 2 shows the diagram of the neural network used in this study.

Before training the neural network, the data needs to be normalized. The normalization depends on the type of transfer function used in the input and hidden layers of the neural network. A tan-sigmoid transfer function was selected; therefore, the data will be normalized according to formula (1) with values ranging from -1 to 1.

In Figure 4, the Transfer functions of the outputs of neurons in the inner and outer layers of the neural network are shown:

$$a = \text{tansig}(n) = \frac{2}{1+e^{-2n}} - 1 \quad (1)$$

When creating the neural network—after setting its parameters and input data, which should ideally be pre-normalized—the neural network randomly arranges the datasets and divides them into three parts:

- 70% for training
- 15% for validating the neural network
- 15% for testing

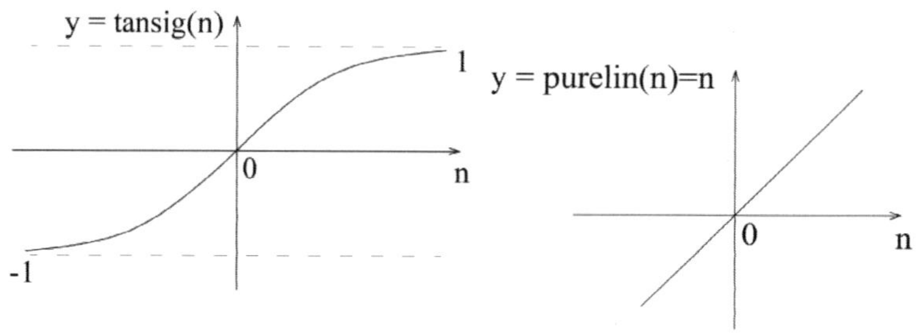

Figure 4. Transfer functions of the outputs of neurons in the inner and outer layers of the neural network.

Training begins by sequentially comparing input data with corresponding output data while adjusting the parameters of the neural network. Such a cycle of training is called an epoch. Validation data are used as criteria for the accuracy of the training—they serve only for comparison. The neural network "sees" them for the first time and does not change the values of the weights and biases of its neurons. Such a cycle of training is called an epoch. Figure 1 shows the error function of the training (validation value). The number of epochs for training the neural network is usually predetermined. The training process can be halted if the Mean Square Error from the validation increases during 6 consecutive epochs. The final assessment of training accuracy is obtained from the last 15% of the test data, which the neural network has not seen even once.

If the validation loss doesn't improve for 6 consecutive epochs, training will stop early. Early stopping is used to prevent overfitting by stopping the training process when the validation loss has not improved for a specified number of epochs.

A large part of the sized and constructed elements are analytically generated with the corresponding steps of the main parameters. This can explain the high values of the correlation coefficients.

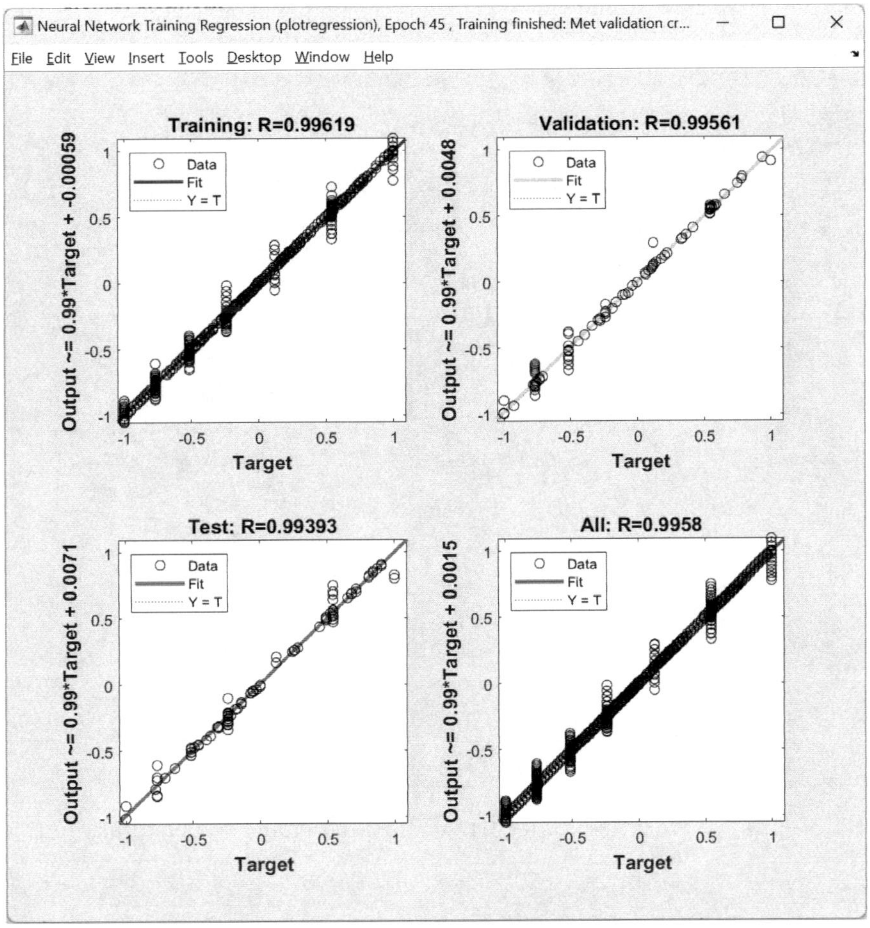

Figure 5 . Neural Network Training correlation coefficients with normalized values.

In Figure 5, the results of the neural network training are presented - the correlation coefficients for the predicted and actual values of the output variables represented by the two neurons, As1 and As2. In Figure 6 are presented the correlation coefficients for actual values of model.

The results show, as one might expect, that the correlation coefficient for the normalized and denormalized values of the input and output variables is almost identical, with a difference of less than one percent.

As an additional check for the correctness of the obtained results, a slab-beam model, which the neural network has not "seen," is used – the combination of the beam height and slab thickness was not used for training. A beam height of 60 centimeters and a slab thickness of 10 centimeters are assumed.

An appropriate load is assumed, and the beam is designed according to EC2 and the Bulgarian National Annex.

IOP Conf. Series: Materials Science and Engineering **1323** (2024) 012019 doi:10.1088/1757-899X/1323/1/012019

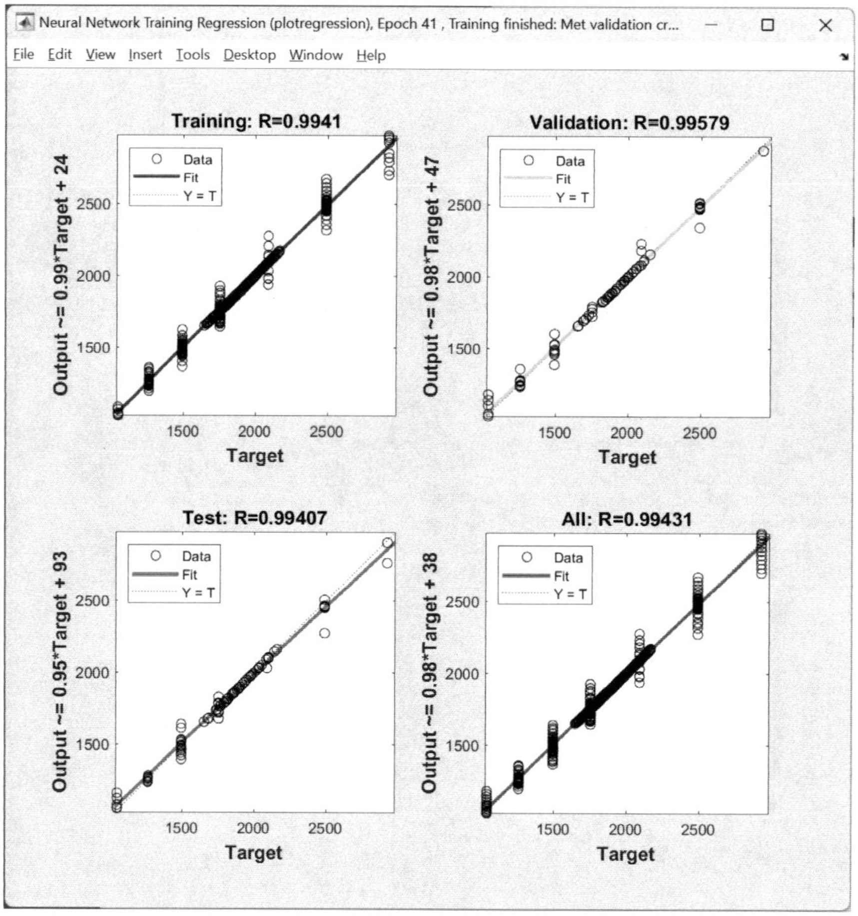

Figure 6. Neural Network Training correlation coefficients with actual values.

The corresponding output parameters are compared with the results already obtained from the analytical solution. The difference is around 1 percent, which confirms the accuracy of the model.

Conclusion

A similar approach could also be used for other structural elements, including the entire set of a building's structure. Each of these elements has specific features and factors that complicate the implementation of such a neural network—it would grow in volume and structural complexity, but in principle, the approach and its implementation remain the same. It is a matter of time and accumulated work and results for the future adoption of such methods in assessing the parameters of reinforced concrete structures at the stage of investment planning and conceptual design.

Besides the stages of preliminary planning and sketch design, the approach of training a neural network could find other practical applications. For example, in assessing the available amount of reinforcement and concrete in old buildings and structures due for demolition, for the purposes of managing construction waste.

References

[1] Kouteva-Guentcheva M. Boshnakov Kr. Engineering reading of the international BIM standards, Annuals of UACEG, 51 (5) 213-224, (2018) ISSN 1310-814X; ISSN 2534-9759 (in Bulgarian)

[2] Kouteva-Guentcheva M. Georgiev B. Boshnakov Kr. Towards a strategic plan for the introduction of building andinformation modeling in the construction industry in Bulgaria, Annuals of UACEG, 51 (12) 43-52, Sofia (2018) ISSN 1310-814X; ISSN 2534-9759 (in Bulgarian)

[3] Boshnakov Kr. Belev B. Mollova G. Seismic performance assessment and approaches for strengthening of existing flat slab reinforced concrete building, Proceedings of the 14 th International Scientific Conference VSU'14, 5-6 June 2014, Vol. IV, 262-267, Sofia (2014) (in Bulgarian)

[4] Boshnakov Kr. Kouteva-Guentcheva M. Applicability of damage indicators for assessment of test frame structure performance under seismic excitation, , Proceedings of the First scientific-applied conference with international participation "Rehabilitation and strengthening of buildings and facilities structures (RSBS2015) Nov. 5-6 2015, University of Architecture, Civil Engineering and Geodesy (UACEG), pp. 142-152, Sofia (2015) (in Bulgarian)

[5] Dimitrov B. Reinforced concrete structures. Tehnika, Sofia (1993) ISBN 954-03-0284-6 (in Bulgarian)

[6] Minev M. Optimization problems in the design of prestressed reinforced concrete girder road bridges.. Tehnika, Sofia (2013) ISBN 978-954-92814-4-6 (in Bulgarian)

[7] Boshnakov Kr., Yakov Vl., Neural Networks Applicability for Design of Reinforced Concrete Sections for Bending. // TEM Journal. Volume 12, Issue 3: 1294-1299, ISSN 2217-8309, DOI: 10.18421/TEM123-08, August 2023

[8] Kapania R.K. and Liu Y. Applications of artificial neural networks in structural engineering with emphasis on continuum models, Langley Research Center (1998) https://ntrs.nasa.gov/citations/20000101656

[9] Kaveh A. Applications of artificial neural networks and machine learning in civil engineering, Springer Cham (2024) DOI: 10.1007/978-3-031-66051-1

AUTHOR INDEX

Institute of Physics
Dirac House, Temple Back
Bristol BS1 6BE UK

ISSN: 1757-8981
ISBN 979-8-3313-1145-2

CURRAN ASSOCIATES INC.
proceedings
.com